Creating the Producti Workplace

In an increasingly competitive environment, companies are being forced to think harder than ever about the way they work, and how they can improve profitability. *Creating the Productive Workplace* provides a critical, multidisciplinary review of the factors affecting workplace productivity.

Productivity is a key issue for individual companies as well as the national economy as a whole. With 70–90% of the costs of running an organisation consisting of the salaries of the workforce, small increases in worker productivity can reap high financial returns. Many studies have shown that productivity at work bears a close relationship to the work environment. This book sets out the most important factors and evidence behind this phenomenon, and offers solutions to providing creative work environments which are conducive to productivity. Contributions are made by an international team of experts:

Bill Bordass, Derek Clements-Croome, Michel Cabanac, Cary Cooper, Roy Davis, Charles E. Dorgan, John Doggart, Francis Duffy, Mathab Farshchi, Max Fordham, Volker Hartkopf, Marshall Hemphill, John Jukes, Stephen Kaplan, Walter Kroner, Adrian Leaman, David Mudarri, Jean Neumann, David A. Schwartz, Valerie Sutherland, Hidetoshi Takenoya, Jackie Townsend, David Wyon, Jennifer Veitch

This book is essential reading for company directors, facilities and estates managers, interior designers, architects and building environmental engineers. It is also a text for undergraduates and postgraduates studying these disciplines and related subjects, particularly those related to intelligent buildings.

Derek Clements-Croome is Professor of Construction Engineering in the Department of Construction Management and Engineering at the University of Reading, UK.

Creating the Productive Workplace

Edited by Derek Clements-Croome

London and New York

First published 2000 by E & FN Spon, an imprint of Routledge
11 New Fetter Lane, London EC4P 4EE

Simultaneously published in the USA and Canada
by Routledge
29 West 35th Street, New York, NY 10001

E & FN Spon is an imprint of the Taylor & Francis Group

Typeset in Sabon by RefineCatch Limited, Bungay, Suffolk
Printed and bound in Great Britain by
St Edmundsbury Press, Bury St Edmunds, Suffolk

British Library Cataloguing in Publication Data
A catalogue record for this book is available from the British Library

Library of Congress Cataloguing in Publication Data
A catalogue record for this book has been requested

ISBN 0–419–23690–2

Contents

Figures

Tables

Contributors

Bill Bordass Principal, William Bordass Associates, London

Professor Michel Cabanac Department of Physiology, Université Laval Quebec

Derek Clements-Croome Professor of Construction Engineering, University of Reading

Cary Cooper BUPA Professor of Organizational Psychology and Health, Manchester School of Management, University of Manchester Institute of Science and Technology

Roy Davis Emeritus Professor of Psychology, University of Reading

John Doggart Director of ECD Energy and Environment Ltd, London

Professor Charles E. Dorgan PE Director, HVAC&R Center, University of Wisconsin, Madison

Dr Francis Duffy Founder, DEGW Plc, London

Mathab Farshchi Research Fellow, Advanced Construction Technology, University of Reading

Norman Fisher Professor of Project Management, University of Reading

Professor Max Fordham Max Fordham & Partners, London

Professor Dr. Eng. Volker Hartkopf Centre for Building Performance & Diagnostics, Carnegie Mellon University, Pittsburgh

Marshall Hemphill Hemphill Interior Technologies, Lancaster, Pennsylvania

John Jukes Optimum Workplace Environments Ltd, Old Coulsdon, Surrey

Stephen Kaplan Professor of Psychology and of Computer Science & Engineering, University of Michigan

Walter Kroner Professor of Architectural Research, Rensselaer School of Architecture, New York

Adrian Leaman Principal, Building Use Studies Ltd, London

Dr David H. Mudarri Indoor Environment Division, US Environmental Protection Agency, Washington

Dr Jean Neumann Senior Consulting Social Scientist, Organisational Change and Technological Innovation Programme, The Tavistock Institute of Human Relations, London

David A. Schwartz, Department of Psychology, University of Michigan

Dr Valerie Sutherland Sutherland-Bradley Associates, Glasgow

Hidetoshi Takenoya Chief Engineer, Kajima Design Europe Ltd, London

Jackie Townsend Director, Greystone International, Brighton

Dr Jennifer Veitch Research Officer, National Research Council of Canada, Institute for Research in Construction, Ottawa, Ontario

Professor David P. Wyon Research Fellow, Johnson Controls Inc, Milwaukee and International Centre for Indoor Environment and Energy, Technical University of Denmark

Foreword

by Nick Raynsford MP, Parliamentary Under Secretary of State
Minister for Construction and London

Buildings provide the environment in which the business community operates, and their procurement and operation represent a considerable business cost. Getting this environment right clearly holds the potential for better performance of both human and physical resources, and thus significant rewards to business and the wider economy. On an individual level, our working environment can have a major impact on our attitudes to our work, and in turn on our personal efficiency and productivity.

To realise this potential, we have to understand better the whole-life performance and economics of buildings. We must also gain a greater understanding of the potentially significant – but currently poorly understood – linkages between the productivity of building occupants and the operation of the internal building environment.

My Department has made improving the productivity of non-domestic buildings one of the four key themes of its new Sustainable Construction research and innovation programme. We have called for proposals to improve understanding of the business benefits of effective building design, and to generate more practical guidance and exemplars, so that clients and occupiers can make informed decisions about improved design and management of indoor environments. We are looking particularly at collaborative, multi-disciplinary, ways of tackling the issues. Our aim is to make sure that the overall performance of buildings is enhanced.

That is why this book is so timely and why I am delighted to be associated with it. The book highlights the issues employers and everyone who designs and commissions buildings need to address to create more productive and, above all, more *sustainable* workplaces. This will be good for individual employees, businesses, communities and the natural environment now and in the future. I very much hope that readers of the observations and suggestions contained in the following pages will be inspired to make or seek positive changes to their workplace.

Preface

The culture of living and working is undergoing accelerating social and technological changes. The release of creative energy in people is vital for them as individuals as well as for communities and society at large. Creative self-fulfilment leads to vitality in nations. There is recognition now that the daily rhythms of life need to be appreciated; there is also a need to understand how we think and concentrate and under what conditions our performance diminishes or improves. A high level of sustained focused concentration is necessary for a high level of productivity. Environments for working in are becoming more fluid to meet changing work patterns and to deal with all the different ways we need to think and communicate during any single working day. This book opens up some solutions but also poses many problems which, it is hoped, will provoke some thought and ideas for the future.

This book is partly based on the proceedings of a conference entitled 'Creating the Productive Workplace' held in London in October 1997. Over 200 people attended. Delegates came from a wide range of disciplines, including those who are interested in acquiring knowledge about human behaviour in buildings, and those who have to design and manage buildings in practice. I would like to thank the team that helped me to compose this conference, which included John Doggart from Energy Conscious Design (ECD), Nigel Oseland from British Research Establishment Ltd, and Aubrey Rogers from Bournemouth University. The administration was admirably executed by Peter Russell and Lilian Slowe, of Workplace Comfort Forum, together with their team of helpers. I would also like to pay tribute and give thanks to the sponsors and associates who have supported the event financially and with time and advice.

Sponsors: The Department of the Environment, Transport and the Regions (DETR); Energy Efficiency Best Practice programme; Clearvision Lighting Limited; RMC Group plc; Trigon Limited.

In Association with: Association of Consulting Engineers; British Council for Offices; British Institute of Facilities Management (BIFM); Building Research Establishment Limited (BRE); Building Services Research &

Information Association (BSRIA); Construction Industry Council; Construction Industry Environmental Forum; De Montfort University (Department of Design Management); ECD Energy & Environment Limited; Ecological Design Association; Energy Design Advice Scheme (EDAS); European Intelligent Building Group; University of Bournemouth (Department of Product Design & Manufacture); University of Reading (Department of Construction Management & Engineering).

I would also like to thank Maureen Taylor and John Jewell for their usual diligence in helping to prepare the manuscript.

Derek Clements-Croome
Reading, October 1998

Part 1

Creativity, environment and people

Chapter 1

Indoor environment and productivity

Derek Clements-Croome

In the journal of the British Council for Offices entitled *Office* (Summer 1997) it is reckoned that 'advanced building intelligence' should increase the productivity of occupants by 10 per cent annually as well as improving efficiency to satisfy owner occupiers. In contrast, 'standard intelligence' can improve efficiency by 8 per cent annually and improve efficiency which results in a payback within two to four years. The argument is that in an intelligent building there is less illness and absenteeism. Intelligent buildings do not mean that masses of technology are necessary. Simple adaptable building forms combined with appropriately specified building services and technologies should result in a high-quality business value intelligent building. Three major UK studies were carried out in 1997–98 on productivity: by SBS Business Solutions and the Building Research Establishment; by the Post Office Property Holdings Policy Planning and Development Group; and by the University of Reading in conjunction with several industrial partners.

At a fundamental level we need to understand the nature of concentration. It is now possible to measure brain rhythm patterns very easily (Dowson, 1997) and to diagnose these scans; feedback frequencies can be used to correct deficiencies. Table 1.1 gives the characteristics of the various brain rhythms.

Table 1.1 Characteristics of brain rhythms

Brain rhythm	*Frequency (Hz)*	*Characteristics*
Delta	0.5–3	Deep sleep
Theta	3–7	Dreaming; artistic, creative, intellectual thought; meditative concentration
Alpha	7–12	Conscious relaxation
Beta	12–30	Concentration; autonomic, processes; emotional states
Gamma	30–60	Unknown but possibly linked to various psychological states
Lambda	60–120	

The beta and theta rhythms are related to various states of concentration. As people tire, the beta rhythm reduces and the alpha rhythm increases.

We need to see if brain rhythm patterns are influenced by the environment. For example fragrances do have effects on the central nervous system but brainwave patterns can be affected by the subject's beliefs and thoughts about the stimulus (Klemm *et al.*, 1991; Torii *et al.*, 1988; Lorig and Roberts, 1990).

Traditionally, thermal comfort has been emphasised as being necessary in buildings, but is comfort compatible with health and well-being? The mind and body need to be in a state of health and well-being for work and concentration. This is a prime prerequisite for productivity. High productivity brings a sense of achievement for the individual as well as increased profits for the work organisation. The holistic nature of our existence – and productivity is one example of this – has been neglected because knowledge acquisition by the classical scientific method has dominated research. This method is controlled but limited; the world of reality is uncontrolled, subjective and anecdotal but nevertheless is vitally important if we are to understand systems behaviour. Personal experiences count. On this basis it is possible to reconsider comfort in terms of the quality of the indoor environment and employee productivity.

Bedford (1949) interprets health as meaning bodily efficiency, well-being and safety. For hundreds of years empirical observations have suggested that the quality of the air has a profound influence on health. There are many examples of seamen, people working in cotton factories, spinners and many other industrial workers being affected by the environment (Croome and Roberts, 198; Bedford, 1949). Most of this work makes deductions about the effect of the environment on productivity based on absenteeism rates, sickness records and incidents of accidents. In most heavy industries the relationship between work output and temperature is very clear. This early work also indicates that exceptionally competent workers are less affected by difficult atmospheric conditions than those of ordinary capacity. There are abundant studies that show a general pattern where work performance, as well as the likelihood of accidents, depends on mental alertness. This can be disturbed by fatigue, lethargy, alcohol, drugs and the environment. Lethargy, for example, is one feature found to be prevalent in buildings exhibiting sick building syndrome (SBS). Sickness records are difficult to analyse because sickness may be due to several causes, some of which have nothing to do with the environment in the workplace. Nevertheless, comparative trends between places which have good or not so good environments are worth while.

Dorgan (1994) analysed some 50,000 offices in the USA to see how they met ASHRAE (American Society of Heating, Refrigeration and Airconditioning Engineers) Standards 62-1989 and 55-1992. About 20 per cent could be classified as healthy buildings and always met the standards during occupied periods; about 40 per cent were generally healthy buildings and met the standards during most occupied periods; 30 per cent were unhealthy buildings and failed to meet the standards during most occupied periods; 10 per cent of the buildings had positive SBS in which more than 20 per cent of the occupants complained of more than two SBS symptoms, and frequently as many as six of the more common SBS symptoms.

Health is the outcome of a complex interaction between the physiological, personal and organisational resources available to the individual and the stress placed upon them by their physical environment, work, and home life. A deficiency in any one of these factors increases the stress and decreases human performance.

Illness symptoms occur when the stress on a person exceeds their ability to cope and where resources and stress both vary with time so that it is difficult to predict outcomes from **single causes**. SBS is more likely with warmer room conditions, and this also means wasteful energy consumption. When temperatures reach uncomfortable levels, output is reduced. On the other hand, output improves when high temperatures are reduced by air-conditioning. When temperatures are either too high or too low, error rates and accident rates increase. While most people maintain high productivity for a short time under adverse environmental conditions, there is a temperature threshold beyond which productivity rapidly decreases (Lorsch and Abdou, 1993). Mackworth (1946) found that overall the average number of errors made per subject per hour increased at higher temperatures, and that the average number of mistakes per subject per hour under the various conditions of heat and high humidity was increased at higher temperatures, especially above 32°C. Vernon (1936) demonstrated relative accident frequencies for British munitions plant workers at different temperatures, and found that the accident frequency was a minimum at about 20°C in three factories.

Pepler (1963) showed that variations in productivity in a non-air-conditioned mill were influenced by temperature changes, although absenteeism was apparently not related to the thermal conditioning; on average an 8 per cent productivity increase occurred with a 5K decrease in temperature. All these aspects help to promote a well-designed building. The importance of various factors is summarised in Table 1.2: it can be seen that natural daylight and ventilation are rated highly, but green issues and the use of atria are also significant.

Clearly, any building that does not maximise its natural daylighting is likely to be unpopular with office occupiers. The high value attributed to the use of windows rather than air-conditioning partly reflects the generally low level of effectiveness achieved by air-conditioning in many buildings, but also, more fundamentally, the inherent need for natural light and good views out of the building.

Wilkins (1993) reports that good lighting design practice, particularly the

Table 1.2 Importance ratings of various designs factors (Ellis, 1994)

	Importance			
Feature	*Very*	*Quite*	*Not very*	*Not at all*
Best use of natural daylight	57%	31%	10%	1%
Ventilation using windows	30%	4%	25%	3%
Thermal design for building	12%	40%	36%	6%
Energy-saving green design	15%	36%	37%	8%
Use of atria & glazed streets	4%	20%	52%	18%

use of daylight, can improve health without compromising efficiency. Concerns about the detrimental effects of uneven spectral power distribution and low-frequency magnetic fields are not as yet substantiated. Wilkins (1993) states that several aspects of lighting may affect health, including (i) low-frequency magnetic fields; (ii) ultraviolet emissions; (iii) glare; and (iv) variation in luminous intensity. The effects of low-frequency magnetic fields on human health are uncertain. The epidemiological evidence of a possible contribution to certain cancers cannot now be ignored, but neither can it be regarded as conclusive. The ultraviolet light from daylight exceeds that from most sources of artificial light. Its role in diseases of the eye is controversial, but its effects on skin have been relatively well documented. The luminous intensity of a light source, the angle it subtends at the eye, and its position in the observer's visual field combine to determine the extent to which the source will induce a sensation of discomfort or impair vision. Glare can occur from the use of some of the lower intensity sources, such as the small, low-voltage, tungsten–halogen lamps. It is reasonable to suppose that in the long term, glare can have secondary effects on health and that visible flickering can have profound effects on the human nervous system. At frequencies below about 60 Hz flickering can trigger epileptic seizures in those who are susceptible; in others it can cause headaches and eye-strain. Wilkins (1993) concludes that the trend towards brighter high-efficiency sources is unlikely to affect health adversely, and may indeed be advantageous. The trend could have negative consequences for health if it were shown that the increasing levels of ambient light at night affect circadian rhythms. Improvements in brightness and the evenness of spectral power may be beneficial. In particular, the move towards a greater use of daylight is likely to be good for both health and performance.

In many buildings, users report most dissatisfaction with temperature and ventilation, while noise, lighting and smoking feature less strongly. The causes lie in the way in which temperature and ventilation can be affected by changes at all levels in the building hierarchy, and, most fundamentally, by changes to the shell and services. In comparison, noise, lighting and smoking are affected mainly by changes in internal layout and workstation arrangements, which can often be partly controlled by users.

There are some indications that giving occupants greater local control over their environmental conditions improves their work performance, their work commitment and morale, which all have positive implications for improving overall productivity within an organisation. Building users are demanding more control of fresh air, natural light, noise and smoke. A lack of control can be significantly related to the prevalence of ill-health symptoms in the office environment; there is a widespread agreement that providing more individual control is beneficial. Work by Burge *et al.* (1987) demonstrated the relationship between self-reports of productivity and levels of control over temperature, ventilation and lighting, and showed that productivity increases as individual control rises across all the variables.

Intervention to ensure a healthy working environment should always be the first step towards improving productivity. There are very large individual differences in the tolerance of sub-optimal thermal and environmental conditions. Even if the average level of a given environmental parameter is appropriate for the average worker, large decrements in productivity may still be taking place among the least tolerant. Environmental changes which permit more individual adjustment will reduce this problem. Productivity is probably reduced more when large numbers work at reduced efficiency than when a few hypersensitive individuals are on sick leave. Wyon (1993) states that commonly occurring thermal conditions, within the 80 per cent thermal comfort zone, can reduce key aspects of human efficiency such as reading, thinking logically and performing arithmetic by 5–15 per cent.

Lorsch and Abdou (1994a) summarise the results of a survey undertaken for industry on the impact of the building indoor environment on occupant productivity, particularly with respect to temperature and indoor air quality. They also describe three large studies of office worker productivity with respect to environmental measurements, and discuss the relationship between productivity and building costs.

It is felt in general that improving the work environment increases productivity. Any quantitative proof of this statement is sparse and controversial. A number of interacting factors affect productivity, including privacy, communications, social relationships, office system organisation, management, and environmental issues. It is a much higher cost to employ people who work than it is to maintain and operate the building, hence spending money on improving the work environment may be the most cost-effective way of improving productivity. In other words, if more money is spent on design, construction and maintenance and even if this results in only small decreased absenteeism rates or increased concentration in the workplace, then the increase in investment is highly cost-effective (Woods, 1989).

According to a report by the National Electrical Manufacturers

Association in Washington, DC (NEMA, 1989), increased productivity occurs when people can perform tasks more accurately and quickly over a long period of time. It also means that people can learn more effectively and work more creatively, and hence sustain stress more effectively. Ability to work together harmoniously, or cope with unforeseen circumstances, points towards people feeling healthy, having a sense of well-being, having high morale and being able to accept more responsibility. In general people will respond to work situations more positively. At an ASHRAE Workshop on 'Indoor Quality' held in Baltimore in September 1992 the following productivity measures were recommended as being significant:

- absence from work, or workstation
- health costs including sick leave, accidents and injuries
- interruptions to work
- controlled independent judgements of work quality
- self-assessments of productivity
- speed and accuracy of work
- output from pre-existing work groups
- cost for the product or service
- exchanging output in response to graded reward
- volunteer overtime
- cycle time from initiation to completion of process
- multiple measures at all organisational levels
- visual measures of performance, health and well-being at work
- development of measures and patterns of change over time.

Rosenfeld (1989) describes how when air-conditioning was first conceived it was expected that the initial cost of the system would be recovered by an increased volume of business. He quotes an example where the initial cost of the air-conditioning system for an office building is about £100 per m^2, so that if the average salary is £3,000 per m^2 and there is an occupancy of 10 m^2 per person, then adding 10 per cent to the cost of the system is justified if it increases productivity by as little as 0.33 per cent. Such small differences are difficult to measure in practice. Rosenfeld (1989) shows a relationship between the savings in working hours and the incremental cost of the system for a range of salaries, and concludes that an improvement in indoor air quality can be more than offset by modest increases in productivity. This leads to the general conclusion that high-quality systems which will have higher capital costs can generate a high rate of return in terms of productivity. In addition, systems will be efficient, be effective, have low energy consumptions and consequently achieve healthier working environments with lower CO_2 emissions.

Holcomb and Pedelty (1994) attempt to quantify the costs of potential savings that may accrue by improving the ventilation system. The increase in cost can be offset by the gain in productivity resulting from an increase in employee work time. Higher ventilation rates generally result in improved indoor air quality. Collins (1989) reported that 50.1 per cent of all acute health conditions were caused by respiratory conditions due to poor air quality. Cyfracki (1990) reported that a productivity increase of 0.125 per cent would be sufficient to offset the costs of improved ventilation. It should be mentioned again that some studies have shown a decrease in SBS symptoms with increased ventilation rates (Samimi and Seltzer, 1992) while others have not (Moray, 1989). Holcomb and Pedelty (1994) conclude that although there is some inconsistency there is sufficient evidence to suggest an association between ventilation rates, indoor air quality, SBS symptoms and employee productivity.

In one case study reported by Lorsch and Abdou (1994b) it is not clear whether the drop in productivity was due to a reduction in comfort, the loss of individual control or frustration due to being inconvenienced. According to Pepler and Warner (1968) people work best when they are slightly cool, but perhaps not sufficiently cool for it to be termed discomfort, and should not be too cool for too long.

Lorsch and Abdou (1994b) conclude that temperatures which provide optimum comfort may not necessarily give rise to maximum efficiency in terms of work output. The difficulty here is that this may be true for relatively short periods of time, but if a person is feeling uncomfortable over a long period of time it may lead to a decrement in work performance. However, there is a need for more research in this area. It almost seems that for optimum work performance a keen sharp environment is needed which fluctuates between comfort and slightly cool discomfort.

It is much more difficult to test the effects of temperature on mental than on physical performance. For example, the lowest industrial accident rate occurs around 20°C and rises significantly above or below this temperature. The other problem is the interaction with other factors which contribute towards the productivity. Motivated workers can sustain high levels of productivity even under adverse environmental conditions for a length of time which will depend on the individual.

Lorsch and Abdou (1994b) analyse several independent surveys which show that when office workers find the work space environment comfortable, productivity tends to increase by as much as 5–15 per cent when air-conditioning is introduced, in the opinion of some managers and researchers. These are, however, only general trends; there are few hard data and some findings are contradictory. Kobrick and Fine (1983) conclude that it is difficult to predict the capabilities of groups of people, never mind individuals, in performing different tasks under given sets of climatic conditions.

A study for the Westinghouse Furniture Systems Company in Buffalo, New York (BOSTI, 1982) suggested that the physical environment for office work may account for a 5–15 per cent variation in employees' productivity. The general conclusion was that people would do more work on an average work day if they were physically comfortable.

Woods *et al.* (1987) reported that satisfaction and productivity vary with the type of heating, ventilation or air-conditioning system. Central systems appeared to be more satisfactory than local ones, the most important factor being whether there is cooling or not. In one study on user-controlled environmental systems by Drake (1990), the ability to have local control was important in maintaining or improving job satisfaction, work performance and group productivity, while reducing distractions from work. For example, some users reported that they wasted less time taking informal breaks compared to times when environmental conditions were uncomfortable. They were also able to concentrate more intensively on their work. The gain in group productivity from the user-controlled environmental system amounted to 9 per cent. A number of studies suggest that a small degree of discomfort is acceptable, but it has to be confined to a level which does not cause distraction.

Work by Kamon (1978) and others shows that heat can cause lethargy which not only increases the rate of accidents but can also seriously affect productivity. Bedford (1949) concluded that there was a close relationship between the external temperature and the output of workers. Deteriorating performance is partially attributable to somnolence due to heat. Schweisheimer (1962) carried out some surveys concerned with establishing the effect of air-conditioning on productivity at a leather factory in Massachusetts, an electrical manufacturing company in Chicago and a manufacturing company in Pennsylvania. In all cases after the installation of air-conditioning the production increased by between 3 and 8.5 per cent during the summer months. On the basis of these investigations Schweisheimer (1966) concluded that the average performance of workers dropped by 10 per cent at an internal room temperature of 30°C, by 22 per cent at 32°C and by 38 per cent at 35°C.

Konz and Gupta (1969) investigated the effects of local cooling of the head on mental performance in hot working environments. The subject had to create words in ten minutes from one of two sets of eight letters, which were printed on a blank form. Poor conditions without cooling resulted in the creation of words dropping by some 20 per cent, whereas with cooling the reduction was about 12 per cent.

Abdou and Lorsch (1994) studied the effects of indoor air quality on productivity. It was concluded that productivity in the office environment is sensitive to those conditions which lead to poor indoor air quality, and this is linked to sick building syndrome. It is recognised that any stress is also influenced by management and other factors in the workplace. Occupants with local control over their environment generally have an improvement in their work effort, but in a more general way there is a synergistic effect of a multitude of factors which affect the physical and mental performance of people. Abdou and Lorsch (1994) concluded that **in many case studies occupants have been highly dissatisfied with their environment, even though measurements have indicated that current standards were being met.** This highlights the need to review standards and the basis on which they are made. Exactly the same conclusion is drawn by Donnini *et al.* (1994).

Productivity depends on four cardinal aspects: **personal, social, organisational** and **environmental** (Fig. 1.1). Although it is difficult to collect hard data which would give a precise relationship between the various individual environmental factors and productivity, there is sufficient evidence to show that there are preferred environmental settings that decrease people's complaints and absenteeism, thus indirectly enhancing productivity. The assessment of problems at the workplace using complaints is unreliable, because there is little mention of issues that are working well, and also the complaints may be attributable to entirely different factors. Abdou and Lorsch (1994) contend that the productivity of 20 per cent of the office workforce in the USA could be increased simply by improving the air quality of the office, and this would be worth approximately $60 billion per year.

Work by Vernon *et al.* (1926, 1930) shows a clear relationship between absenteeism and the average ventilation grading for a space, which was judged by the amount of glazing on various walls; windows on three sides had the highest grading and windows on one side had the lowest. Abdou

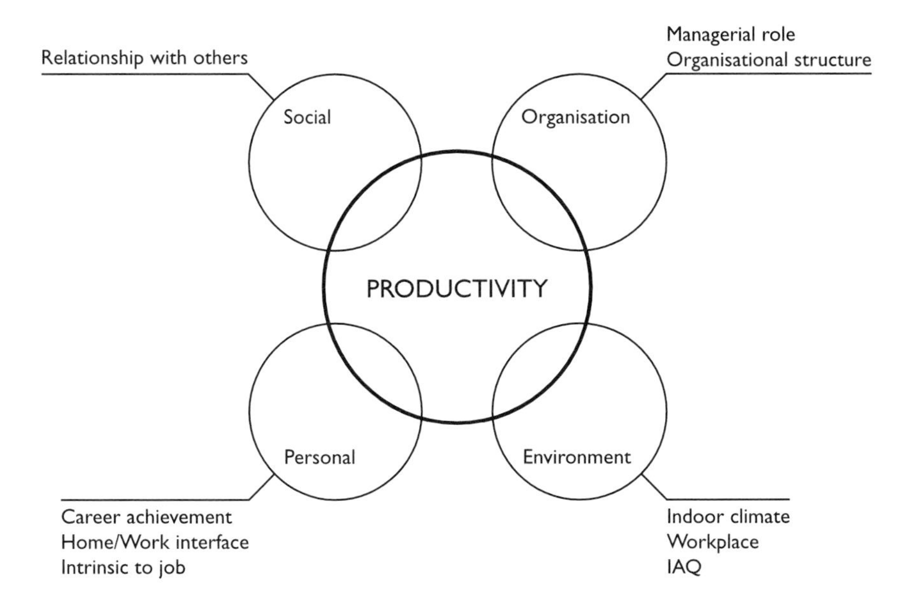

Figure 1.1 Factors which affect productivity

and Lorsch (1994) give the following causes as being the principal ones contributing to sick building syndrome:

- building occupancy being higher than intended
- low efficiency of ventilation
- renovation using the wrong materials
- low level of facilities management
- condensation or water leakage
- low morale and lack of recognition.

In this case lower efficiency of ventilation means that the supply air is not reaching the space where the occupants are, hence the nose is breathing in recirculated stale air. It is important to realise that even if the design criteria are correct for ventilation, the complete design team are responsible for ensuring that the systems can be easily maintained; the owner and the facilities manager also need to ensure that maintenance is carried out effectively. The tenant and occupants should use the building as intended. When new pollutant sources are introduced, such as new materials or a higher occupancy density, then the ventilation will become inadequate.

Burge *et al.* (1987) conducted a study of building sickness among 4,373 office workers in 42 UK office buildings having 47 different ventilation conditions. The data were further analysed by Raw *et al.* (1990). The principal conclusions were that as individuals reported more than two symptoms, the subjects reported a decrease in productivity; none of the best buildings in this survey were air-conditioned and these had fewer than two symptoms per worker on average, whereas the best air-conditioned buildings had between two and three symptoms; women recorded more symptoms than men, but there was no overall difference in productivity; individual control of the environment had a positive effect on productivity; the productivity was increased by perceived air quality; productivity, however, increased with perceived humidity only up to a certain point and then appeared to decrease again. Evidence supporting the importance of individual control of environment is again provided by Preller *et al.* (1990). It should be said that some contrary evidence exists concerning some of these factors, which points to the need for a systems approach to studying the effects of environment in buildings such as that proposed by Jones (1995).

Indoor environment is a dynamic interaction of spatial, social and physical factors, which affects productivity, health and comfort. The indoor environment and consequently the health of people can be affected by building materials and construction as well as the services systems (Rudnai, 1994). A complete analysis of the indoor environmental quality should take into consideration more than indoor air quality and thermal comfort. It should include the quality of lighting, sound levels, layout of individual work spaces, colour schemes, building materials, indoor CO_2 concentrations, radiation and electromagnetic fields, dust levels, and biological contaminants.

Productivity can be related to quality and satisfaction of the work performed. Studies have shown (Clements-Croome *et al.*, 1997; Lorsch and Abdou, 1994a, 1994b; Raw *et al.*, 1994; Woods, 1989; Wyon, 1982; Mackworth, 1946) that productivity at work bears a close relationship to the work environment. Burge *et al.* (1987) demonstrate a strong relationship between self-reports of productivity and ill-health symptoms related to buildings: productivity decreases as ill-health symptoms increase. There is a slightly less marked trend relating productivity and air quality, but it is a significant effect.

According to a Trade Union Congress survey of union health and safety for representatives at more than 7,000 workplaces (*Guardian*, 7 October 1996), work-related stress has become the most serious health hazard faced by British employees. The *Financial Times* (5/6 October 1996) stated that stress at work is costing British industries £79bn at a loss of 40 million working days per year. The concept of the sick building reflects the users' dissatisfaction, which can be as high as 40–50 per cent (Brundrett, 1994). Consumers are increasingly conscious of their rights to have a safe and pleasant environment and also believe that their productivity is impaired in unhealthy conditions. Concerns for productivity have accelerated as salaries have increased. In a typical commercial organisation salaries amount to about 90 per cent of the total costs, so even a seemingly small percentage increase in productivity of 0.1 to 2 per cent can have dramatic effects on the profitability of a company.

Dorgan (1994) defines productivity as the increased functional and organisational output including quality. This increase can be the result of direct measurable decreases in absenteeism, decreases in employees leaving work early, or reductions of extra long breaks and lunches. The increase can also be the result of an increase in the quantity and quality of production while employees were active; improved indoor air quality is an important consideration in this respect. There is general agreement that improved working conditions – and the office environment is certainly one of the more important working conditions – tend to increase productivity. However, determining a quantitative relationship between environment and productivity proves to be highly controversial. While some researchers claimed to have reliably measured improvements of 10 per cent or more, others presented data showing that no such relationships exist. Since the cost of the people in an office is an order of magnitude higher than the cost of maintaining and operating the building, spending money on improving the work environment may be the most cost-effective way to improve worker productivity (Lorsch and Ossama, 1993).

In 1994, the energy use in an average commercial office building in the

USA had an annual cost of approximately \$20 per m^2, whereas the functional cost was approximately \$3,000 per m^2. Functional costs include the salaries of employees, the retail sales in a store, or the equivalent production value of a hotel, hospital, or school. This means that a 1 per cent gain in annual productivity (\$30 per m^2) has a larger economic benefit than a 100 per cent annual reduction (\$20 per m^2) in energy use. In addition, the productivity gains will increase the benefits such as repeat business in hotels, faster recovery times in hospitals, and attainment of better jobs due to better education in schools. A small gain in worker productivity has significant economic impacts, and it makes sense to invest in improving the indoor environment to achieve productivity benefits. Dorgan (1994) states that productivity gains of at least 1.5 per cent in generally healthy buildings, and 6 per cent in sick buildings, can easily be achieved. As costs of improvement in the indoor air are typically paid back over a period ranging from less than nine months to two years, the benefits clearly offset the renewal cost, resulting in a very favourable cost-to-benefit ratio. Some literature indicates that productivity gains may be as high as 5 to 10 per cent. However, achieving such productivity gains may require the use of advanced active or passive environmental control as well as personal controls. Examples of productivity gains in the order of 1 to 3 per cent are found in several studies (Kroner, 1992). Informal (unpublished) and anecdotal reports on productivity gains have been researched in supermarkets, fast food outlets, retail department stores, schools, and office buildings. Estimated gains in sales ranging from 4 to 15 per cent in retail stores during summer months were reported by Dorgan (1994).

By focusing on the productivity benefits, projects which improve the indoor environment are increasingly moved away from an **energy-saving** scenario to one emphasising **productivity increase**. Even if a proposed project improves the indoor environment but increases the energy cost by 5 per cent, the project may still be economically feasible if the productivity increase is greater than 0.04 per cent. Wyon (1993) states that the 'leverage' of environmental improvements on productivity is such that a 50 per cent increase in energy costs of improved ventilation would be paid for by a gain of only 0.25–0.5 per cent in productivity, and capital investments of \$50 per m^2 would be paid for by a gain of only 0.5 per cent in productivity. The payback time for improved ventilation is estimated to be as low as 1.6 years on average, and to be well under one year for buildings in which the ventilation is below currently recommended standards.

An increase in productivity can be achieved with either no increase in energy use (or even a decrease), or with an increase in funding for the given level of technology. The use of energy recovery systems, and the increased use of such technologies as advanced filtration, dehumidification, thermal storage and natural energy, are examples of energy improving technologies whose cost can be offset by increases in productivity. The increased building services budget should allow for the introduction of the best system, not the cheapest. Any indoor environment productivity management programme should be able to include reducing energy consumption as one of the design objectives. Improving indoor environment will provide a high return on investment through productivity gains, health saving and reduced energy use. The benefits of improved indoor environment are improved productivity, increased profits, greater employee–customer–visitor health and satisfaction, and reduced health costs. The potential productivity benefits of improved indoor environment are so large that this opportunity cannot be ignored. There are indirect, long-term, and social benefits.

References

Abdou, O.A. and Lorsch, H.G. (1994) The impact of the building indoor environment on occupant productivity: Part 3: Effects of indoor air quality, *ASHRAE Trans. 1*, **100**(2), 902–913.

Bedford, T. (1949) Airconditioning and the health of the industrial worker, *Journal of Institution of Heating and Ventilating Engineers*, **17**, 112–146.

BOSTI (Buffalo Organisation for Social and Technological Innovation) (1982) *The Impact of the Office Environment on Productivity and the Quality of Working Life*. Buffalo, NY: Westinghouse Furniture Systems.

Brundrett, G. (1994) Health and the built environment, *Renewable Energy*, 5, II, 967–973.

Burge, S.A., Hedge, A., Wilson, S., Harris-Bas, J. and Robertson, A. (1987) Sick building syndrome: a study of 4373 office workers, *Ann. Occ. Hygiene*, **31**, 493–504.

Clements-Croome, D.J., Kaluarachchi, Y. and Li, B. (1997) What do we mean by productivity, *Workplace Comfort Forum*, October, London.

Clements-Croome, D.J. and Roberts, M. (1981) *Airconditioning and Ventilation of Buildings* (UK: Pergamon).

Collins, J.G. (1989) Health characteristics by occupation and industry: United States 1983–1985. *Vital Health and Statistics* **10**(170) Hyattsville, MD: National Center for Health Statistics.

Cyfracki, L. (1990) Could upscale ventilation benefit occupants and owners alike? *Indoor Air '90, 5th International Conference on Indoor Air Quality and Climate*, 5, 135–141. Aurora, ON: Inglewood Printing Plus.

Dorgan, C.E. *et al.* (1994) Productivity link to the indoor environment estimated relative to ASHRAE 62-1989, *Proceedings of Healthy Buildings '94*, Budapest, 461–472.

Dowson, D. (1997) Personal communication, 27 August.

Drake, P. (1990) *Summary of Findings from the Advanced Office Design Impact Assessments*, Report to Johnson Controls Inc., Milwaukee, WI.

Ellis, R. (1994) *Tomorrow's Workplace*, a Survey for Richard Ellis by The Harris Research Centre.

Holcomb, L.C. and Pedelty, J.F. (1994) Comparison of employee upper respiratary absenteeism costs with costs associated with improved ventilation, *ASHRAE Trans.*, **100**(2), 914–920.

Jones, P. (1995) Health and comfort in offices, *Arch. J.* 8 June, 33–36.

Kamon, E. (1978), Physiological and behavioural responses to the stresses of desert climate, in *Urban Planning for Arid Zones*, ed. G. Golany (New York: Wiley).

Klemm, W. *et al.* (1991) *Topographical EEG Maps of Human Responses to Odours*, Texas A & M University, unpublished report.

Kobrick, D.J. and Fine, M.M. (1983) Climate and human performance, in *The Physical Environment at Work*, eds Oborne and Gruneberg, 69–197 (New York: John Wiley and Sons).

Konz, S. and Gupta, V.K. (1969) Water-cooled hood affects creative productivity, *ASHRAE J.* 40–43.

Kroner, W.M. and Stark-Martin, J.A. (1992) Environmentally Responsive Work Stations and Office Worker Productivity. *Workshop on Environment and Productivity*, June (contact Department of Architecture, Rensselaer University, Troy, NY).

Lorig, T. and Roberts, M. (1990) *Chem. Senses*, **15**, 537–545.

Lorsch, H.G. and Abdou, O.A. (1994a) The impact of the building indoor environment on occupant productivity: Part I: Recent studies, measures and costs, *ASHRAE Trans.*, **100**(2), 741–749.

Lorsch, H.G. and Abdou, O.A. (1994b) The impact of the building indoor environment on occupant productivity: Part II: Effects of temperature, *ASHRAE Trans.*, **100**(2), 895–901.

Lorsch, H.G. and Abdou, O.A. (1993) The impact of the building indoor environment on occupant productivity, *ASHRAE Trans.*, **99**, Part 2.

Mackworth, N.H. (1946) Effects of heat on wireless operators hearing and recording morse code messages, *Br. J. Ind. Med.*, **3**, 143–158.

Moray, N. *et al.* (1979) *Final report of the Experimental Psychology Group. Mental Workload: Its Theory and Measurement*, ed. Moray, N., 101–114 (New York: Plenum).

NEMA (1989), *Lighting and Human Performance: a Review*, a report sponsored by the Lighting Equipment Division of the National Electrical Manufacturers Association, Washington, DC, and the Lighting Research Institute, New York.

Pepler, R.D. (1963) *Temperature: Its Measurement and Control in Science and Industry*, **3**, ed. Hardy, 319 (Rheinhold).

Pepler, R.D. and Warner, C.G. (1968) *ASHRAE Trans.*, **74**, Part II.

Preller, L., Zweeis, T., Brunekeef, B. and Boleij, J.S. (1990) Sick leave due to work related health complaints among office workers in the Netherlands. *Indoor Air '90, 5th International Conference on Indoor Air Quality and Climate*, 227–230.

Raw, G.J., Roys, M.S. and Leaman, A. (1990) Further findings from the Office Environment Survey: Productivity. *Indoor Air '90, 5th International Conference on Indoor Air Quality and Climate*, **1**, 231–236.

Raw, G.J., Roys, M.S. and Leaman, A. (1994) *Further Findings from the Office Environment Survey, Part 1: Productivity*, Building Research Establishment, Garston, Note N79/89.

Rosenfeld, S. (1989) Worker productivity: hidden HVAC cost, *Heating/Piping/Air Conditioning*, September, 69–70.

Rudnai, P. (1994) *Proceedings of Healthy Buildings 1994*, Budapest, **1**, 487–497.

Samimi, B.S. and Seltzer, J.M. (1992) Sick building syndrome due to insufficient and/or nonuniform fresh air supply in a multi-storey office building, *IAQ '92: Environments for People*, 319–322.

Schweisheimer, W. (1966) Erhaehung und Leistung und Produktion, *Wärme, Luftungs und Gesundheitstechnik*, Nov., 278–285.

Torii, S. *et al.* (1988) Chapter in *Perfumery: the Psychology and Biology of Fragrance*, ed. van Toller and Dodd (New York: Chapman Hall).

Vernon, H.M. (1936) *Accidents and their Presentation* (Cambridge University Press).

Vernon, H.M., Bedford, T. and Warner, C.G. (1926) A physiological study of the ventilation and heating in certain factories, *Rep. Ind. Fatigue Res. Bd*, no. 58, London.

Vernon, H.M., Bedford, T. and Warner, C.G. (1930) A study of heating and ventilation in schools, *Rep. Ind. Health Res. Bd*, no. 35, London.

Wilkins, A.J. (1993) Health and efficiency in lighting practice, *Energy*, **18**(2), 123–129.

Woods, J.E., Drewry, G.M. and Morey, P. (1987) Office worker perceptions of indoor air quality effects on discomfort and performance, *4th International Conference on Indoor Air and Climate*, Berlin, 464–468.

Woods, J.E. (1989) Cost avoidance and productivity in owning and operating buildings, *Occupational Medicine, State of the Art Reviews* **4**(4), eds J.E. Cone and M.J. Hodgson; also *Problem Buildings: Buildings Associated Illness and the Sick Building Syndrome*, 753–770 (Philadelphia, PA: Hanley & Belfus).

Wyon, D.P. (1982) The effect of moderate thermal stress on the potential work performance of factory worker, *Energy and Buildings*, April.

Wyon, D.P. (1993) The economic benefits of a healthy indoor environment, *Healthy Air '94*, Budapest, 405–416.

Chapter 2

Creativity in the workplace

Jackie Townsend

This is a book about well-being; the well-being of any building and of the people working inside it and how the combination of the building and the occupants can be enhanced to produce an environment which encourages creative, productive work and pleasure: pleasure in the environment and pleasure in the work.

It occurs to me that this desire to create an environment which is conducive to creative and productive work indicates quite a radical shift in the whole philosophy of work and the workplace. I would suggest that work and the workplace, for the great majority of people, have not been instigated, designed, begun and built with the workers themselves in mind. Most office buildings are lumps of grey concrete with bits of glass in, not particularly beautiful or inviting to any of the senses, but purely functional. They speak of power and money rather than creativity and pleasure.

It is entirely possible to build buildings which enhance the environment and make us feel good. Most cathedrals do this, especially the great ones such as St Paul's and Chartres. Because people visit them with a feeling of respect, religious or not, and mostly treat them in a respectful way, this energy builds up so that the building performs its function in a more and more fulfilling way. It gets better and better. Of course, the Church speaks of a different kind of power: its own, which is also not entirely altruistic or benevolent except sometimes on an individual level, and that of the Other, the Divine, whatever you might perceive that to be. So it can be done and it has been done.

Therapists who work with the body talk about function and form. Form implies the function, and function guides the form. For instance, your two feet are not particularly big but they can carry the weight of your whole body, without you falling over when you stand still. They do this because the bone in your lower leg goes down into your heel, so that it is the bones which are supporting your weight, but then you have maybe around nine inches of progressively smaller bones with tendons, muscles and skin, blood vessels and nerves, which spread out from your heel and then become five toes. They are ideally designed for their purpose. Function and form.

There are basically three types of buildings. There are buildings built for worship, buildings for living in and buildings for working in, although there might be some overlap, for instance in hospitals, shops and leisure centres. I feel it is pertinent to ask, with respect to creativity and creating a productive workplace, which of these are designed and put together with the most respect for human life and human possibility and endeavour.

At the conference on 'Creating the Productive Workplace' (London, 1997), Liam Hudson spoke about boundaries, both physical and non-physical, as in, for example, what is familiar and what is not, what is accepted, perhaps because it is part of our conditioning, and what is not, yet. In suggesting these comparisons maybe I am myself treading the boundary between the sacred and the profane, which the Japanese architect, Shigeru Uchida, considers to be the original boundary of all boundaries, in the avenue of thought I am offering here. Perhaps the relevance of any boundary, real or imagined, is our willingness to push it to a different place or even remove it altogether, since boundaries of any sort create a separation, and that usually takes the form of a separation between people, a segregation, which causes pain. So, to consider the boundaries between these three types of building, beginning with the boundary between the sacred and the profane, if buildings built for worship go into the sacred box, by the same criteria the buildings built for working in should go into the profane box, since by the standards we usually set money and power are profanities. They are not of the Divine. 'Render therefore unto Caesar the things which are Caesar's; and unto God the things that are God's.' Buildings for living in could, I suppose, have a foot in both camps, depending on the circumstances.

But maybe we could take a look at the intention that goes into the construction of the three main types of buildings. So for buildings that are for living in, we think about the things we do in our homes and the functions that they need to perform for us. We sleep, eat, keep ourselves and our belongings clean, entertain ourselves, our families and our guests and relax, and we are protected from the elements. So we have to allow for all of that when building and buying homes, and of course the extent to which all these functions are made more or less comfortable depends on the amount of money we have available to spend on the structure and the fixtures and fittings. Buildings for working in are usually very functional with a little bit of comfort. So there are lots of workspaces, whether group or individual; some eating areas, rest areas and washing and toilet facilities, all usually fairly basic, especially if they are used only by the staff. Areas that are used by the public, any of whom may become clients or customers, are usually more luxurious. This is not because the company cares about the comfort of the general public more than it cares about the comfort of its staff; it is part of its public relations, the image that it is anxious to project about who it is and what it does.

Buildings for worship are different. The word 'cathedral' derives from the Latin *cathedra* which itself comes from the Greek *kathedra*, meaning 'a seat'. A cathedral is a church in a diocese containing the bishop's throne. They are built on sacred ground, incorporating the ley lines, which are the meridians of the earth. Their function is to give us a location to try to reach something larger or higher than our individual selves.

But a cathedral takes an average of eighty years to build. That's quite a long time, more time than any company would have available if, for instance, it wanted a new headquarters, or to open a new branch.

Cathedrals also earn a lot of money for the Church and, when they were originally built, for the local community, in times when wealth was invested in productive assets, rather then bundles of paper made from trees which have to be cut down, sitting in bank vaults and building societies and gambled in the money markets. Even today, in Chartres, the bulk of the city's businesses live from the tourists who visit the cathedral 800 years after its completion. So sometimes the sacred and the profane can be put in the same box. There is some affinity.

Apart from their physical design and construction, the difference between these three types of buildings is the energy contained within them, which itself derives from the intention behind them and the purpose for which they were built. Let's take a look at that. By 'intention' I mean the thought that created any building in the first place. A community in which the Church is powerful says, 'Let's build a cathedral. It will give people work to do; our craftsmen can do their best work there and employ apprentices to continue their skills. People will come and bring us wealth, in one form or another, which we can put into the cathedral to keep it beautiful or we can use in the community.' (Recent studies have revealed that the quality of life for the common labourer in Europe was the highest in the twelfth to thirteenth centuries, when most of the cathedrals were built. It was possibly even higher than it is today.)

A modern housebuilder says, 'I'm going to build some houses. People will buy them to live in and that will earn me a living. I can build each house to my own specifications, make sure the building regulations are covered, and cut corners wherever possible. I'm not going to live in it so I will not bother to make it as good and as pleasing as if I were. I will waste the space in the attic by leaving it unlit and without electricity, gas and water connections. They can have tiny gardens because they are going to be sitting in front of the television instead of growing vegetables.' A look at any modern housing estate bears out this assumption, cynical as it is. I live in a town where there are lots of streets of Victorian houses; in fact I live in one myself. They are mostly terraced and they do have tiny gardens (we do grow vegetables in ours) and they were built for working people, but they are also spacious, elegant and very solid. You can put large items of furniture in them and still have room to walk around. They are gentle houses and you can't help feeling that they have been designed, built and put together with more care, more thought, more time, more creativity and more pleasure than the houses on the modern estate.

Another builder, a bit higher up the financial scale, decides to build an office block. If he gets his timing right and he has a ready buyer he can make quite a lot of money, which goes either into another office block or into the bank and the money markets. There's nothing inherently wrong with that, but these buildings are less than they could be. There is less room for more care, more thought, more time, more creativity and more pleasure. They are not appropriate because they rarely fit with the intention. These buildings, then, are less likely to become workplaces that are creative as well as productive.

All of this is part of the purpose and is included in the intention, and it makes a difference in the energy which is inherent in the building, whatever kind of building it is, and the energy that we bring to it. A church is a place of permanence and stability, even while the Church itself waxes and wanes, and even though some churches have been converted into homes and even offices, because that is what it represents. People visit cathedrals to appreciate the architecture but also to experience the peace and the restorative energy that exudes from these buildings. It helps us to find a space in ourselves, the same kind of space that is found in meditation, and which is not the same as thinking. The body enjoys this space and there is a resonance in buildings which have the kind of energy that allows this space to be found. I would suggest that we try to find ways to incorporate these kinds of intentions into our modern buildings. The energy will then come because it has been invited, so to speak. And this will encourage creative endeavour.

The energy in and around any building comes from the intention that was there before its construction, the structure itself (both the design and the actual building of it), the feelings of the people who spend their time there (whether living, working or praying), and the energy that other people bring to it. You can tell a house that has a good atmosphere while you are still standing on the front doorstep; a cathedral is a place of peace and harmony because that is what people bring to it; a large building such as a block of offices or a university can give a feeling of pleasure if that intention is part of it, and such buildings do exist and make very strong statements about themselves and their owners because of the contrast they provide to all their neighbours. One such building is the headquarters of the NMB Bank just outside Amsterdam (as described in a corporate PR publication, *NMB Bank's Head Office*). It is made up of several medium-sized towers, all linked by walkways on the inside. The aim of the bank in its design and construction was to achieve a better balance between the organisational requirements of the bank and the needs of the staff as individuals as well as bankers. It was designed by Ton Alberts, an Amsterdam architect who is known as working on a consciously human scale. It has sloping walls (to deflect the sound of passing traffic upwards and create better acoustics; also the sun's heat is used more efficiently) and there are no 90° angles. It is very light and spacious, with wide staircases going only short distances so that staff are encouraged to use the stairs rather than the lifts, which are incorporated into the walls in such a way as to be easily overlooked. It has four restaurants, each very different to the other three and each used by all members of staff. The colours everywhere are light and soft, and every wall is slightly different to the previous one. The architect says that the aim is to allow people to go about their work in a relaxed manner. 'They shouldn't feel they're working; they should just have the idea that they're "getting on with things".' It is very low on energy consumption, using natural ventilation and cooling systems and uses daylight to maximum advantage. The overall feeling is welcoming and refreshing and one of lightness. When you first see it, it looks quite amazing and rather bizarre, but it makes you smile and you keep on smiling as you enter because the feeling of it is pleasurable and the people you see going about their business and 'getting on with things' are a part of it. The building is not just to keep them warm and dry and meet their physical needs while they sit at their desks and shuffle and exchange bits of paper – it is the expression of a philosophy, and I feel that the Bank is to be commended for taking such a progressive stance and literally putting its money where its mouth is.

If we want to have workplaces that, of themselves, allow and encourage the possibility of creative and productive work, then we have to look at why that is not happening already (if indeed that is the case). I would suggest that there are two things going on here which are relevant to well-being in the workplace. It is generally agreed that we are currently going through a time of enormous change. Most people do not like change and resist it with all their might. Most people also have a tendency, when faced with a problem, to want somebody else, or something else, to take responsibility for it, and so absolve them of their responsibility, and of their opportunity to do something about it. As a species we are creative. We are bundles of energy that is creative. We have ourselves created, and are creating, this period of change that we are experiencing. If there is a problem, then the problem is our resistance and our expectation that things will stay the same while political, social and business structures, and all their ramifications, reorganise themselves around us. Albert Einstein observed that problems cannot be solved at the same level at which they were created. What we need is a truly creative response to change, and to learn to be creative in a completely new way; the buildings that we construct and work in can be part of that. But, there again, it is no bad thing to be reminded that hugely creative endeavours have been accomplished in a shed at the bottom of the garden. It really depends on how we choose to use the energy that is at our disposal, that in fact is us.

The other thing that is going on is also due partly to a reluctance, as groups and as individuals, to take responsibility and a tendency to go instead for the easy option of letting other people tell us what to do. Millions of people go to work every day and are told what to do. The natural urge to creative self-expression is minimised and the result is negativity and resentment, or dependence on what is seen as a higher authority. In this way we limit ourselves. We can also limit ourselves by taking too specific an orientation. We live in a society based on economics and business is the dominant institution. The orientation of business is making money, and the orientation of education at this time is moulding people to fit into business and economics. The move towards more intellectual knowledge and greater specialisation can mean that other kinds of knowledge and experiential skills are neglected, to the detriment of all parts of our society and our lives. We are rapidly establishing a knowledge society, as opposed to a society of practical and experiential skills. It seems pertinent to ask a few questions about knowledge and also to look at the role of creativity. On one level life is a process of learning and it can come to us or we can seek it out. So we go to school and college and evening classes and workshops to learn about things and find out about what is going on in the world around us and what we can do to be a part of it. And I think it is a good idea to pursue actively this acquisition of knowledge. It helps us to make a contribution and to participate in society. But I think it is also important to realise that this is only one kind of knowledge. Essentially it is passed-down knowledge. Finding out from other people what other people know; from their experience, their experiments and their collection of passed-down knowledge. There is also the knowledge that we have from our own experience and experiments. So we each have a pool of inherited or passed-down knowledge, mostly to do with the intellect because it is not our own experience, and then we have our own experience and things we have tried ourselves, which brings in emotion, so now we have knowledge on an intellectual level and knowledge on an emotional level: about interacting with other people, experiencing how other people respond to us and our ideas, experiencing relationships, children, families, rejection and acceptance in many different circumstances, and learning how to handle all of that. In this way we start to mature through our emotions and we gain insight, and this can be classed as knowledge, and it is valuable and usable. Because of the emphasis on intellectual or passed-down knowledge in the education system and especially in the National Curriculum, emotional growth and maturity lag behind intellectual growth, and I think it is probably true to say that most people never get anywhere near emotional maturity, although they may have several letters after their names to show how much intellectual knowledge they have amassed.

Then there is intuitive knowledge, which comes from within, and which is dismissed only by those who don't have it. I think it was Jung who said that the only real knowledge comes from within. And this is also of value and can be learned, but it is a slow process and often a painful one, because it demands a great deal of emotional growth and finding out about oneself. For this reason it is avoided by most people most of the time.

And then there is instinct, part of the primitive brain. It is difficult to quantify instinct because it's so tied up with our emotions and our physicality. Small children are probably happiest with their instincts, because they haven't yet been educated out of them. Michel Odent, one of the pioneers of water birth, believes that instinct is a vital part of our being which has historically been much maligned and neglected but which we need to resurrect and nurture if we are to learn to care for ourselves and our planet in an enlightened way (Odent, 1990). It is also a resource for creativity, which, I have found, is like an underground stream into which all the other streams which are the parts of yourself feed; they all flow along together quite nicely, lying dormant and maturing, and waiting, because all the best ideas have to await their time and place, or be dismissed as irrelevant or cranky. Then creativity develops what might be likened to hot springs, which bubble up to the surface, into consciousness, and then it can be put to use.

The essence of creativity is that it brings into being something new. The intellectual knowledge and the level of emotional maturity that you have attained are obviously substantial parts of this creativity, but there needs also to be a freedom of expression and the opportunity to use all resources.

My feeling is that what is required is a much deeper perception than saying that if you put people in nice friendly buildings then they will do good work. For a while they probably will, but I suspect the effect will wear off as the new environment becomes the norm. Abraham Maslow's hierarchy of human needs (in the 1940s) bears this out. Once basic needs for physical and emotional survival and safety are met, people are more likely to be motivated by psychological and emotional factors of a higher order such as job satisfaction, relationships with others, self-esteem and fulfilment. Professor Clements-Croome says that productivity is related to the morale of the people working for the organisation. I think that says it all, really. I think work is important. Whatever kind of work it is and whether it is paid or unpaid, we spend a lot of time doing it and it is the largest avenue for self-expression that we have, and whether you're cleaning the public lavatories at Piccadilly Circus or running a multinational company, that still applies.

However, it has been estimated that only 27 per cent of us enjoy what we do as work. Many people experience work as a major source of anxiety, unhappiness, insecurity, frustration and turmoil, which is not only a shame in human terms but a huge waste of resources, talent and creativity. Real joy comes from creativity, or the expression of one's own creative energy, which emanates from deep within a person, from the soul. Everyone has it and there are no exceptions. However, what the great majority of us also have is huge amounts of habit, and habit kills creativity. We brush our teeth in the same way; we clean the bath in the same way; we put our clothes on in the same way – right sock then left sock, or possibly the other way round.

It's hard to remember, and yet I presume you put something on your feet every day. Is it before or after underwear? Does it matter? What matters is the lack of conscious awareness because our responses to these kinds of tasks, or doings, are habitual. Habit is death to creativity because it is mindless and unconscious. Constant conscious awareness is hard work but it brings enormous rewards, one of which is access to your own creativity.

I believe that this creativity, or ability to be creative, is potentially immensely satisfying and very relevant to the work area, especially if we want that to be creative and productive.

So how can we tap into that? Apart from cleaning the bath in a different way and remembering the order in which you dress yourself, how do you start to be creative and what happens when you are?

The dictionary definition of creative is 'having the ability to create, characterised by originality of thought; having or showing imagination; designed to or tending to stimulate the imagination'.

Creative energy is life energy. And life energy is on a level of feeling and emotion. The life force is an energy of feeling and it directs through feelings and emotions. The things that you have the most energy for are the things that you want to do. So life energy and therefore creative energy come from desire, which itself is a feeling for some kind of satisfaction. It moves us in a specific direction. If you watch a flower opening towards the sun in the morning, that is the energy of desire, and through that, the flower is able to go through its life cycle. There is a biological blueprint and desire may simply be a part of that, but without the desire, nothing happens. If we had no desire for a mate, no children would be born, and it is the same for the flower.

So, feeling is where your life energy is. Feeling is what the life force comes from. Maybe feeling even **is** the life force. You do something because you want to. You buy something because you want it. You talk to someone because you are attracted to them. You travel abroad because you want to see and experience different countries and cultures. This is the energy you use to access creativity, and you use it with awareness.

As I understand it, we have two brains. One of these brains we have in common with all mammals and it is called the sub-cortical nervous system or the primitive brain. It is associated with the basic adaptive systems, that is to say, the hormonal and immune systems. Emotions and instincts are linked to the activity of the primitive brain. The other brain can be called the rational brain or the neocortex. It is a kind of supercomputer able to collect, put together, associate and store information. The neocortex has two hemispheres which are joined by the corpus callosum, and it seems that the right side of the neocortex is usually more directly in touch with the primitive brain. The left hemisphere is mainly concerned with analytical, logical thinking, language and mathematical work. The right is responsible for spatial relations, for arts, recognising faces and patterns. Its information handling is diffuse, simultaneous, pulling together many facets at once.

What we also have is a masculine and a feminine aspect of the psyche, which ties in with the right and left sides of the neocortex. Masculine qualities are thinking, knowledge, analysis, discipline, focus, determination, accomplishment and endeavour. Feminine qualities are to do with Eros, the principle of relatedness, intuition, feeling, emotion and instinct, myth, symbol and metaphor. If the female part of you has a creative idea or concept, it is the male in you which takes it out into the world and brings it into form. As an analogy, think of a couple who live together and love each other very much. The role of the female is to nourish and nurture. She's the one who stays at home and looks after what you have and sees what you need. Speaking metaphorically, she goes through the cupboards to see what is there and what needs to be there. She does that by feeling and being aware and by investigating. The role of the male is to protect, and to find ways to go out into the world and get what is needed. In turn, she supports him in doing this.

The seeds from a man and a woman need to come together to create new life. In the same way the inner male and female come together to create something new. The Chinese might categorise them as yin and yang, which are the opposite ends of the same whole, forced apart but always seeking to come together. So you have water and fire, both necessary for life, and opposite but complementary; sun and moon, light and shade, day and night, north and south, yes and no, and so on.

If you were given a project, for instance, you would ask the female 'how do you feel about this, do you have some ideas, can you relate it to any previous experience?'. She would then set to work and her antennae would bring in information, ideas and concepts, from within herself and from her environment; from the air, so to speak, where the ideas are floating around. It is the role of the male to put all of this diffuse information into a structure and make it concrete, and present it. To think of the words and put them on paper, of the physical structure and construct it. They work together and sometimes, when you are doing something, and you're really feeling good about it and the energy is flowing and everything is happening just as you want it to, and you feel complete – just for a while you don't need anything or anyone – then the two brains are functioning as one, and if you have an awareness of that it brings a feeling of satisfaction.

One of the things we can do with our lives is try to bring these two into balance and harmony, but this is difficult when one is so much more favoured than the other. At the moment our society and our education system and even our world view – the belief system that most of us have inherited and carry around with us – favour almost exclusively the functions of the left hemisphere.

Ever since Descartes said 'I think, therefore I am', logic, fact and analytical thought have been pursued, to the detriment of the feminine which says 'I feel, therefore I am'. It is essential to realise that we are whole beings and we need to acknowledge and use all the parts of ourselves in order to achieve our potential, and to access our creativity.

In order for people to be creative in the workplace, they have to have a certain amount of autonomy; they have to have some freedom to express themselves in what they do. This means being willing to take personal responsibility and it means being given opportunities to solve problems, run projects, have ideas for new ways of doing their tasks, in a creative way. This means dealing with change, learning to relish the challenges that change brings, making choices that contribute to personal growth as well as the growth of the company, and using logic and intuition (right and left brain). In this way companies and the people in them can flourish and prosper, and this is surely the objective. This also means the loss of some security, for companies and the people they comprise.

I spent some time in the local office of a large insurance company. Most of the staff were either in front of a PC all day or on the telephone. Management felt that they were well catered for. They had vending machines for a constant supply of drinks, a subsidised canteen, a coffee shop, and there were regular staff outings and social events. But while they were in the building they were unable to give themselves a break for relaxation. It took only a few seconds to go to the vending machine and back to the desk. The lights were on all day regardless of the weather because large filing cabinets were in front of the windows, blocking natural light. The food in the canteen was over-cooked, over-salted and over-sugared, while the coffee shop was in the basement – again with no natural light – with tables and chairs bolted to the floor, plastic flowers on the tables, with coffee and tea from more vending machines and in plastic cups. There was an attractive garden in an open square in the middle of the building, but it wasn't open for the staff except on hot summer days, to let some air into the building. You had the feeling of being trapped. There were so many things that company could have done, but the first and most important would be to ask its staff what they wanted. In that way people could think about what they wanted in the way of facilities for work and for refreshment, opening the way for innovation and change happening from within rather than imposed from without.

There are companies that function in a creative and productive way. At one American ice-cream company which has recently set up business in the UK, for instance, all company workers regularly swap jobs, so that every person can do any task that needs to be done and so, of course, no-one is indispensable. There is a large manufacturing company in São Paulo, Brazil where there are no secretaries, receptionists or PAs so everyone, including top managers, types letters, stands over photocopiers, fetches their own guests and makes their own coffee. Employees choose their own working hours and set their own salaries. There is a waiting list of people wanting to join this company, in any capacity, because it is understood that everyone has a contribution to make and every opportunity is given to create an avenue for each person to do that. In twelve years this company has multiplied itself six times, productivity has increased by 700 per cent and profits by 500 per cent, in a country with high inflation and a chaotic national economic policy.

A start can be made in very simple, small but fundamental ways. The potential in terms of the growth of the individual and the company is, quite literally, unlimited, because creative energy, expressed and channelled into activities that encourage freedom of choice, participation and some degree of autonomy, gives personal satisfaction in a way that nothing else does. It is unlimited because it flows, quite naturally, out of every person into the environment around. It also generates a lot of energy, both physical and emotional, and excitement. People start to enjoy themselves at work, they have ideas and think of new ways of doing things and new things to do, they communicate more honestly and openly and start to function on a higher level. They have fun. Maybe this is where seeing with the heart can come in, as people feel less threatened and have less need to compete and be the best. They are more expansive, more kind and more loving.

Imagine a company staffed from top to bottom (or inside to outside) by individuals whose creative energy is being directed into exercising their power of choice in the opportunities they have been given, taking personal responsibility, participating at every level and creating a kind of career dialogue, whether as a cook in the staff restaurant or a sales executive, with the company they work for. A dynamic is being created that is palpable, it is a pleasure to be there, and that is what makes it work.

Reference

Odent, M. (1990) *Water and Sexuality*, Harmondsworth: Arkana.

Chapter 3

Consciousness, well-being and the senses

Derek Clements-Croome

Practice powerfully affects performance for simple and complex tasks
(Pashler, 1998)

For an organisation to be successful and to meet the necessary targets, the performance expressed by the productivity of its employees is of vital importance. In many occupations people work closely with computers within an organisation which is usually housed in a building. Today, technology allows people to work easily while they are travelling, or at home, and this goes some way to improving productivity. There are still, however, many people who have a regular workplace which demarcates the volume of space for private work but is linked to other workplaces as well as to social and public spaces. People produce less when they are tired; have personal worries; suffer stress from dissatisfaction with the job or the organisation. The physical environment can enhance one's work, but an unsatisfactory environment can hinder work output.

Concentration of the mind is vital for good work performance. Absolute alertness and attention are essential if one is to concentrate. There is some personal discipline involved in attaining and maintaining concentration, but again the environment can be conducive to this by affecting one's mood or frame of mind; however, it can also be distracting and can contribute to a loss of concentration.

A number of personal factors which depend on the physical and mental health of an individual, and a number of external factors which depend on the environment and work-related systems, influence the level of productivity. Experimental work on comfort often looks at responses of a **group** as a whole and this tends to mask **individuals**' need for sympathetic surroundings to work and live in. People also need to have a fair degree of personal control about various factors in their environment. They react to the environment as a whole, not in discrete parts, unless a particular aspect is taxing the sensory system.

Productivity can be measured in **absolute** or **direct** terms by measuring the speed of working and the accuracy of outputs by designing very controlled experiments with well-focused tests. **Comparative** measures use scales and questionnaires to assess the individual opinions of people concerning their work and environment. **Combined** measures can also be employed, using for example some physiological measure such as brain rhythms to see whether variations in the patterns of the brain responses correlate with the responses assessed by questionnaires.

The nature of consciousness

> 'We do not understand how the mind works – not nearly as well as we understand how the body works, and certainly not well enough to design utopia or to cure unhappiness.' (Pinker, 1997)

How do the neural processes occurring in our brains while we think and act in the world relate to our subjective sensations? Crick and Koch (1997) believe that this is a central mystery of human life. The fundamental question that needs to be understood is the relationship between the mind and the brain. We are conscious or aware of events central to our attention or concentration at any one time. Often there are peripheral events which feature in only a fleeting way in our consciousness unless they manage to distract us. The ability to focus the concentration or alertness for a particular event, such as the work we are undertaking, is an important issue when discussing productivity. For high productivity we need high and sustained levels of concentration centred on the task being carried out. There are many short-term, medium-term and long-term factors which can contribute towards lowering productivity and these include low self-esteem, low morale, an inefficient work organisation, poor social atmosphere or environmental aspects such as excessive heat or noise. Factors which lower productivity, by distracting our attention and diluting concentration, include lethargy, headaches and physical ailments. These factors all feature in surveys carried out on building sickness syndrome.

Crick and Koch (1997) discuss visual consciousness in trying to reach an understanding about how the brain interprets the visual world based on the information perceived by the visual system. Past experience evolved through living, or from our genes, features strongly in our responses to the environment around us. The stimuli from the environment trigger this system and arouse our consciousness to various levels of concentration. The human perceptual sensory systems process information from visual, sound, touch, smell and taste sensations. Our surroundings create a sensory experience and hence must have some effect on the way we work. Conditions external to the body can disturb these systems; internal disturbances due to drugs or alcohol, for example, can also upset the response to the world around us.

Greenfield (1997) believes that consciousness is impossible to define. She goes on to state that neuronal connectivity is a very important feature of the brain, which means that it is the connections rather than the neurons themselves that are established as a result of experience. It is the pattern of these experience-related connections that largely distinguishes the individuality of a human being. The pattern of these connections can change quite rapidly. It appears that the brain remains adaptable and sensitive to experience throughout a person's life. Consciousness changes as biorhythms and flows of hormones alter through daily cycles. Greenfield (1997) concludes that a critical factor could be the number of neurons that are gathered up at any one time, and it is this which determines one's consciousness.

The state of knowledge about environmental factors is uneven. It is probably fair to say that there is a high level of knowledge about how heat, light and sound affect our thermal, visual or auditory responses. There is much less information about how we react to **combinations** of these stimuli and also about how electromagnetism, geomagnetism and chemical fields affect the sensory system. An added complication is that human responses are partly physiological and partly psychological. This makes the measurement of responses difficult because objective measures of lighting or temperature levels are comparatively easy, but assessing people's judgements about preferred or acceptable levels of light and heat is much more difficult. Yet another complication is that reception of information from visual images, music or speech, smells or touch interact with one another. The sense organs extend beyond the eyes, ears, nose, mouth and skin and include the vestibular organs concerned with orientation, posture and locomotion, as well as a variety of respiratory and thermo receptors which respond to air quality, pressure and temperature.

Our response to the world around us occurs at various levels. For example, a cartoon can depict recognisable people from the skimpiest of outlines. The outline form and a few added clues about detail are all that is required to recognise the person being represented. Contrast this with a portrait by Rembrandt, for example, in which colour, texture and shading give much more detailed information which triggers higher orders of aesthetic and emotional response. Likewise, the quality of the environment has a basic structure upon which is superimposed more detail. For example, air movement can be represented at a basic level by a mean velocity, but a more complete picture would refer to the degree of turbulence, the peak as well as the mean velocity, and the periodicity of the air-flow. Crick and Koch (1997) describe various levels of representations that occur in the visual field. They suggest that there may be a very transient form or fleeting awareness that represents simple features and does not require an attention–awareness mechanism. The renowned psychologist William James believed that consciousness was not **a thing** but was **processed thought**, which involved attention and short-term memory. From brief awareness the brain constructs a view-centred representation and the visual inputs awaken a greater level of attention. Crick and Koch go on to suggest that this in turn probably leads to a three-dimensional object representation and thence to more cognitive ones.

In the design of the productive workplace an attempt is being made to set conditions which allow selected information to be perceived and transmitted quickly through the human perceptual sensory system. This pathway must not be trammelled by extraneous information from peripheral stimuli. Efficient work processes and organisation, and controllable environmental conditions can help this process given that the person is healthy in mind and body and there is no interference at a social level from any other person.

Chalmers (1996) asks the following questions about consciousness.

- How can a human subject discriminate sensory stimuli and react to them appropriately?
- How does the brain integrate information from many different sources and use this information to control behaviour?
- How is it that subjects can verbalise their internal states?
- How do physical processes in the brain give rise to subjective experience?

Crick and Koch (1997) suggest that **meaning** is derived from the linkages among the various representations of the neuron firing fields which are spread through the cortical system in a vast network, equivalent to a huge database which is changing as the experience of the individual increases throughout life. Changes bring about the process of learning. However, many questions remain unanswered. The existence of consciousness does not seem to be derivable from physical laws, and because consciousness is strongly subjective, there is no direct way to monitor it, although questionnaires and semi-structured interviews are techniques which are employed. Chalmers (1996) goes on to say that it is possible to use people's descriptions of their own experiences. There have been several surveys of productivity using self-assessment techniques. This is a valid procedure as long as the subject attempts to structure the information output in an objective fashion.

Recent work at Rochester University in New York suggests that the mind can affect the immune system. Stress can decrease the body's defences and increase the likelihood of illness, resulting in a lowering of well-being. Stresses come from a variety of sources: the organisation, the job, the person and the environment. It is likely that building sickness syndrome is triggered by unfavourable combinations of environmental conditions which stress the mind and body and lower the immune system, leaving the body more sensitive to environmental conditions. The biological chain seems to be that stress acts on mind and brain, to which the hypothalamus reacts and the hormone ACTH is released, and then the hormone cortisol in the blood increases to a damaging level. This chain of events interferes with human performance and consequently productivity is lowered.

Architecture and the senses

Buildings should be a multi-sensory experience. This section is strongly influenced by the book entitled *The Eyes of the Skin: Architecture and the Senses* by Juhani Pallasmaa (1996), who elegantly describes this belief in great detail. All students of building design, whether architects or engineers, should read this book.

During the Renaissance the five senses were understood to form a hierarchical system from the highest sense of vision down to touch. This reflects the image of the cosmos in which vision is correlated with fire and light, hearing with air, smell with vapour, taste with water and touch with earth. It is by vision and hearing that we acquire most of our information from the world around us. But one should not underestimate the importance of the other senses, such as the olfactory enjoyment of a meal or the fragrance of flowers, and responses to temperature, which all provide a bank of sensory experience that helps to mould our attitudes and expectancies about the environment. The senses not only mediate information for the judgement of the intellect; they are also channels which ignite the imagination. This aspect of thought and experience through the senses, which trigger the body and mind, is stimulated by the environment and people around us but, when we are inside a building, it is the architecture of the space which sculpts the outline of our reactions. Merleau-Ponty said that the task of architecture was **to make visible how the world touches us.**

In Buddhism there are nine levels of consciousness (Allwright 1998):

- the **five senses** felt by the eyes, ears, nose, mouth and skin
- the **integration of senses** using reason and logic
- **rational thought** expressed via self-awareness and intuition
- the **stores of experience** in the short- and long-term memories
- **pure consciousness** within the inner self; this also involves emotion.

At the heart of architecture is the fundamental question of how buildings in their design and use can confront the questions of human existence in space and time and thus express and relate to humans being in the world. If this question is ignored the result is soulless architecture which is a disservice to humanity. There is a danger, for example, that the ever-increasing pace of technology is distorting natural sociological change and this makes it difficult for modern architecture to be coherent in human terms.

Buildings must relate to the language and wisdom of the body. If they do not they become isolated in the cool and distant realm of vision. For example, people passing by a building gain a visual impression which they like, dislike or have no particular feeling about. Buildings are a vital part of a nation's heritage and so they are historically important. This is in stark contrast to the sculptor, whose work can be selected and located according to individual choice. But in assessing the value of building, how much attention is paid to the quality of the environment inside the building and its effects on the occupants? The qualities of the environment affect human performance inside a building and these should always be given a high priority. This can be considered as an **invisible aesthetic**, which together with the **visual impact** makes up a **total aesthetic**.

Multi-sensory experience

Buildings should provide a multi-sensory experience for people and uplift the spirit. A walk through a forest is invigorating and healing due to the interaction of all sense modalities; this has been referred to as the **polyphony of the senses.** One's sense of reality is strengthened and articulated by the interaction of the senses. Architecture is an extension of nature into the man-made realm and provides the ground for perception, and the horizon to experience from which one can learn to understand the world. Buildings filter the passage of light, air and sound between the inside and outdoor environments; they also mark out the passage of time by the views and shadows they offer to the occupants. Pallasmaa (1996) gives an example to illustrate this point. He believes that the Council Chamber in Alvar Aalto's Säynätsalo Town Hall recreates a mystical and mythological sense of community where darkness strengthens the power of the spoken word. This demonstrates the very subtle interplay between the senses and how environmental design can heighten the expression of human needs within a particular context.

Although the five basic senses are often studied as individual systems covering visual, auditory, taste, smell, orientation and the haptic sensations, there is an interplay between the senses. For example, eyes want to collaborate with the other senses. All the senses can be regarded as extensions of the sense of touch, because the senses as a whole define the interface between the skin and the world. The combination of sight and touch allows the person to get a scale of space, distance or solidity.

Qualitative attributes in building design are often only considered at a superficial level. For example, in the case of light the level of illuminance, the glare index and the daylight factor are normally taken into account. But in great spaces of architecture there is a constant deep breathing of shadow and light; shadow inhales, whereas illumination exhales light. The light in Le Corbusier's Chapel at Ronchamps, for example, gives the atmosphere of sanctity and peace. How should we consider hue, saturation and chroma in lighting design, for example?

Buildings provide contrast between interiors and exteriors. The link between them is provided by windows. The need for windows is complex but it includes the need for an interesting view, contact with the outside world and, at a fundamental level, it provides contrast for people carrying out work in buildings. Much work today is done by sitting at computers at close quarters and requires eye muscles to be constrained to provide the appropriate focal length of vision, whereas when one looks outside towards the horizon the eyes are focused on infinity and the muscles are relaxed. There are all kinds of other subtleties, such as the need to recreate the wavelength profile of natural light in artificial light sources, which need to be taken into account. Light affects mood. How can this be taken into account in design?

The surfaces of the building set the boundaries for sound. The shape of the interior spaces and the texture of surfaces determine the pattern of sound rays throughout the space. Every building has its characteristic sound of intimacy or monumentality, invitation or rejection, hospitality or hostility. A space is conceived and appreciated through its echo as much as through its visual shape, but the acoustic concept usually remains an unconscious background experience. At midnight on 31 December 1999, it is intended that the sound of church bells will ring and echo throughout the land. The sound will be a powerful uniting experience for the nation. It will make us aware of our citizenship and awaken any patriotic feelings that we have within us. Recall what you feel when hearing an organ in a cathedral, or the burst of applause at the end of a concert, or the cries of seagulls in a harbour.

It is said that buildings are composed as the architecture of space, whereas music represents the architecture of time. The sense of sound combines the threads of this notion. Without people and machines, buildings are silent. Buildings can provide sanctuary or peace and isolate people from a noisy, fast-moving world. The ever-increasing speed of change can temporarily be reduced by the atmosphere created in a building. The opposite is true when working with computers or watching television, for example. Architecture emancipates us from the embrace of the present and allows us to experience the slow healing flow of time. Again, buildings provide the contrast between the passing of history and the time-scales of life today.

The most persistent memory of any space is often its odour. 'Walking through the gardens of memory, I discover that my recollections are associated with the senses', wrote the Chilean writer Isabel Allende (*The Times*, April 1998). Every building has its individual scent. Our sense of smell is acutely sensitive. Strong emotional and past experiences are awakened by the olfactory sense. Again think of the varying olfactory experiences such as in a leather shop, a cheese stall, an Indian restaurant, a cosmetic department or a flower shop; all awaken our memories and give, or do not give, pleasure. Wine and whisky connoisseurs know that flavour is best sensed using the nose, whereas texture is sensed in the mouth. Odours can influence cognitive processes which affect creative task performance as well as personal memories. These tasks are influenced by moods, and odours can affect these also (Warren and Warrenburg, 1993; Erlichman and Bastone, 1991; Baron, 1990).

Various parts of the human body are particularly sensitive to touch. The hands are not normally clothed and act as our touch sensors. But the skin of the body reads the texture, weight, density and temperature of our surroundings. Proust gives a poetic description of a space of intimate warmth next to a fireplace sensed by the skin: 'it is like an immaterial alcove, a warm cave carved into the room itself, a zone of hot weather with floating boundaries'.

There is a subtle transference between tactile, taste and temperature experiences. Vision can be transferred to taste or temperature senses; certain odours, for example, may evoke oral or temperature sensations. The remarkable world-famous percussionist Evelyn Glennie is deaf but senses sound through her hands and feet and other parts of her body. Marble evokes a cool and fresh sensation. Architectural experience brings the world into a most intimate contact with the body.

The body knows and remembers. The essential knowledge and skill of the ancient hunter, fisherman and farmer, for example, can be learnt at a particular time but, more importantly, the embodied traditions of these trades have been stored in the muscular and tactile senses. Architecture has to respond to ways of behaviour that have been passed down by the genes. Sensations of comfort, protection and home are rooted in the primordial experiences of countless generations. The word 'habit' is too casual and passes over the sequels of history embedded within us. Isabel Allende describes the idea for her book *Aphrodite* (1998) as being 'a mapless journey through the regions of sensual memory'.

From early times the fire has been a symbol of human multi-sensory experience. The fire gave light in darkness; it produced warmth for the body and heat for cooking; it was a protection from hungry animals of prey; it was a social focus. A campfire today, as throughout history, enhances our well-being and uplifts the spirits.

Well-being and productivity

Myers and Diener (1997) have been carrying out systematic studies about awareness and satisfaction with life among populations. Psychologists often refer to this as **subjective well-being**. Findings from these studies indicate broadly that those that report well-being have happy social and family relationships; are less self-focused; less hostile and abusive; and less susceptible to disease. It appears also that happy people typically feel a satisfactory degree of personal control over their lives, whether in the workplace or at home. It is probably fair to assume that it is more likely that the work output of a person will be high if their well-being is high. Jamison (1997) reviews research which links manic-depressive illness and creativity. Many artists such as the poets Blake, Byron and Tennyson, the painter Van Gogh and the composer Robert Schumann are well-known examples of manic-depressives. The work output of such people is distinguished but lacks continuity. Mozart and Schubert were not classified as manic-depressives and their work output was consistently high throughout their short lifespans. In contrast, Robert Schumann was very prolific during 1839–41 and 1845–53, whilst suffering hypomania throughout 1840 and 1849. Between these timespans he had suffered from severe depression, and before 1838 and after 1853 made suicide attempts.

In the workplace one is not expecting creativity at this level of genius; rather, it is hoped that there will be a consistently high standard of work performance. It is interesting to consider some case studies of the competitors at the Mind Sports Olympiad held in London during August 1997 (Henderson, 1997). One competitor could memorise a pack of 52 cards in just over 30 seconds. His daily routine involves running four miles a day; no alcohol for six weeks before a tournament, during which he eats a lot of pasta and other high-carbohydrate food to keep the blood sugar level high; he also takes regular doses of a Chinese herb called *Ginkgo biloba* to improve blood circulation; he practises the trance-like state which is needed to perform his memory feats and has regular brain scans. This competitor believes in lowering his brain activity to the optimum concentration level for this type of feat. This means that his brain activity rate is reduced to 5–7 Hz, which allows a higher degree of meditative-type concentration than the normal brain activity of 12–14 Hz. Another competitor stated that he had a meat-free diet and a fitness programme, ran marathons and played tennis matches and dived to improve concentration. Diving is about poise and balance, and requires the same sort of mental rigour that is needed for competition draughts.

Tony Buzan is one of the organisers of the Olympiad, and believes that mind training techniques can open up a new sphere of mental fitness, which needs to be integrated with a physical fitness programme. The brain uses 40 per cent of the body's oxygen, and a healthy body promotes brain activity. Buzan goes on to say that the imagination can do for the mind what weight training can do for the body. Everyone can do concentration exercises almost anywhere by, for example, watching a vase of flowers and concentrating on every detail, then closing his or her eyes and imagining it.

Well-being reflects one's feelings about oneself in relation to one's world. Warr (1998) proposes a view of well-being which comprises three scales: pleasure to displeasure; comfort to anxiety; enthusiasm to depression. There are job and outside-work attributes which characterise one's state of well-being at any point in time and these can overlap with one another. Well-being is only one aspect of mental health; other factors include personal feelings about one's competence, aspirations and degree of personal control.

People do not have to be Olympiad competitors to get more out of their work. Townsend (1997) states that 25 per cent of us enjoy our work but the rest of us do not. Productivity suffers as a consequence, due to the workplace being more a place of conflict and dissatisfaction. Lack of productivity shows up in many ways, such as absenteeism, arriving late and leaving early, over-long lunch breaks, careless mistakes, overwork, boredom, frustration with the management and the environment. In the same way as the Olympiad competitors aim to focus their mind completely on the task in hand, Townsend (1997) believes we can all try to do this and, when we succeed, the whole body feels different.

Townsend (1997) goes on to say that people in the workplace can be encouraged to use both halves of their brain. The left-hand part is concerned with logic, whereas the right-hand side is concerned with feeling, intuition and imagination (Ornstein, 1973). If logic and imagination work together, problem-solving becomes more enjoyable and more creative. Of course some people thrive on change while others prefer to do repetitive types of work. There seems to be no doubt that the industrial and commercial worlds can play a leading role in increasing the awareness of their workforce of all of these possibilities. It is also important to start this way of thinking in school children.

References

Allwright, P. (1998) *Basics of Buddhism*. UK: Taplow Press.

Baron, R. (1990) Environmentally Induced Positive Effect; Its Impact on Self-efficacy, Task Performance, Negotiation and Conflict, *J. Appl. Soc. Psychol.*, **16**, 16–28.

Chalmers, D.J. (1996) *The Conscious Mind; In Search of a Fundamental Theory*. Oxford: Oxford University Press.

Crick, F. and Koch, C. (1997) The problem of consciousness, *Scientific American*, Special Issue 'Mysteries of the Mind', Jan. 19–26.

Erlichman, H. and Bastone, L. (1991) *Odour Experience as an Affective State*, Report to the Fragrance Research Fund, New York.

Greenfield, S. (1997) How might the Brain Generate Consciousness? *Communication Cognition*, **30**, 3–4, 285–300.

Henderson, M. (1997) Mental athletes tone their bodies to keep their minds in shape, *The Times*, 19 August, 6.

Jamison, K.R. (1997) Manic Depressive Illness and Creativity, *Scientific American*, Special Issue, 'Mysteries of the Mind', Jan., 44–49.

Myers, D.G. and Diener, E. (1997) The Pursuit of Happiness, *Scientific American*, Special Issue, 'Mysteries of the Mind', Jan., 40–49.

Ornstein, R.E. (1973) *The Nature of Consciousness*. London: Viking.

Pallasmaa, J. (1996) *The Eyes of the Skin: Architecture and the Senses*. London: Academy Editions.

Pashler, H.E. (1998) *The Psychology of Attention*. Cambridge, MA: MIT Press.

Pinker, S. (1997) *How the Mind Works*. London: Allen Lane.

Townsend, J. (1997), How to draw out all the talents, *The Independent*, tabloid section, 24 July, 17.

Warr, P. (1998) *What is our Current Understanding of the Relationships between*

Well-Being and Work?, Economics and Social Sciences Research Council Seminar Series at Department of Organisational Psychology, Birkbeck College, London (ed. R. Briner), 22 Sept. and *Journal Occ. Psychol.* (1990) **63**, 193–210.
Warren, C. and Warrenburg, S. (1993) Mood benefits of fragrance, *Perfumer and Flavourist*, **18**, Mar./Apr., 9–16.

Chapter 4

Pleasure and joy, and their role in human life

Michel Cabanac

Experiments on human subjects showed that the perception of sensory pleasure can serve as a common currency to allow the trade-off among various motivations for access to behaviour. The trade-offs between various motivations would thus be accomplished by simple maximisation of pleasure. A common currency for motivations as different from one another as physiological, ludic, social, aesthetic, moral, and religious is necessary to permit competition for access to behaviour. Therefore, all motivations can be compared to one another from the amount of pleasure and displeasure they arouse. It follows that the main properties of sensory pleasure should belong also to joy. Indeed, joy and sensory pleasure share identical properties; they are contingent, transient, and they index useful behaviours.

The behavioural final common path

One basic postulate of ethology is that behaviour tends to satisfy the most urgent need of the behaving subject (Tinbergen, 1950; Baerends, 1956). One shortcoming of the theories of the optimisation of behaviour proposed by ethologists and behavioural ecologists is that the mechanism by which behaviour is optimised is never mentioned. In other words, they do not explain **how** the subject 'decides'. The notion of **behavioural final common path** is a first step on the way leading to an answer to that question. Paraphrasing Sherrington's image of the motoneuron final common path of all motor responses, McFarland and Sibly (1975) pointed out that behaviour is also a final common path on which all motivations converge. This image incorporates all motivations into a unique category since behaviour must satisfy not only physiological needs but also social, moral, aesthetic, playful motivations. Indeed, it is often the case that behaviours are mutually exclusive; one cannot work and sleep at the same time. Therefore, the brain, responsible for the behavioural response, must rank priorities and determine trade-offs in the decisions concerned with allocating time among competing behaviours. The brain can be expected to operate this ranking by using a **common currency** (McFarland and Sibly, 1975; McNamara and Houston,

1986). In the following pages pleasure and joy will be proposed as this common currency.

Sensory pleasure

Sensory pleasure possesses several characteristics: pleasure is contingent, pleasure is the sign of a useful stimulus, pleasure is transient, pleasure motivates behaviour. In the commerce of a subject with stimuli, it has been shown experimentally that the wisdom of the body leads the organism to seek pleasure and avoid displeasure, and thus achieve behaviours which are beneficial to the subject's physiology (Cabanac, 1971). Relations exist between pleasure and usefulness and between displeasure and harm or danger. For example, when subjects are invited to report verbally, the pleasure aroused by a skin thermal stimulus can be predicted knowing deep body temperature (Cabanac *et al.*, 1972; Attia, 1984) (Fig. 4.1).

A hypothermic subject will report pleasure when stimulated with moderate heat, and displeasure with cold. The opposite takes place in a hyperthermic subject. Pleasure is actually observable only in transient states, when the stimulus helps the subject to return to normothermia. As soon as the subject returns to normothermia, all stimuli lose their strong pleasure component and tend to become indifferent. Sensory pleasure and displeasure thus appear especially suited to being a guide for thermoregulatory behaviour (Fig. 4.2).

The case of pleasure aroused by eating shows an identical pattern (Fig. 4.3). A given alimentary flavour is described as pleasant during hunger and

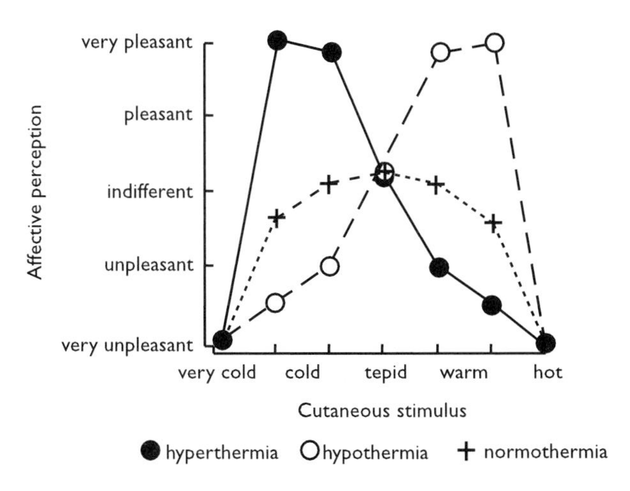

Figure 4.1 Pleasure (positive ratings) and displeasure (negative ratings) reported by a subject in response to thermal stimuli presented for 30 s on the left hand

Source: Cabanac (1986).

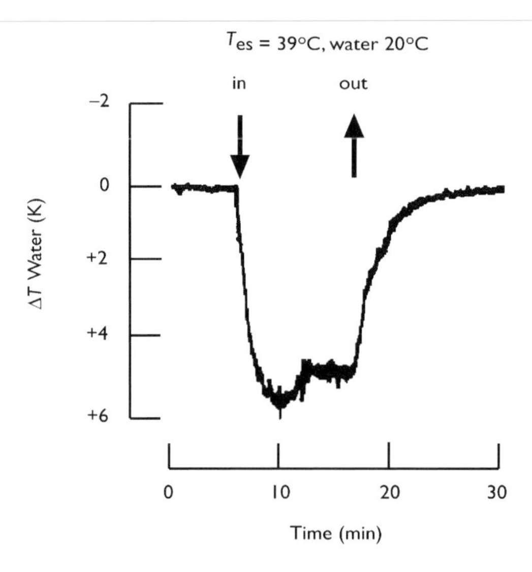

Figure 4.2 Direct calorimetry of the heat loss by a hypothermic subject's hand when dipped into highly pleasant cold water at 20°C: the plateau around min 12 is a steady state, after initial deflection, when heat flow taken from the hand by the cold water is equal to heat flow brought to the hand by arterial blood (T_{es}, oesophageal temperature). Knowing the flow of water running round the hand, it is easy to calculate that the heat lost by the hand corresponds to ~73 W

Source: Cabanac *et al.* (1972).

becomes unpleasant or indifferent during satiety. Measurement of human ingestive behaviour confirms the above relationship of behaviour with pleasure; it has been repeatedly demonstrated in the case of food intake (Fantino, 1984, 1995), that human subjects tend to consume foods that they report to be pleasant and to avoid foods that they report to be unpleasant. Pleasure also shows a quantitative influence: the amount of pleasurable food eaten is a function of alimentary restrictions and increases after dieting. The result is that pleasure scales can be used to judge the acceptability of food.

Thus, in the cases of temperature and taste, the affective dimension of sensation depends directly on the biological usefulness of the stimulus to the subject. This was already noticed by Aristotle (quoted by Pfaffmann, 1960). The word 'alliesthesia' was coined to describe the fact that the affective dimension of sensation is contingent, and to underline the importance of this contingency in relation to behaviour (Cabanac, 1971): a given stimulus will arouse either pleasure or displeasure according to the internal state of the stimulated subject. The seeking of pleasure and the avoidance of displeasure lead to behaviours with useful homeostatic consequences. Garcia *et al.*

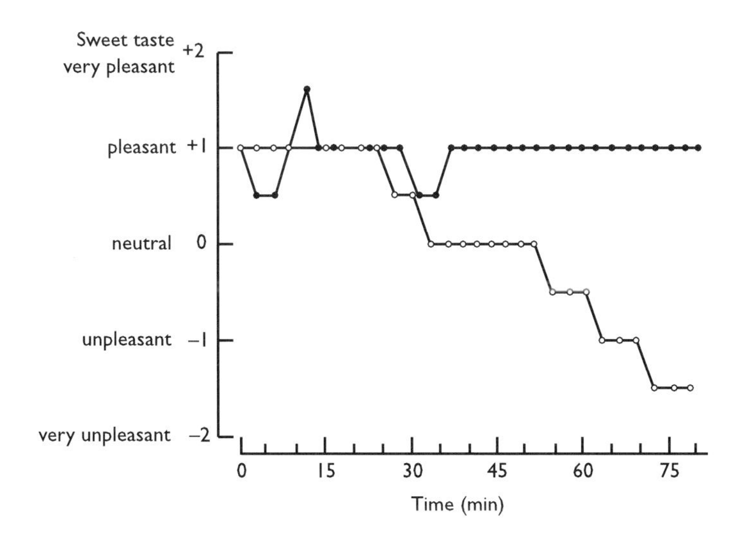

Figure 4.3 Pleasure (positive ratings) and displeasure (negative ratings) reported by a fasted subject in response to the same gustatory stimulus, a sample of sweet water presented repeatedly every third minute. Solid symbols, the subject expectorated the samples after tasting; open symbols, the subject swallowed the samples and thus accumulated a heavy sucrose load in his stomach. It can be seen that the same sweet taste that first aroused pleasure in the subject aroused displeasure once the subject was satiated

Source: Cabanac (1971).

(1985) have shown how past history, such as illness induced in association with the taste of an ingested substance, can 'stamp in' a change in the affective quality of that taste. The behaviour of subjects instructed to seek their most pleasurable skin temperature could be described and predicted from their body temperatures, and the equations describing their behaviour were practically the same as those describing autonomic responses such as shivering and sweating (Cabanac *et al.*, 1972; Bleichert *et al.*, 1973; Marks and Gonzalez, 1974; Attia and Engel, 1981).

It is possible therefore from verbal reports to dissociate pleasure from behaviour and to show thus that the seeking of sensory pleasure and the avoidance of sensory displeasure lead to behaviours with beneficial homeostatic consequences. Pleasure thus indicates a useful stimulus and simultaneously motivates the subject to approach the stimulus. Pleasure serves both to reward behaviour and to provide the motivation for eliciting behaviour that optimises physiological processes. One great advantage of this mechanism is that it does not take rationality or a high level of cognition to produce a behaviour adapted to biological goals. ('Rational' is understood, here, in its philosophical acceptation (i.e. reason), and not in its narrower economical sense.) Indeed, conditioned food aversion can be induced during sleep and under anaesthesia (Garcia, 1990). As soon as a stimulus is discriminated, the affective dimension of the sensation aroused tells the subject, animal or human, that the stimulus should be sought or avoided.

Pleasure and comfort

Let us keep the examples taken from temperature to understand the difference between comfort and pleasure. Comfort is a general feeling, whereas sensory pleasure applies to the sensation aroused by a precise stimulus. Thermal comfort used to be defined as the 'subjective satisfaction with the thermal environment' (Bligh and Johnson, 1975). However, Fig. 4.4 shows that this definition is inadequate. This figure simplifies the experimental results presented in Fig. 4.1. It can be seen that the cases of 'subjective satisfaction with the thermal environment', boxes P and I, represent a heterogeneous category including pleasure and indifference. Pleasure, P boxes, occurs when there is an internal trouble, hypothermia or hyperthermia. In these cases a pleasant stimulus, e.g. warm skin in a hypothermic subject, tends to correct the trouble, i.e. results in normothermia, and then turns unpleasant, i.e. to follow up with our example, warm skin in a normothermic subject. Pleasure provides 'subjective satisfaction with the thermal environment', but the situation is highly unstable. This led to the new definition of thermal comfort as 'subjective indifference to the thermal environment' (IUPS Commission for Thermal Biology, 1987). Defined this way, comfort is stable and can last and is clearly different from pleasure whose characteristic is to be transient.

Internal state \ Stimulus	cold	neutral	warm
hypothermia	U	I	P
normothermia	U	I	U
hyperthermia	P	I	U

Figure 4.4 Fig. 4.1 data simplified to a 3 × 3 matrix: the affective dimension of thermal sensation depends on the subject's internal state. A thermal stimulus feels unpleasant (U), indifferent (I) or pleasant (P) depending on body core temperature. Pleasure occurs only in dynamic situations when a stimulus tends to correct an internal trouble

Source: Cabanac (1986).

A feeling of comfort indicates therefore that everything is right, but this is not a very exciting feeling, whereas pleasure indicates, in a troubled situation, a useful stimulus that should be consumed but will not last once the trouble is corrected.

Conflicts of motivations

If pleasure indicates usefulness, it would be of interest to explore situations with simultaneous and possibly conflicting multiple motivations. Several experiments were conducted where one motivation was pitted against another (e.g. sweet *v.* sour, temperature *v.* fatigue, chest *v.* legs). In all these experiments the subjects' behaviours were repeatedly coherent. In the bidimensional sensory situations imposed by the experimenters the subjects described maps of bidimensional pleasure in sessions where their pleasure was explored. They tended to move to the areas of maximal pleasure in these maps, in sessions where their behaviour was explored.

Thus, subjects in conflict situations tend to maximise their sensory pleasure as perceived simultaneously in both dimensions explored. In these experiments the subjects tended to maximise the algebraic sum of their sensory pleasure, or to minimise their displeasure. As a corollary of this observation, it can be stated that, in a situation of conflict of motivations, one can predict the future choice of the subject from the algebraic sum of affective ratings of pleasure and displeasure given by the subject to the conflicting motivations. This theoretical situation is presented in Fig. 4.5.

Such a prediction is not surprising if one considers that, at each instant, all motivations are ranked to satisfy only the most urgent and that there must exist a common currency to actuate the behavioural final common path (McFarland and Sibly, 1975; McNamara and Houston, 1986; Cabanac, 1992). The results of the above experiments show that sensory pleasure fulfilled the conditions required of a common motivational currency, at least in the case of the behaviours selected which have clear physiological implications.

From pleasure and comfort to joy and happiness

Let us now see another implication of the behavioural final common path and the common currency. In everyday life, physiological motivations must compete with other motivations: social, ludic, religious, etc. Since pleasure was the common currency allowing trade-offs among various physiological motivations, the same currency must also allow comparison of all motivations in order to rank them by order of urgency. In turn, the transience of pleasure can be found also in other aspects of consciousness than sensation. Happiness is considered generally as the aim of life. Yet, the pursuit of happiness is fallacious if one does not know what happiness is. In the same

	Resulting affective experience	Action
Behaviour 1	a → A	yes
Behaviour 2	B → b	no
Behaviour 1 + Behaviour 2	a + B → A + b	yes

with $a < A$, and $B > b$, and with $a + B < A + b$

Figure 4.5 Mechanism by which a behaviour (behaviour 1) that produces displeasure can be chosen by a subject if another behaviour (behaviour 2) that produces pleasure is simultaneously chosen. The necessary and sufficient condition for behaviour 2 to occur (action) is that the algebraic sum of affective experience (pleasure) of the yoked behaviours is positive ($a + B < b + A$). Capital letters indicate larger pleasure than lower-case letters

Source: Cabanac (1992).

way as there are two different elements in sensation–sensory pleasure (highly positive but transient) and comfort (indifferent but stable) – it is possible to recognise two elements in the affectivity of global consciousness: positive and transient joy, and indifferent but stable happiness (Cabanac, 1986, 1995). Happiness is to joy what comfort is to pleasure (Table 4.1). This duality explains the disappointment expressed by many a writer and some philosophers when they deal with happiness. They use the word 'happiness' to describe a pleasant experience which they expect to last and, when they see that it is transient, erroneously conclude that happiness does not exist. They should have used the unambiguous word 'joy' (for additional discussion on the nature of happiness, see Cabanac (1986, 1995)). Thus, the transience of pleasure can be found also with joy.

Table 4.1 Joy is to happiness what pleasure is to comfort

	Sensation	*Consciousness*
pleasant	pleasure	joy
indifferent	comfort	happiness

Optimisation of behaviour

The word 'optimality' applied to behaviour can be ambiguous because it bears somewhat different meanings when used by ethologists, economists and physiologists (Lea *et al.*, 1987). Ethologists differentiate between goal and cost. Economists differentiate between utility and cost. The goal of a subject, as well as utility, is some entity that an optimal behaviour will tend to maximise, which may appear tautological to the physiologist. The cost is a characteristic of the environment that the optimal behaviour will tend to minimise. All would agree that an optimal behaviour gives the maximal net benefit (or fitness) to the behaving individual. Specialists diverge in their definition of benefit (or fitness). The benefit can be defined in terms of reproductive efficacy (Krebs and Davies, 1981) as well as financial profit and physiological function (McFarland, 1985). We are concerned here with this last aspect: physiological benefit.

To the physiologist a behaviour is optimal when it leads to homeostasis. Optimal behaviour could be recognised easily when subjects instructed to seek their most pleasurable skin temperature selected stimuli which, after data analysis, could be described by mathematical models identical to the models describing the autonomic responses (Cabanac *et al.*, 1972; Bleichert *et al.*, 1973; Marks and Gonzalez, 1994), or when the best performance coincided with minimal discomfort (Fig. 4.6). One may wonder whether

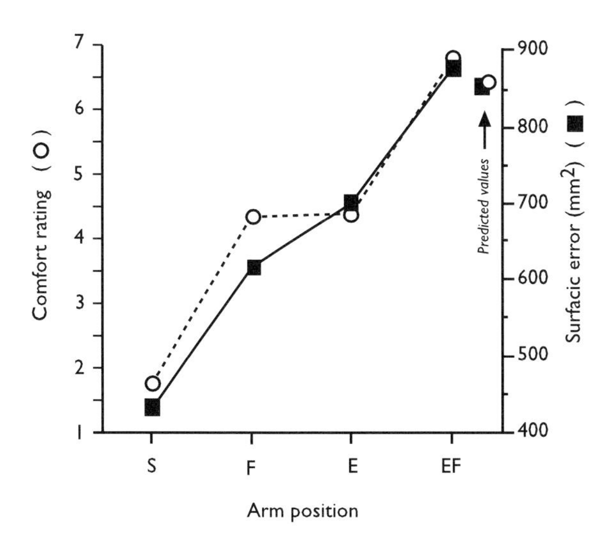

Figure 4.6 Increase in discomfort rate and in the error of performance observed over various arm positions: in S, the subject adopted the most comfortable natural position; in F, the wrist was maximally flexed; in E, the arm was maximally extended vertically; in EF, the subject adopted both F and E

Source: Rossetti *et al.* (1994).

optimisation, as seen from the physiological point of view, was also achieved in the experiments quoted above where subjects maximised the algebraic sum of two modalities of sensory pleasure. Physiological criteria of optimisation showed that maximisation of pleasure was the key to optimal behaviour in the experimental conflicts of motivations studied.

The hypothesis according to which pleasure would also signal optimal mental activity was tested empirically (Fig. 4.7). Ten subjects played video-golf on a Macintosh computer. After each hole they were invited to rate their pleasure or displeasure on a magnitude estimation scale. Their ratings of pleasure correlated negatively with the difference (par minus their performance), i.e. the better the performance, the more pleasure was reported. This result would indicate that pleasure is aroused by the same mechanisms, follows the same laws in physiological and cognitive mental tasks, and leads to the optimisation of performance.

Conclusions

Given that pleasure is an index of the physiological usefulness of a stimulus and of optimal function in a mental activity, the law of the common currency renders it inevitable that joy is the index of a useful conscious event. The relationship of joy with usefulness may become non-univocal with the increased complexity of the mind process.

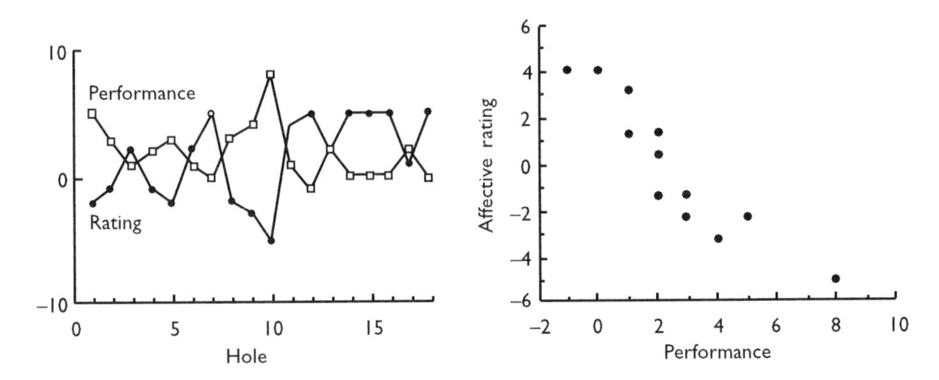

Figure 4.7 Results obtained in the golf video-game for the subject with the highest correlation between performance and affective experience. (Left) Performance and rating of affective experience plotted over the successive holes of the golf game: the amplitude of the rating scale was open and left to the subject's estimation; positive rating for pleasure and negative rating for displeasure. (Right) Rating of affective experience (dependent variable) plotted against performance (independent variable): it can be seen that when performance improved (lower scores), pleasure increased. This correlation was significant in nine subjects out of ten

Source: Cabanac *et al.* (1997).

It is obvious that among those events or behaviours that arouse joy there are some whose usefulness we cannot foresee. The case of drug addiction comes immediately to mind. The answer to this argument is twofold. First, from a Darwinian point of view it is not necessary that joy be useful on each mental event. To be passed on to future generations, it is sufficient that joy gives some advantage to the subjects who possess it. From this point of view, we can compare pleasure and joy to curiosity. On some occasions the outcome of curiosity may be noxious. Yet everybody will agree that curiosity gives an evolutionary advantage to the subjects who possess it. Second, usefulness of sensory pleasure is mostly proximate, usefulness being judged from its immediate survival value. However, sensory pleasure can be associated also to long-term usefulness. Sexual pleasure is a powerful reward of reproductive behaviour. Its usefulness may be assigned to the species, rather than the individual. Similarly, joy can be the sign of an integrative behaviour, useful in the short term for the subject or in the long term for the species. The joy of love may have no immediate survival usefulness (arguably the opposite), but finds its usefulness in the outcoming reproductive behaviour. Pleasure as an index of useful sensation can be innate or acquired. Similarly, one may easily accept that homologous joy can be acquired. The hormic joy associated with effort has to be taught and learnt.

Since no behaviour can escape the law of pleasure maximisation, one may question what remains of liberty in such a situation. Human liberty is often misunderstood as the freedom to do anything. Actually it is to be understood as the freedom to choose one's own way to maximise pleasure and joy. Among the motivations sorted by Sulzer (1751) as sensory, intellectual, and moral, the last has always been considered by the philosophers as the most rewarding.

References

Attia, M. (1984) Thermal pleasantness and temperature regulation in man. *Neurosci. Biobehav. Rev*, 8, 335–343.

Attia, M. and Engel, P. (1981) Thermal alliesthesial response in man is independent of skin location stimulated. *Physiol. Behav.*, 27, 439–444.

Baerends, G.P. (1956) Aufbau des tierischen Verhaltens, in *Handbuch der Zoologie* (ed. Kükenthal, W., Krumbach, T.). Teil 10 (Lfg 7). Berlin: De Gruyter & Co.

Bleichert, A., Behling, K., Scarperi, M. and Scarperi, S. (1973) Thermoregulatory behavior of man during rest and exercise. *Pflüg. Arch.*, 338, 303–312.

Bligh, J. and Johnson, K.G. (1973) Glossary of terms for thermal physiology. *J. Appl. Physiol.*, 35, 941–961.

Cabanac, M. (1971) Physiological role of pleasure. *Science*, 173, 1103–1107.

Cabanac, M. (1986) Du confort au bonheur. *Psychiatr. Fr.*, 17, 9–15.

Cabanac, M. (1992) Pleasure: the common currency. *J. Theoret. Biol.*, 155, 173–200.

Cabanac, M. (1995). *La quète du plaisir*, 169. Montréal: Liber.

Cabanac, M., Massonnet, B. and Belaiche, R. (1972) Preferred hand temperature as a function of internal and mean skin temperatures. *J. Appl. Physiol.*, **33**, 699–703.
Cabanac, M., Pouliot, C. and Everett, J. (1997) Pleasure as a sign of efficacy of mental activity. *Eur. Psychol.*, **2**, 226–234.
Fantino, M. (1984) Role of sensory input in the control of food intake. *J. Auton. Nerv. Syst.*, **10**, 326–347.
Fantino, M. (1995) Nutriments et alliesthésie alimentaire. *Cah. Nutrit. Diétét.*, **30**, 14–18.
Garcia, J. (1990) Learning without memory. *J. Cognit. Neurosci.*, **2**, 287–305.
Garcia, J., Lasiter, P.S., Bermudez-Rattoni, F. and Deems, D.A. (1985) A general theory of aversion learning. *Ann. NY Acad. Sci.*, **443**, 8–21.
IUPS Commission for Thermal Biology (1987) Glossary of terms for thermal physiology. *Pflüg. Arch.*, **410**, 567–587.
Krebs, J.R. and Davies, N.B. (1981) *An Introduction to Behavioural Ecology*, 292. Sunderland, MA: Sinauer Associates.
Lea, S.E.G., Tarpy, R.M. and Webley, P. (1987) *The Individual in the Economy*, 627. Cambridge: Cambridge University Press.
McFarland, D.J. (1985) *Animal Behaviour*, 576. London: Pitman.
McFarland, D.J. and Sibly, R.M. (1975) The behavioural final common path. *Phil. Trans. R. Soc. Lond.*, **270**, 265–293.
McNamara, J.M. and Houston, A.I. (1986) The common currency for behavioural decisions. *Am. Nat.*, **127**, 358–378.
Marks, L.E. and Gonzalez, R.R. (1974) Skin temperature modifies the plesantness of thermal stimuli. *Nature*, **247**, 473–475.
Pfaffmann, C. (1960) The pleasures of sensation. *Psychol. Rev.*, **67**, 253–268.
Rossetti, Y., Meckler, C. and Prablanc, C. (1994) Is there an optimal arm posture? Deterioration of finger localization precision and comfort sensation in extreme arm-joint postures. *Exp. Brain Res.*, **99**, 131–136.
Sulzer, M. (1751) Recherches sur l'origine des sentimens agréables et désagréables. *Mém. Acad. R. Sci. Belles Let. Berlin*, **7**, 57–100.
Tinbergen, N. (1950) The hierarchical organization of mechanisms underlying instinctive behaviour. *Symp. Soc. Exp. Biol.*, **4**, 305–312.

Chapter 5

Emotion and the environment: the forgotten dimension

Mahtab Akhavan Farshchi and Norman Fisher

Introduction

In the creation of architectural spaces that meet the dynamic, conflicting and complex multifaceted social and physical requirements, the design discipline has to be informed of how spaces are perceived, judged and evaluated by their users. For a long time, despite the multiplicity of place appraisal, there has been the seemingly **objective** evaluation of buildings by financial criteria, which has overwhelmingly driven the process. This approach strongly resembles the Cartesian view of a place as purely a geometrical space. In the late twentieth century, however, many philosophers, geographers and social psychologists have expanded the notion of space beyond its geometrical connotations.

For Lefebvre (1994) the search was to find a unitary theory of the physical, mental and social space aimed to reconcile the **mental** space, 'the space of the philosopher', and **real** space, 'the physical and social spheres in which we live'. Among social psychologists, Canter (1983) emphasised the social, spatial and services **facets** of a place. His approach to place evaluation, empirically supported using facet theory, aimed to explain the interrelationship between these facets. Other researchers have since applied his approach to investigate environmental meaning (Groat, 1985), and building users' evaluation of places (Hacket and Foxall, 1995). Among contemporary geographers Massey (1994, 1995) and Harvey (1989) have critically analysed the social relations and spatial organisation of the space created under the ruling of a capitalist system of production.

Generally, advances in the theory of the place on the one hand, and the empirical evidence in support of the multifaceted impact of the place on the individual's social behaviour on the other, have induced the emergence of a new approach to building appraisal which focuses not only on the geometry of space, but also on its social, cultural and emotive aspects. Place appraisal is partly an attitudinal response. An attitude is expressed as 'an evaluative response to an individual, a group or a thing' (Hewstone *et al.*, 1997). Evaluative responses have cognitive, affective, and behavioural components, and studies in social and cognitive psychology have suggested that these three components are correlated with one another. A positive thought is associated with a positive feeling and hence behaviour, although thoughts, actions and feelings may vary in their degree of positivity or negativity. This tripartite model of attitude has been tested empirically and has been accepted as a valid framework. The model presented in this chapter assimilates a conceptual framework for building appraisal, taking account of the above principles.

Purpose, intention and goal-directed behaviour

The theory of the 'place' was first introduced as a non-design theory of place appraisal by Canter (1983) and originated from environmental and social psychology research. The essence of Canter's contribution is in his emphasis on the **purposive** nature of human activity and the dependence of behaviour on place. Other theories, such as the theory of reasoned action (Ajzen and Madden, 1986), also predict the actual behaviour of the individual as a function of their intention to perform a particular action. Behavioural intentions are determined by three factors: (a) the attitude towards performing; (b) other subjective norms of the individual; and (c) perceived behavioural control. The last of these reflects the individual's belief about how easy or difficult performance of a given behaviour is likely to be. Since individuals are different, their purposes can also vary according to their particular social and individual circumstances. Evidence suggests that age, gender, education, status or social role can affect attitudes. On the whole, individual or group purposes are inspired by a number of social or economic motivators. People get engaged in varied activities to fulfil their social or economic roles. Architectural spaces are designed for human activities and so depend on technology for their operations. The character of a space affects human emotions and behaviour. Thus success in design depends on how well the space satisfies the range of human needs of the occupants.

Environmental meaning and evaluation

The appraisal and the expression of satisfaction owes much to the structure of our thinking processes and thus to the languages used as the communication medium. In the evolutionary process of brain development, languages have become the vehicles for further advances and have enabled communication of concepts (although not all communication is through language). Sternberg (1996) has briefly described the process by which we allocate meaning to concepts:

> We encode meanings into memory through concepts–ideas (mental representations) to which we may attach various characteristics and with

which we may connect various other ideas, such as through propositions – as well as through images and perhaps also motor pattern for implementing particular procedures.

So the question is, how can attributing meaning affect appraisal? The answer may be found in the hidden cognitive responses that mediate between external stimuli and overt behavioural responses (Fiedler, 1996, p. 136). To explain cognitive responses one needs to understand stages of the cognitive process as well as comprehending the social nature of information processing. Fiedler (1996) describes a process starting from perceptual inputs to behavioural outputs, characterised by various feedback loops. There is a logical order in the cognitive process: perception results in earlier stages, and categorisation takes place in later stages (see Fig. 5.1). This explanation suggests that we cannot categorise a building as belonging to some architectural style unless we have perceived the building, but we can perceive building attributes independently of style categorisation. Cultural and subjective norms affect cognitive processes and their representations in language via meanings that are attributed to them.

The social aspect of attitude

To understand and explain social behaviour and attitude formation (i.e. productivity, satisfaction at work, or affective appraisal of places), we need to emphasise the social nature of the cognitive process. Research on place evaluation is enriched by the inclusion of the cognitive processes and also by emphasising the social nature of information processing.

Sensory experience

Within our nervous system receptors receive sensory information (as sensations via the eyes, ears, nose and skin) while effectors transmit motor information (e.g. movements of the large and small muscles). The body acts in response to the information that it receives, although little of this enters our consciousness. Sensations are the data of perception, though sensation and perception of an object are not the same thing. Errors may occur through

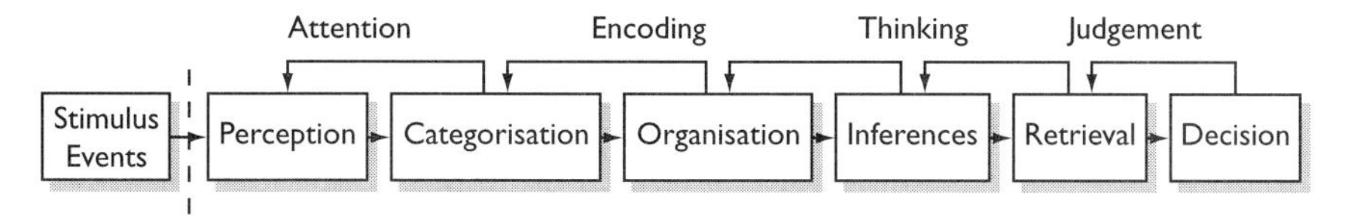

Figure 5.1 Conceptual framework for cognitive stages in information processing

Source: Fiedler (1996).

'intelligent leaps of the mind' leading to illusion (Gregory, 1997), which raises the question of what is 'objective' and what is 'subjective'. Locke (1690), the precursor of modern psychology, suggested that there are two kinds of characteristics: the **primary** characteristics, such as hardness, mass and extension of objects in space and time, are free from the mind and are objective; the **secondary** characteristics do not exist in the world but are created within us, and are therefore subjective. The secondary characteristics are affected by the state of the sensing organism. For example, colours can change if we look through haze or wine.

Our senses operate independently from each other but can naturally reinforce one another. We can distinguish the colour of a flower despite its smell, or its touch. However, when remembering that flower, our experience is based on our memory stored as a series of representations with particular characteristics. Particular features of the object may influence this representation; such as shape, smell or colour. Canter (1977) explains the role of human memory in its ability to 'discover ways of escaping from the complete sway of immediate circumstances' by building up a residue of experience which it can utilise later. Bartlett (1932) first introduced the notion of 'schemata':

> The sensory cortex is a storehouse of past experiences. They may rise into consciousness as images but more often, as in the case of spatial impressions, remain outside central consciousness. Here, they form organised models of ourselves which may be called schemata. Such schemata modify the impressions produced by incoming sensory impulses in such a way that the final sensations of position or locality rise into consciousness charged with a relation to something that has gone before.

The schemata, however, are not only concerned with the real and physical characteristics of an object or a thing itself. The schemata of a building can be of the image of the building in its own right or as part of a bigger reality such as a town, or a community. People usually form the representations of reality that are in accord with their purposes, motivation and interest. Memory retrieval ability is affected by our emotions, moods, state of consciousness, schemata, and other internal contextual features (Sternberg, 1996, p. 273).

Imperceptibility

In addition to the stimuli that can be processed by our sensory system, the environment affects us in other ways, which are not recognisable to us. Such stimuli can cause changes in our psychological state, which apparently lack any conscious experience of their cause. Harmful imperceptible stimuli are invisible lights, gases, chemical compounds, radiation, etc., which may be harmful to our well-being. Our bodies, for example, have not developed mechanisms to detect carbon monoxide. Small doses of this gas can result in feelings of exhaustion, fatigue, lassitude and drowsiness. Nitrous oxide, or 'laughing gas', has a distinctive odour, but it is not as famous for its smelling quality as it is for its laughing (Russell and Snodgrass, 1987).

Perception

Perception consists of a set of complex processes by which we recognise, organise, and make sense of the sensations we receive from environmental stimuli (Sternberg, 1996). Theories of perception can be divided into two main groupings. **Bottom-up** theories (sometimes called data-driven theories), such as the Gestalt school or Gibson's ecological model, start from the bottom, or the physical stimulus, such as the observable pattern or form that is being perceived and is then categorised and organised into concepts. **Top-down** theories focus on the high-level cognitive processes, existing knowledge, and prior expectations. Empirical evidence suggests that these theories are not as incompatible as they may sound. Generally, it is accepted that perception depends on active physiological processes. Marr (1982, p. 127) has proposed a computational theory of visual perception that combines the richness of the sensory information and the value of prior knowledge and experience in perception. This theory has been used as a basis for the application of artificial intelligence on visual perception. In Marr's view, shapes are derived from images via three essential stages: (a) the primal sketches which describe intensity changes, locations of critical features such as terminal points, and local geometrical relations; (b) the $2\frac{1}{2}$-D sketch, giving a preliminary analysis of depth, surface discontinuities, and so on, in a frame that is centred on the viewer; (c) a 3-D model representation, in an object-centred co-ordinate system, so that we see objects as much as they are in 3-D space though they are presented from just one viewpoint.

The nature of emotion

The functions of emotion are regarded as being to probe the nervous system into play to prepare the organism to cope with threatening or stressful situations. The presence of this activity motivates an organism to make a response to decrease it. Emotion is a crucial mechanism for survival, which provides an adaptive response to stimuli in the environment. The physiological response of the organism to the stimuli is the production of emotional behaviour. Speculations about the nature of emotions have engaged philosophers from as far back as the ancient Greek civilisation to the present day. The dominant theory up to the 1920s was the James–Lange theory, which proposed that bodily changes follow directly from perception of the existing facts. Our feeling about these changes is what they regarded as emotions. Cannon's (1929) critique of the James–Lange theory offered an alternative by suggesting that emotions were the results of concurrent brain stem and cortical events. The main question however, remained:

> Do we experience emotion because we perceive our bodies in a particular way, or are there specific emotional neutral patterns which respond to environmental events and then release bodily and visceral expressions?

The behaviourist school in the decades after Cannon insisted on dealing with only the objective and the observable as the basic psychological data, which implied that behaviour and action are determined by the knowledge and thought of the organism. A change of direction in cognitive theories of emotion was made by Schachter and Singer (1962). The main contribution of this theory was the acknowledgement of visceral arousal as a necessary condition for emotional experience, yet explaining the quality of emotion as depending on cognitive and perceptual evaluations of the external world and the internal state. The recent theories of emotion have tried to explain the role of emotions in adaptive behaviour. Gregory (1997) explains this development:

> Rather than emotion appearing as an interfering, irrelevant, and chaotic state of affairs, it seems that different kinds of situations (and cognitions) become especially marked if they occur in the 'emotional' visceral context. This notion corresponds with the common experience that emotionally tinged events occupy a special place in our memories. The visceral component of the emotion may well serve as an additional cue for retrieval from memory of specific events, and sets them apart from the run-of-the-mill catalogue of events. The 'emotional' memory of a visit to a theatre is selected from among all the plays we have seen; it is 'special'. . . . Current wisdom would suggest that any discrepancy, any interruption of expectations or of intended actions, produces undifferentiated visceral (automatic) arousal. . . . The quale of the subsequent emotion will then depend on the ongoing cognitive evaluation (meaning, analysis, appraisal) of the current state of affairs.

Mandler (1984) provided a contrasting view of the affective experience to that of Schachter and Singer's (1962). This view regarded the affective experience as a result of the arousal of the automatic nervous system and the cognitive processing of information about that environment and stored representations of the previous experience with that environment. However, Purcell (1986, 1992) argued that in this process the schema-based processing of environmental experience can be interrupted by the discrepancy between aspects of the environment and the prototypical basis for the default values in the schema. He further stressed the cognitive components of the emotion (1992).

In recent decades, the role of emotions in directing human behaviour has also received a lot of attention from managers (Locke, 1996). The cross-country studies of emotions reveal that people hold norms of emotion perception in which the socially desirable emotions are those which are perceived to be positive and moderate (Paez *et al.*, 1996). Regulation of emotions (i.e. increasing positive feelings rather than negative ones) therefore could help managers within organisations. Emotions are essential parts of the fabric of human experience and play a crucial role in determining the nature and quality of a person's day-to-day functioning. Earlier research also revealed that despite the seemingly apparent nature of emotions as discrete and separate aspects of human functioning, these are in reality linked to cognition and actions, which together are the interdependent aspects of the information-processing–acting system mediating the internal and external interaction of humans with their environments. Hebb (1949) suggested that we often notice those emotions which interrupt the ongoing behaviour, hence prompting different reactions. Western's (1994) model introduced an integrative model of **effect regulation**, which converges aspects of behavioural, cognitive, social-cognitive and psychodynamic perspectives. This model describes effect regulation as a mechanism by which a selective retention of behavioural and mental processes is maintained.

It has been suggested that emotions have three main components (Mandler, 1984; Schachter and Singer, 1962). The behavioural component could be 'expressive (such as frowns, smiles, weeping, gesture, or tone of voice) or instrumental (such as fight or aggression)'. The physiological components as argued by Russell and Snodgrass include 'changes in, or activity clearly inverted by, the automatic nervous system'. The mental component of emotions is less well specified than the other two named above. It is, however, implied that the person who is undergoing an emotion is well aware that he or she is afraid, angry or happy.

Types of emotion

One view classifies emotions into a limited number of key emotions. Some emotion theorists believe that any type of emotion can be defined in terms of one or more basic emotions. Ekman (1972), for example, argued that there are at least six basic emotions: sadness, disgust, anger, fear, surprise, and happiness. Two more emotions, shame and interest, have been added to the above list by Izard (1977). This is notwithstanding 2000 emotion-denoting words that exist in the English language. Russell and Snodgrass (1987) have referred to five types of emotional response, i.e. emotion, mood, affective appraisal, emotional episodes and emotional dispositions.

Emotions

Generally, emotions refer to a **heterogeneous class of different phenomena** (Russell and Snodgrass, 1987). The narrow definition of emotion refers only to such prototypical episodes as falling in love. Emotions can be broadly defined to include vague feelings of mood, attitudes, preferences, or just about anything that is not coldly rational (Russell and Snodgrass, 1987). Zajonc (1980) has defined emotions as **bodily feelings that reveal preferences.** Clore and Parrott (1991) identified emotions as the means of conveying information to individuals about the nature of the psychological situation.

Moods

These are **the core emotion-tinged feelings of a person's subjective state at any given moment.** As Russell and Snodgrass argue, to be in a certain mood is to feel calm, upset, depressed, excited, unhappy, or neutral. By this definition, a person is always in a mood. Moods are usually measured in self-report formats in which research begins with a series of factor-analysis studies. In the study by Nowlis and Nowlis (1956), several bipolar factors such as pleasure–displeasure and activation–deactivation summarised the mood domain. There is still much controversy over the proper number of dimensions in the description of mood. The core of these arguments is whether moods are unipolar or bipolar. More recent evidence favours bipolarity.

Russell (1980) showed a circumplex model of mood (Fig. 5.2) in which individual mood descriptors fall in a roughly circular order in a space defined by two underlying bipolar dimensions of feeling: the first, they argue, is pleasure–displeasure or happiness–unhappiness, and the second is arousal–sleepiness. In this model pleasure and arousal are identified as the key dimensions of mood. This circumplex model was developed using 40 descriptors, which were drawn systematically from a set of 105 adjectives. Mehrabian and Russell (1974) have concluded that: 'Sensory dimensions are consistently appraised as mood altering, although this does not necessarily guarantee that they are in fact mood altering'.

Some aspects of our perception do not need to reach our consciousness to affect us. This is seen in the effects of lighting in our moods. It has been indicated that sunlight has a positive effect on humans' mood (Cunningham, 1979), and artificial light has shown to have a negative effect on our moods (Thornington, 1975). Hellman (1982) has suggested that lack of natural sunlight could be a factor in depression, jetlag and sleep disorders in psychiatric cases. The amount and type of light also seems to affect the migration of birds, and egg production on poultry farms (Wortman *et al.*, 1964).

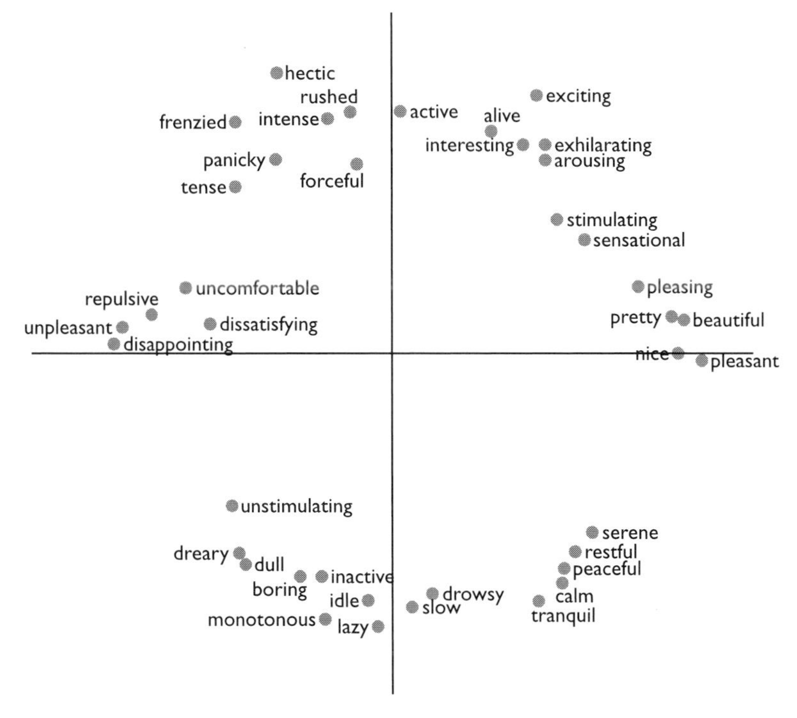

Figure 5.2 Circular ordering of mood descriptors

Source: Russell (1980).

Affective appraisal

Affective appraisal refers to **our judgements of things as pleasant, attractive, valuable, likeable, preferable, repulsive and so on.** Affective appraisal is always directed towards a thing, i.e. a quality of the object is appraised. It is the object that appears pleasant or disgusting. It is distinguished from moods, in which affective appraisal could occur with no inner emotional feelings. For example, the appraiser may evaluate a subject while experiencing a certain mood, which is not related to the appraisal event.

Emotional dispositions

These relate to long-term emotions: **a tendency to do or think or feel particular things when the right circumstance occurs.** For example, feeling for our parents is an emotional disposition. Emotional disposition is said to be a disposition because it exists even during the times when we are not thinking or feeling anything about our parents, but it is manifested on certain occasions.

Emotional episodes

These are **emotional reaction to something, with the reaction typically involving co-ordinated and distinctive physiological, behavioural and mental changes like someone, suffering a grief at death, getting angry at someone and being frightened by a bear in the woods** (Russell and Snodgrass, 1987). Emotional episodes have common characteristics with both affective appraisals and mood. They are about something, as in affective appraisal, and refer to a core subjective feeling, as in mood. Emotional episodes are prototypical examples of what is usually meant by 'emotion'.

Environment and behaviour

Environmental psychology, a growing field of research, has not yet formulated any unified theory of environment and behaviour. In general there are six theoretical perspectives, which have been supported by empirical evidence (Bell *et al.*, 1996). These are briefly explained here.

The arousal approach

In this approach the environmental stimulus affects people by increasing their arousal as measured physiologically. Arousal here is regarded as an intervening variable with distinct effects on behaviour. Arousal moves along a continuum of low arousal (towards sleep) to high arousal. Both pleasant and unpleasant stimuli can heighten arousal. The change in arousal level causes people to seek information about the their internal states, to find out whether there is any kind of threat to their well-being. Arousal can therefore affect the level of performance, since some tasks require higher level of arousal than others. The relationship between arousal and performance is dependent on the complexity of the task concerned. Noise, heat and crowding are the most notable environmental factors affecting arousal.

The stimulation load approach

When the information provided by the environment exceeds the individual's capacity to process, overload occurs (Bell *et al.*, 1996). The level of interacting environmental stimuli is important in behaviour. This concept is regarded as environmental load or overstimulation, which can affect behaviour through its demand on attention and information processing. Kaplan and Kaplan (1989) have identified four environmental factors contributing to information overload: (a) a new environment or '**being away**'; (b) an experience that is extended in time and space; (d) an interesting or engaging environment; and (d) the ability of the environment to facilitate the achievement of the intended purpose.

Understimulation can also lead to severe anxiety or other psychological anomalies. Research suggests that environments should provide more complexity and stimulation in order to restore excitement and a sense of belonging. In order to deal with the problem of understimulation, Wohlwill (1966) advocated scaling the environment along a number of dimensions of stimulation, including intensity, novelty, complexity, temporal changes or variations, surprisingness and incongruity (Bell *et al.*, 1996, p. 122).

The adaptation level

This theory suggests that in order to deal with the problems of overstimulation or understimulation the environment should provide a mechanism by which we can regulate it according to our own acceptable levels.

The behaviour constraint approach

The loss of perceived control over environmental stimulation can also lead to arousal or strain on our capacity for information processing. When individuals perceive that they are losing control over the environment, they first experience discomfort (negative effect) and then try to gain some control. Bell *et al.* (1996) give examples of different types of control: (a) behavioural (e.g. turning off the noise); (b) cognitive (e.g. deciding that a contaminant in water is not toxic); and (c) decisional (e.g. choosing to live in a quiet neighbourhood).

The stress approach

Current models regard stress as the outcome of an unbalanced interaction of the person with its environment (Salvendy, 1997). Elements of the environment such as noise and crowding are viewed as stressors, although other social factors can also cause stress, such as job pressures, family discord, or moving to a new home. Stress comprises emotional, behavioural and physiological components. Generally, stress can stem from three sources: cataclysmic events, personal stressors and background stressors (Bell *et al.*, 1996). In any case the probability of an event becoming stressful is determined by how the stress is appraised by the individual. This cognitive process of stress appraisal is a function of personal factors (e.g. intellectual resources, knowledge and past experience, and motivation) and cognitive aspects of the specific stimulus situation (e.g. control over the stimulus, predictability of the stimulus, and immediacy of the stimulus) (Bell *et al.*, 1996, p. 133).

Stress will occur if the environmental demands are greater than the person's capabilities, and/or if the person's expectations are greater than the environment supplies (Salvendy, 1997, p. 1061). In psychological terms, stress is the response to the individual's interpretation of the meaning of environmental events against the individual's appraisal of her/his coping resources.

The ecological approach

This approach is primarily concerned with the specific effects of the environment on behaviour. Barker (1968), the principal advocate of the ecological psychology, believes that environment and behaviour have ecological interdependencies. According to Barker's model, the physical setting of the environment can suggest to us what behaviours we can expect within that environment.

Emotions as a crucial component of environmental meaning

Emotions can be regarded as part of a cognitive process revealing the individual's psycho-physiological state. Such reactions can be measured using verbal descriptors such as the studies carried out by Vielhauer (1965) and Russell *et al.* (1981), or physiological measurements of the bodily changes. Vielhauer developed a semantic scale for the description of the physical environment. Using bipolar adjectives, she identified the degree of appropriateness of each descriptor in the context of six specific environments presented as pictures to her subjects. She identified four main factors and five sub-factors, which could – independently of the person or the room – be regarded as the idiosyncratic characteristics of the room itself. These are: (1) aesthetic appeal; (2) physical organisation; (3) size, and (4) temperature–ventilation as the main factors and (1a) style; (1b) functional ugliness; (1c) colouring; (2a) organisation; (2b) cleanliness; (3a) phenomenological size; (3b) physical size; and (5) lighting as the sub-factors. Russell *et al.* (1981) tried to separate effect from perception/cognition in their study of environmental meaning. They reviewed the use of semantic scaling and argued that the discrepancy between studies in environmental meaning has emerged because they have measured the different – but highly correlated – components of meaning, i.e. perceptual/cognitive and affective. Russell *et al.* (1981) identified various effect-denoting concepts which could be empirically tested and used as a single system. For example, they hypothesised a stressful situation as one which is both unpleasant and arousing – a bipolar opposite of a relaxing situation, which is both pleasant and unarousing.

In design terms the most referred to positive and negative emotions for the description of the environment are the concepts originating from comfort and stress. Slater (1985) has given a scientific interpretation of comfort as **a pleasant state of physiological, psychological and physical harmony between a human being and the environment.** Comfort is considered as a multidimensional construct influenced by many factors and, as Zhang (1996) demonstrated, it does not relate only to physical aspects of things. For example, in her study, seating comfort was positively correlated with the appearance of the chair. Activities are the individuals' responses to environmental interpretations, i.e. the meanings that they attribute to their environment. Therefore, human activity is as much about individuals as it is about the social aspect of their interactions.

Generally, stress relates to negative effects such as nervousness, tension and anxiety and can cause a deficit in task performance and interpersonal behaviour. Evans and Cohen (1987) have argued that stress is a person- or situation-based concept. For example, crowding causes stress by excessive stimulation; to avoid that an individual maintains a preferred interaction distance from others in order to avoid excessive arousal, stimulation, and a variety of stressors associated with proximity that is too close. Evidence suggests that individual differences play a role in perception of density. Other studies have shown different responses from female and male respondents to density. Ross (1973), for example, found that female groups responded positively to high occupation density while for male groups it was negative. This demonstrates that both genders follow norms, which belong to their specific sex. Further, Taylor (1981) has argued that the consequences of density can be mediated by perceived control, social structure, and type of activity and physical features of the environment.

Stress can have both positive and negative consequences. Different activities require different levels of stress.

Can we assign affective appraisals for places?

Most of our activities are conducted within built spaces (e.g. we buy our food at supermarkets, walk and relax in our gardens, or dine in restaurants). Our appraisals depend on an array of factors, some derived from our sensory experience and some based on our past experiences mediated by culture and our subjective norms. For example, we may like a restaurant for its tranquil lighting, or comfortable temperature, the convenience of its location, aesthetic quality, tasty food or hospitable service. The affective qualities of the space such as liveliness, attraction, calmness or excitement can help us store and then retrieve information from our long-term memory which otherwise is forgotten. Thus in the evaluation of places we should allow for measuring emotive as well as cognitive components of appraisal. Considering that environmental setting can affect behaviour (Barker, 1968), we may conclude that given a specific emotional content for a space we can expect certain behaviour to occur.

Development of a conceptual model for affective appraisal of spaces

Architectural design can benefit from broadening its scopes by consideration of achievements in social sciences, especially environmental and social psychology. Design, a problem-solving activity, aims to create spatial solutions to respond to human habitual needs, which may appear in the form of physical, technical, social, managerial or functional constraints. Achieving a balance between these factors requires a systematic approach in order to create optimal solutions in the form of habitable geometrical spaces, yet with identity and inspiration, and to complement human interaction and well-being. Thus, understanding human needs and aspirations can be considered as one of the most crucial parts of any design conceptualisation.

Design is generally an incremental and hierarchical process. Some decisions, such as the spatial relationships between different spaces, are part of early design decisions, while others, such as reconfiguration of the internal organisation of the space, can be modified by the changing need of the space (Farshchi and Fisher, 1997; Farshchi, 1998).

To distinguish between different contents and contexts of design information, Farshchi and Fisher (1997) offered a preliminary model, which divided design information into two interrelated levels, i.e. macro and micro. Form, structure and style convey social and cultural meanings. Design starts from development of a core central concept which is based on the type of activities (functions) expected from the building, the needs of users with social and personal motives, and purposes that affect how well they can achieve their individual and social goals. Technology can be used to facilitate the functions and performance of activities.

The conceptualisation of design is a cognitive process. Design can increase its user friendliness by better understanding of the concept of needs and how these are evaluated. Concepts in architectural design are socially rooted. In the Western cultural context there seems to exist commonly held values or beliefs for the concept of 'home' or 'church'. In the case of airports, this new experience for humankind, the question of purpose is left unanswered. With the high rate of technological and sociological change and increasing number of users, the existing concepts of airports are vague and unsatisfactory. Seeking a philosophical justification for a relatively new means of transport (air travel) motivated a group of architects and historians to come together at the 1998 Airport Conference at the Architectural Association in London. It was claimed that in our modern society such narratives are vague and confusing: for example, we are not sure what to expect from our experience in an airport. Do we need to use airports as we use train stations? Should they be spaces of timeless quality, being universal and inviting, or places of no identity where the notion of 'place' is diluted? Should airports be regarded as crossroads to our global cultures, where interaction is carried at levels of leisure, arts, and trade? Thus, fulfilling a philosophical need is also part of the design conceptualisation.

The model presented in Fig. 5.3 offers a form of generic methodology that can be adopted for different design circumstances. The concept of satisfaction here is regarded as a positive or negative response to environmental stimuli. Individual differences such as age, gender, physical abilities, past experience and education, as well as social factors such as goal/purpose, intention, social role, cultural values, and norms, can affect preference. The model suggests that although individual differences are important in the selection of the incoming information, socially held values and role are useful predictors for emergent behaviours. The individual's past experience can increase or decrease the level of arousal due to their degree of novelty or mere exposure (familiarity). Behaviour in groups can be mediated by the groups' norms.

The incoming information (noise, light, temperature, smell and touch) reaches the individual through the sensory organs. The information is selected and processed which is in line with the individual's purposes or motivations. The level of adaptation to environmental stimulus varies in individuals, and some may require a higher level of environmental control than others.

To provide practical steps for improving designs' user-friendliness and satisfaction, we earlier referred to research on 'place'. In turning a physically

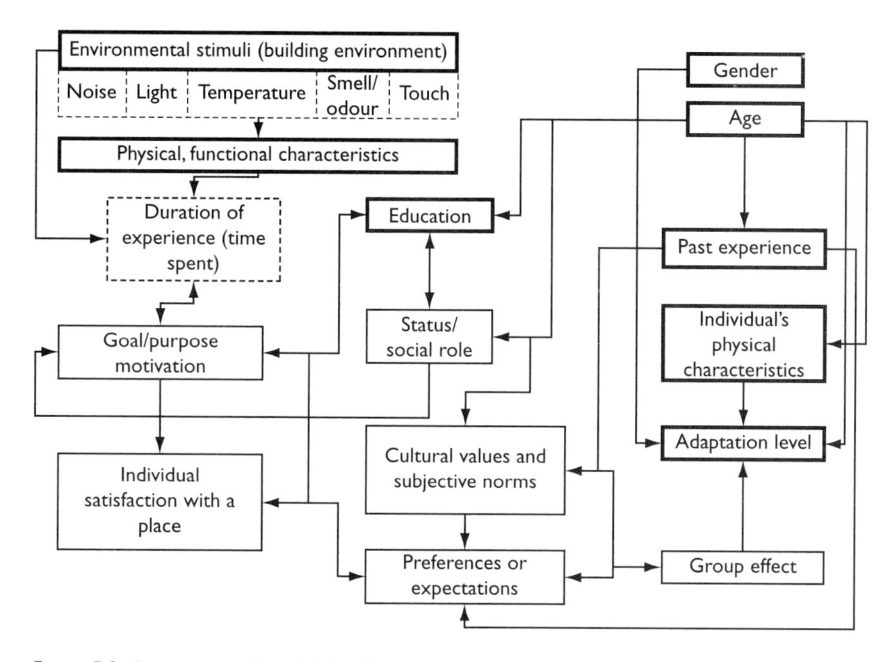

Figure 5.3 A conceptual model for the appraisal of spaces

structured space into a 'place' with philosophical, social, and artistic identity or meaning, we should also explain needs as a holistic concept to help the definition of design attributes. Perceived needs are a social phenomenon. As a control mechanism, culture can play a significant role in mediating needs, norms of behaviour, acceptable values and expectations in any society. Such norms of behaviour can also provide codes of acceptable concepts in architectural design. Farshchi (1998) gave a description of needs in architectural design, i.e. socio-psychological, economic and philosophical.

Overall there are two worlds of abstraction in which architecture is conceived – the 'real' and the 'experienced'. In creating an architectural space the perceived world of the architect/designer, or the 'experienced', should closely resemble the 'real' one. The journey of the architect/designer between the two realms finally resolves in his or her finding a formal solution to the 'problem' or 'design context' via architectural forms. None the less, the dialectic process between form and context may fail to reflect or embrace all intricacies or complexities of the real world. To emancipate design from the tangle of designers' subjectivity, the design problem needs to be objectively defined, i.e. the collective representation of the experienced world by members of society which become the building's beneficiaries. Yet, objective thinking is contingent on a number of presuppositions as follows. First, design is in effect a social phenomenon, with the aims of enclosing for the purpose of exclusion, partitioning, spatial fixation, and maintaining spatial distances between people, as it is also to facilitate communication and traffic (Ankerl, 1981). Second, social spaces are affected by the social needs for territorial sovereignty of not only the residents, but also the social establishment, which is a need even when the space is empty. Ankerl (1981) described the characteristics of the architectural design as the creation of spaces, which reflect the physical, technical, social and psychological needs. Architectural space in his terminology is expressed as:

> A closed surface with human access into its concavity . . . where the envelope has at least one side defining an inside or interior . . . a volumetric magnitude with particular geometric shape . . . sensed by more than one sensory organ . . . has a physical language . . . has a purposeful creation so to isolate a numerable set of individuals for certain activities . . . form interference from the broader physical and social environment . . . protects the intimacy or privacy of small or large groups . . . stimulates communication.

Conclusions

Architectural spaces can either facilitate or restrict human actions and thus cause positive or negative emotions. Modern buildings are criticised for their incompatibility with their surroundings, unsatisfactory sensory qualities or alienation of people who use them or live in or around them. Current design theorists, such as Gero (1996), have aimed to develop detailed accounts of design by providing a breakdown of design parameters in terms of functions, behaviour and structures. Although this approach has provided a language by which design tasks can be rationalised and formalised, these can be criticised for their disregard of the emotional content of design. In order to develop a taxonomy of the concept of evaluation/satisfaction we need to understand the meaning of concepts such as comfort, convenience and efficiency to the individuals. As the experience of each individual is varied and is based on her or his personal needs and expectations, it is important to know whether there are any commonly held views among building users. Due to the temporal nature of our experience within the built environment, it is important to identify the implications of this variable.

Earlier we referred to the notions of familiarity and novelty. Through the passage of time and repeated exposures, our sensory system makes sense of environmental clues by developing schemata. Appraisal is largely affected by the characteristics of new stimuli causing increased levels of arousal. To avoid boredom and sleepiness, and to generate enthusiasm and motivation in the occupants and other building users, design continuously has to provide new stimulation and motivation to maintain arousal and positive feelings. At the same time, spaces which have a changeable population of users, such as in airports, require information that is communicated in a logical order and also spaces which offer an interesting and engaging experience for those who work there.

The amount of information that we can process through our sensory system is also important for our overall satisfaction, and as our experience is dynamic and multidimensional we need to select from an array of environmental clues those which are pertinent to our purposes and goals. The analogy proposed here is what we call the 'motorway' and the 'countryside' driving experiences. This example refers to movements within the space at various speeds. At higher speeds of movement, due to an information overload, the brain selects only those environmental clues, that are directly important to its sensory needs, while at lower speeds, in addition to the sensory information, attention can also be drawn to more specific clues such as the distance between street lamps or the texture of the road, which relate to specific goals or purposes of the driver, i.e. driving. At the same time, the driver may also process information, which may be somewhat distant from her immediate purpose, i.e. enjoying the scenery at a country lane. Limitations in the information capacity of the brain require rationalisation of the stored information. But retrieval can be improved when an ongoing experience is associated with some emotional response to the external stimulus.

This chapter aimed to highlight the importance of the emotive components of the cognitive process in the appraisal of places, i.e. evaluating and judging places. The underlying cognitive stages such as perception, categorisation, organisation, inferences and retrieval of the stimulus events can be intervened by attention, encoding and thinking. Design as an information system can help stimulate attention and can help the process of encoding and schemata modification. User-friendly design will reduce unnecessary stress and other negative emotions, while maximising positive emotions.

References

Ajzen, I. and Madden, T. (1986) Prediction of goal-directed behaviour: attitudes, intentions, and perceived behavioural control. *J. Exp. Psychol.*, no. 22, 435–474.

Ankerl, G. (1981) *Experimental Sociology of Architecture: A Guide to Theory, Research and Literature*. Mouton Publishers.

Barker, R.G. (1968) *Ecological Psychology: Concepts and Methods for Studying the Environment of Human Behaviour*. Stanford, CA: Stanford University Press.

Bartlett, F.C. (1932) *Remembering: A Study in Experimental and Social Psychology*. Cambridge: Cambridge University Press.

Bell, P.A., Greene, T.C., Fisher, J.D. and Baum, A. (1996) *Environmental Psychology* (4th edn). USA: Harcourt Brace.

Cannon, W.B. (1929) *Bodily Changes in Pain, Hunger, Fear and Rage*, 2nd edn. New York: Appleton.

Canter, D. (1977) *The Psychology of the Place*. London: Architectural Press Ltd.

Canter, D. (1983) The purposive evaluation of places: a facet approach. *Environ. Behav.* **15**, no. 6., Nov.

Clore, G.L. and Parrott, L. (1991) Moods and their vicissitudes: thoughts and feelings as information, in J.P. Forgas (ed.), *Emotion and Social Judgements*, 107–124. Sydney, Australia: Pergamon Press.

Cunningham, M.R. (1979) Weather, mood and helping behaviour. *J. Personality Soc. Psychol.* no. 37, 1947–1956.

Ekman, P. (1972) Universals and cultural differences in facial expression of emotions, in J.R. Cole (ed.) *Nebraska Symposium on Motivation*. Lincoln, NE: University of Nebraska Press.

Evans, G.W. and Cohen, S. (1987) Environmental stress, in D. Stokols and I. Altman (eds), *Handbook of Environmental Psychology*, vol. 1, 571–610.

Farshchi, M. (1998) The theory of design information – a hierarchical approach, *2nd European Conference on Product and Process Modelling in the Building Industry*, BRE, Garston, August.

Farshchi, M. and Fisher, G.N. (1997) The emotional content of the physical space, paper presented to RICS, *COBRA 97 Conference*, Portsmouth, 10–12 Sept.

Fiedler, K. (1996) Processing social information for judgements and decisions, in M. Hewstone, W. Stroebe, and G.M. Stephenson, *Introduction to Social Psychology*. Oxford: Blackwell.

Gero, J. (ed.) (1996) *Advances in Formal Design Methods for CAD*. London: Chapman and Hall.

Gregory, R. (1997) *The Oxford Companion to the Mind*. Oxford: Oxford University Press.

Groat, L. (1985) *Psychological Aspects of Contextual Compatibility in Architecture: A Study of Environmental Meaning*, PhD thesis, University of Surrey.

Hacket, P.M.W. and Foxall, G.R. (1995) The structure of consumers' place evaluation, *Environ. Behav.*, **27**, no. 3.
Harvey, D. (1989) *The Conditions of Post Modernity*. Oxford: Blackwell.
Hebb, D.O. (1949) *The Organisation of Behaviour*. New York: Wiley.
Hellman, H. (1982) Guiding light, *Psychol. Today*, no. 10, 22–28.
Hewstone, M., Mastead, A.S.R. and Stroebe, W. (ed.) (1997) *The Blackwell Reader in Social Psychology*. Oxford: Blackwell.
Izard, C.E. (1977) *Human Emotions*. London: Plenum.
Kaplan, S. and Kaplan, R. (1989) The visual environment: public participation in design and planning, *J. Soc. Issues*, no. 45, 59–86.
Lefebvre, H. (1994) *The Production of Space* (translated by D. Nicholson Smith). Oxford: Blackwell.
Locke, J. (1690) An essay concerning human understanding, in R. Gregory (1996) *The Oxford Companion to the Mind*. Oxford: Oxford University Press.
Locke, K. (1996) A funny thing happened! The management of consumer emotions in service encounters, *Org. Sci.*, 7, no. 1, Jan./Feb., 40–59.
Mandler, G. (1984) *Mind and Body: Psychology of Emotion and Stress*. New York: Norton.
Marr, D.C. (1982) *Vision*, San Francisco, CA: Freeman.
Massey, D. (1994) *Space, Place and Gender*. Oxford: Blackwell.
Massey, D. (1995) *Spatial Division of Labour, Social Structures and the Geography of Production* (2nd edn). London: Macmillan.
Mehrabian, A. and Russell, J.R. (1974) *An Approach to Environmental Psychology*. MIT Press, Cambridge, MA.
Nowlis, V. and Nowlis, H.H. (1956). The description and analysis of mood. *Ann. NY Acad. Sci.*, no. 65, 345–355.
Paez, D., Marques, J. and Insua, P. (1996) The representation of emotions in groups: the relative impact of social norms, positive–negative asymmetry and familiarity on the perception of emotions. *J. Soc. Psychol.*, **26**, 43–59.
Purcell, A.T. (1986) Environmental perception and affect, a schema discrepancy model. *Environ. Behav.*, **18**, no. 1.
Purcell, A.T. (1992) Abstract and specific physical attributes and experience of landscape. *J. Environ. Man.* **34**, 159–177.
Russell, J.A. (1980) A circumplex model of affect, *J. Personality Soc. Psychol.*, no. 39, 1161–1178.
Russell, J.A., Ward, L.M. and Pratt, G. (1981) Affective quality attributed to environments – a factor analysis study. *Environ. Behav.*, **13**, no. 3, 256–288.
Russell, J. and Snodgrass, J. (1987). Emotion and the environment, in D. Stokols and I. Altman (eds) *Handbook of Environmental Psychology*, vol. 1, New York.
Salvendy, G. (1997) *Handbook of Human Factors*, 2nd edn. New York: Wiley.
Schachter, S. (1971) *Emotion, Obesity and Crime. American Psychologist.*
Schachter, S. and Singer, J.E. (1962) Cognitive, social and physiological determinants of emotional state. *Psychol. Rev.*, no. 69, 379–399.
Slater, K. (1985) *Human Comfort*. Springfield Thomas.
Sternberg, R.J. (1996) *Cognitive Psychology*. USA: Harcourt Brace.
Taylor, R.B. (1981) Perception of density. *Environ. Behav.*, **13**, no. 1, Jan.
Thornington, L. (1975) Artificial lighting – what colour and spectrum, *Lighting Des. Appli.*, no. 16, 16–21.

Vielhauer, J.A. (1965) *The Development of a Semantic Scale for the Description of the Physical Environment*, PhD dissertation, Louisiana State University.

Western, D. (1994) Towards an integrative model of affect regulation: applications to social–psychological research, *J. Personality*, **62**, no. 4, Dec.

Wohlwill, J.F. (1966) The physical environment: a problem for a psychology of stimulation, *J. Soc. Issues*, no. 22, 29–38.

Wortman, R.J., Axelord, J. and Fischer, J.E. (1964) Melatonin synthesis in the gland: effect of light mediated by the sympathetic nervous system, *Science*, no. 143, 1328–1329.

Zajonc, R.B. (1980) Feeling and thinking: preferences need no inferences. *Am. Psychol.*, **39**, 117–123.

Zhang, L. (1996) Identifying factors of comfort and discomfort in sitting. *Human Factors*, **38**(3), 377–389.

Chapter 6

A broad definition of comfort as an aid to meeting commitments on carbon dioxide reduction

Max Fordham

This book is about the productive workplace. I am an environmental engineer for buildings and so I am concerned about perceptions based on data from the eyes, the ears, and the thermal sensors of the body. Human beings evolved in the world in a form that was adapted to the wide range of environments occurring naturally. Light levels vary from 0 to 100,000 lux, sound levels from very quiet to enough to break the eardrums, and temperatures characterised by ranges from, let us say, –40°C to +50°C. We have not adapted to meet the extremes of environment, but we are adapted to survive in the most common middle ground. We have developed buildings to relieve the stress caused by extreme environmental conditions and to extend the range of locations in the world where we can survive. We imagine that by reducing the environmental stress we can increase our ability to be productive.

As an engineer, I am not professionally equipped to comment on the impact on people of their neighbours. However, as the son of a psychologist I am very much aware that my ability to function is dependent on my internal state of mind and the impact of my fellow workers. I am writing this before hearing the speakers at a conference, and in case it seems necessary I would like to remind you that a piece of research exists into the effects on productivity of altering the physical environment. Each alteration to the physical environment produces an immediate improvement in productivity. After a while the improvement tails off until the next alteration produces another spurt. Eventually, the environment is restored to the original *status quo* and the productivity increases. My psychological understanding of this phenomenon is that productivity of people is improved if they know they are loved, and special attention, such as special teaching or special changes to the environment, are signs of love which improve productivity.

We cannot form the future without understanding the *status quo*. I think the future is dominated by a conflict. On the one hand, economic growth requires that we invent new things to consume and that we consume more of those things which we have already invented. On the other hand, the success of any species of organism tends to produce an explosion in its consumption and waste so that it exhausts the resources available and is engulfed in its waste product. This is happening to us. It doesn't matter which resource or which sink you focus on, we can see that neither the sources nor the sinks are infinite and we had better think of ways of becoming less greedy. Fossil fuels, fish, tropical rain forests, cobalt, copper, water – you name it, the resource is limited. The atmosphere, the sea, the land, all of these sinks for our waste products are being altered by our activities. The pressures of economic growth are provided by the market and call for:

- air-conditioning as a sign of status in even the most benign climates
- heating to standards to encourage us to wear lightweight clothes in the winter
- electric lighting to supplement adequate natural light.

The forces of the market are very powerful, and I do not care to predict that we will beat them. However, I hope we shall and that we will meet the target of designing buildings for temperate climates that need no heating energy, that need no electricity for lighting if the sun is above the horizon, that need no electrical energy for refrigeration nor electrical energy to circulate air through buildings. We do not know how to meet these aims but we do know how to try. One of the first issues we need to understand is what constitutes comfort. Comfort has been defined by Mike Humphreys of Oxford Brookes University as 'the absence of discomfort; discomfort is alleviated by making adjustments' (1995). This may seem arcane, but it is better than the ASHRAE definition: 'Comfort is a state of mind'.

Ole Fanger ISO (1994) has carried out measurements on the parameters which affect thermal comfort as follows: the amount of clothing, the dry bulb air temperature, the moisture content of the air, long-wave radiation, short-wave radiation, and the rate of metabolism of an individual. These measurements show that with the right combination of the parameters, a person maintains the correct skin temperature with a minimum amount of heat loss by evaporation and resides in a state which does not prompt him or her to want to make adjustments. If one wants to predict whether a person will be able to make suitable adjustments in a particular environment, that can be done using the Fanger equations. We must all be able to envisage what it is like to sit in a building wearing light clothes and shirt sleeves and deciding to put on a light jersey or cardigan. Having put on the extra clothes, you might decide it is too warm and decide to take them off. The light jersey or cardigan makes a difference to the Fanger Comfort Vote of about two points. Thus you can see that the Fanger Comfort Vote Measurement is pretty subtle, and one point would hardly make a person want to make further adjustments. They could be said to be comfortable. A wide range of radiation environments and air temperature environments produce comfortable conditions, and it is absolutely not necessary to try to control each parameter at a centrally defined position. To try to standardise a set of environmental criteria for the workplace which are unnecessarily tightly defined will lead to expensive installations and extravagant energy use.

There is already a European Directive which requires the humidity in rooms to be 'reasonable'. Imagine sitting in a committee and refusing to agree that the humidity should be reasonable! The statement is either meaningless, in which case it should be eliminated from any document, or it is without bound, namely the relative humidity should be between 0 per cent and 100 per cent: is that reasonable? No, we can all agree that 50 per cent is reasonable, and it is. However, to maintain 50 per cent relative humidity in a building in the winter is tantamount to making a requirement for refrigeration and air conditioning. I propose that 'reasonable' for relative humidity is anything within the range 10 per cent to 90 per cent, with the proviso that the individuals in the space can make adjustments to the environment to bring the Fanger Comfort Vote score in the range –2 to +2.

The Fanger conditions cannot be extrapolated to all the conditions that people enjoy. People spend money to lie in the sun on tropical beaches in conditions which are too hot for the Fanger comfort criteria. They swim in cold water, they walk on mountains, and the Fanger system is not adaptable enough to explain their thermal situation. As the temperature rises, one of the involuntary adjustments that people make is to open the capillary vessels in the flesh immediately under the skin and raise the skin temperature. An additional adjustment is that glands which produce sweat are activated so that water collects on the surface of the skin and heat can be lost by evaporation. The blood is at a temperature of 37°C, and if the air outside were at 37°C and saturated then no heat could be lost from the surface of the skin to the air. However, the Fanger equations predict that no heat can be lost at a temperature of 35°C because the skin temperature is implicit in the equations and is set at 35°C. In order to understand extreme conditions we need to modify our model.

The adjustments which people might make are constrained by the social environment. My company was asked to design an office building in Manchester without air-conditioning. We explained that the occupants of the building would have to make certain adjustments. For example, during hot summer weather in the office it would be good for them to take off their coats and loosen their ties. However, as solicitors they said that this was not permitted, and we had to say that if that was the case they needed an air-conditioned building. I am sure we have all experienced the situation of being comfortable in nothing but a bathing suit. The air temperature is likely to have been well above 30°C, possibly with some radiation and a good deal of air movement. It is pretty difficult to design a building in the UK where the summer temperature is as high as that, and we might manage to enjoy the feeling of working dressed in bathing suits. This is not a sensible proposal. Figure 6.1 shows various kinds of temperature associated with

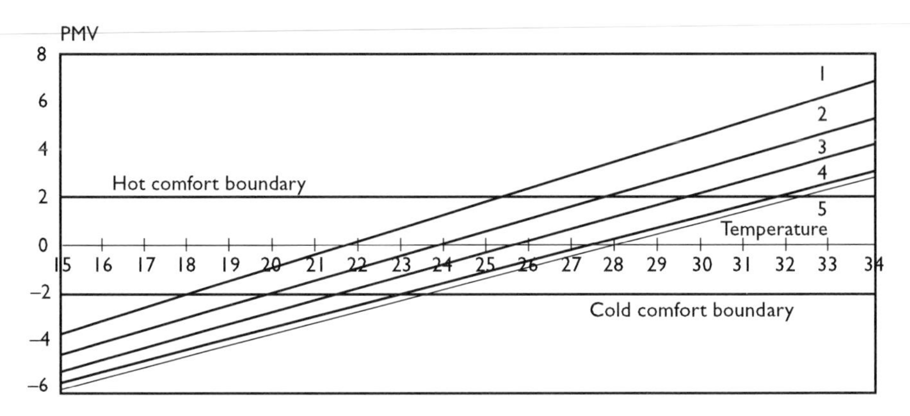

Figure 6.1 Temperature plotted against predicted mean vote (PMV) for five types of clothing: (1) socks, shoes, briefs, light long-sleeved shirt, tie, light sweater, vest, jacket, heavy trousers; (2) socks, shoes, briefs, light short-sleeved shirt, tie, jacket, light trousers; (3) socks, shoes, briefs, light short-sleeved shirt, light trousers; (4) flip-flops, shorts, light short-sleeved shirt; (5) no clothes at all!

different types of clothing and comfort votes. It shows a set of conditions which can easily be achieved during the summer in non-air-conditioned British buildings.

Thus I am advocating that the design summer temperatures in the UK should be modified so that we should specify that on a few days in the year people might want to work dressed in shorts, a short-sleeved shirt and sandals. The temperature of the building during the day would fluctuate so that they would probably want to put on a jersey in the morning and take it off during the afternoon. When we come to winter conditions, a similar set of adjustments about clothes should be adopted; I would like some attention paid to the design of clothes which allow free and easy movement and comfort at quite low air temperatures. We could then design buildings which needed no heating.

In order to prevent buildings from getting too hot in the summer, they need to be cooled down by ventilation at night. There have to be ventilation openings which can be left open without harm from intruders or excessive wind and rain. This is a design problem and it can be solved. The building needs to be cooled down at night to as low a temperature as is tolerable. With clothes – socks, shoes, briefs, a light short-sleeved shirt, sweater, jacket – the building is comfortable in the morning at about 19°C. Then during the day the building and the environment heat up; a person then removes the sweater and jacket and possibly loosens their tie, and is still reasonably comfortable at 19°C to 30°C. You might narrow the range and say 20°C to 28°C. The heat released into the building by computers, people, and light has to be absorbed by the structure. The ventilation air cannot remove heat from the building during the day unless the building is hotter inside than outside, and that is against the comfort strategy. Thus the energy has to be coupled to the thermal capacity of the building. It's no good changing or increasing the thermal capacity of a building if the heat cannot get from the building into the thermal store. In electrical jargon, the capacity must be coupled to the source. The heat in the building flows into the walls, floors and ceiling through the surfaces and the surfaces have a thermal resistance which is not negligible, in fact it is critical. The ultimate coupling from a room through a surface is 8 $W/m^2 K^{-1}$. The degree of coupling to an infinite thickness of concrete is 6 $W/m^2 K^{-1}$ taken over a 24 hour period. Thus it is important to get the surface area up. Tall spaces have a bigger area exposed to a room than low spaces. Small rooms have more surface area per unit floor area than large ones. Once there is enough surface area, the next stage is to ensure that the surfaces are sufficiently heavy to be able to go on absorbing heat for a substantial part of the day, and that means we should avoid lightweight plasterboard or thin metal sheeted partitions and use instead dense concrete or block. We should avoid lightweight false ceilings and think carefully about the floor. Furniture has surfaces and the furniture is thermally coupled to the room. Once one takes the coupling of the air boundary layer into account, there is not much point in providing more thermal capacity than 40 or 50 mm of dense concrete facing onto a room. I am not dealing here with the thermal capacity which enables a basement to maintain a stable temperature with very small temperature fluctuations providing the heat flows are small over a long period. There is an issue here about a potential conflict between the requirements for the thermal capacity and the requirements for acoustic absorption.

Acoustically absorbing surfaces are inherently bad at storing heat. For an office, the acoustic environment needs some thought. Let us consider privacy. In a noisy cocktail party the privacy is almost perfect. It is possible to stand opposite somebody with your nose about 100 mm from their nose and shout at them as loud as you can. They cannot understand a word of what you say. Nobody can hear anybody else's conversation. The reason for this is that the reverberant sound level field in the space is greater than the noise of a single person's voice, even when a few centimetres from the mouth. For an open plan office you want to be able to have a conversation with another person sitting, say, a metre away so that they can hear your voice but the general population cannot. That means the reverberant sound level should be comparable with the direct sound from a person about a metre away. That is a perfectly well defined noise level, and if there is one person per 10 m^2 in an office all talking then it is perfectly possible to calculate the amount of absorption needed in the space. A carpet on the floor will probably do, and then that leaves the ceiling as an exposed dense thermal capacity surface.

I think that deals with the thermal and acoustical problems. We come now to lighting. In buildings where people work, light tends to represent something like half of the CO_2 production of the building. The current consensus is that light levels in buildings should be reduced, and 300 lux is often suggested as a suitable light level. It is common to observe that buildings with apparently good provision for natural lighting are lit with electric lights throughout the day during the winter. One might think that the occupants are being 'naughty' and if the lights were properly controlled by automatic means, then they would be turned off and energy would be saved. Imagine yourself coming to work in the morning in the winter before dawn. The office is dark, the lights are on. After a time it gets light outside and natural light is added to the electric light by which you work. Ideally, the controls would turn the lighting down to maintain a constant 300 lux light level, thus saving energy. You might be satisfied by this strategy, but most people enjoy a little more light and leave the lights on. In our office it is not popular to turn off the lights until the contribution made by the electric light is a small proportion of the total light available, so on bright sunlit days the lights can be turned off and the light level reduces from, say, 1500 to 1200 lux. In my view, buildings should be designed with this in mind so that we really do have generous light. Of course, this brings two dangers: during the winter the heat loss at night is excessive and in summer the heat gain from bright sun is also excessive. The Victorians invented curtains, shutters and blinds to control these disadvantages of windows and I think we should have insulated blinds or shutters which close at night during the winter and close when there is excessive heat gain from light during hot summer days. Roller shutters have the advantage of providing security against intruders.

I have not talked much about mechanical and electrical engineering, because I hope the future does not lie in that sphere. I can design such installations, but I hope they will not be required.

References

Humphreys, M. (1965) What causes thermal discomfort? Proceedings of the Workplace Comfort Forum.

ISO (1994) ISO. 7730: Moderate thermal environments: determination of the PMV and PPD indices and specifications of the conditions for thermal comfort, Geneva. International Standards Organisation.

Chapter 7

Stress and the changing nature of work

Valerie Sutherland and Cary L. Cooper

One consistently emerging theme identified with organisational life in the 1990s is that of constant change, and it is likely that this trend will continue into the next millennium. In addition to predictable life event changes, we face endless modification to our work structure and climate, much of which is fuelled by rapid technological development. Many have raised concerns about the impact of such change on productivity, performance and quality of life. As Richard Hooker (1554–1600) said, 'change is not made without inconvenience, even from worse to better'. The implication is that exposure to change is, in some way, costly to the individual, business and society. Informed organisations are only just beginning to understand the real costs and acknowledge the notion of '**healthy workforce – healthy organisation**'. Thus, is it necessary to understand the relationship between 'change' as a source of stress and our responses to it in both physiological and behavioural terms. However, if change is inevitable, dysfunction and/or distress is not, since it is mismanagement of the change process that is damaging in its consequences. Identification of sources of stress or pressure by diagnosis of potential problems informs action. This proactive approach helps to prevent problems and the need to rely on costly, often ineffective, curative strategies for stress management in the workplace.

Introduction

Research evidence indicates that a wide variety of workplace conditions cause stress, strain or pressure which is associated with a range of physical and psychological ill-health problems. Costs to the individual, business and society are well documented, and informed organisations are only just beginning to understand the real costs of mismanaged stress in terms of business profitability. It is clear that health, well-being and quality of work life are associated with performance and productivity, and so understanding stress and pressure at work is vital if we wish to create a productive workplace. However, for many people at work the changing nature of the work environment is a potential source of stress and pressure which must be managed in a positive way if we are to remain both healthy and productive.

Constant change has been the dominant theme of organisational life in the 1990s, and this pattern seems likely to sustain as we move into the next millennium. While we endeavour to meet the demands associated with predictable life event changes, we must continue to face the endless reshaping of our work structure and climate, embedded in the changing nature of society. Change, it is said, brings progress, improvement to our quality of life, stimulation and variety which relieves us from boredom. Indeed, why would we wish to make changes for the worse? So, why express concerns about the concept of 'change' and use it within the same sentence as the word 'stress'? Therefore, our first objective is to raise awareness about the concept of 'change' as a stressor and the notion that it is mismanaged stress which is damaging in its consequences and costly to the individual, business and society. Next, by considering some of the changes that seem to be ongoing in contemporary work environments we can begin to identify the options available for the effective management of stress which could minimise or eliminate problems that lead to poor productivity and reduced levels of well-being and unhappiness. First, we address the notion that 'change' is a potent source of stress.

'Change' as a source of stress

For many the work environment has become a world of rapid discontinuous change, requiring us to live in a state of transience and impermanence. This situation is potentially damaging because energy is expended by constant adaptation to stimulation from the external environment. Thus, change is a powerful stressor because it necessitates adaptation: whether it is a negative or positive experience, welcomed, feared or resisted, this adaptation or adjustment requires energy (Selye, 1956). Our energy resources are not infinite, and so breakdown of the system, in part or total, will ultimately occur. In Selye's view, impairment of function and structural change are wholly or partly linked to adaptation to stimulation (i.e. arousal). Exposure to a continued state of arousal results in wear and tear on the body which in the extreme results in exhaustion, collapse and finally death of the system. If we accept this theory we can concur with Hans Selye's hypothesis that the only person without stress (i.e. arousal) is a dead person. In Selye's terms, stress should be viewed as, 'stimulation to growth and development . . . challenge and variety . . . the spice of life', and so it is **any** stimulus, event or demand impacting on the sensory nervous system. When our perceived ability to meet a demand exceeds our perceived ability to cope with that demand, the resulting imbalance is acknowledged as a state of stress (that is, we experience unwanted pressure, a lack of control and/or an inability to cope). Stress, therefore, is a subjective experience and is 'in the eye of the beholder'.

This explains why in a given situation one person will be highly distressed, yet another seems to prosper and thrive.

All too often we try to cope with the demands of exposure to change by resorting to maladaptive ways of coping. We drink alcohol because we believe it gives us confidence or helps us to sleep or relax; we drink lots of strong coffee to gain the 'buzz' necessary to sustain long hours of working without a break; we smoke cigarettes to calm our nerves; we use various pills and potions to ensure sleep, or 'pep' ourselves up; or we eat 'comfort' foods, particularly sugars and fats with low or poor nutritional value. These forms of coping render us less fit to cope with the demands of change and in the long term actually become a source of stress when the addiction exacerbates the problem.

In evolutionary terms the stress response was adaptive and vital for survival because various bodily functions/mechanisms were activated in order to prepare the body to respond by either fighting or fleeing from a threat/source of stress. We are physiologically primed to take some form of action which is now denied us in contemporary society, particularly for those in sedentary occupations, because we are not able to release our aggressions by punching someone on the nose when the pressures become intolerable, but neither can we run away from the situation. Societal pressures and job role demands require us to stay and cope without showing sign of weakness. Although we develop some resistance to stress, the energy needed for adaptation becomes depleted if we remain in stressful circumstances for an extended period of time. We cannot maintain a state of resistance indefinitely, because biological activity causes wear and tear on the body which leads to various forms of illness/diseases, and/or weakens our resistance to disease.

Counting the cost of mismanaged stress

Response to stress is not only in physical terms: we can experience psychological, emotional or behavioural reactions. Both directly and indirectly, adaptation to change as a source of stress has potential negative impact which is costly to the individual, the organisation and society. This fact is well documented by a surfeit of studies and illustrates that the effective management of stress in the workplace has tremendous potential benefits, in both humanistic and monetary terms. The costs of mismanaged stress can be high as a result of:

- ill-health – coronary heart disease; certain cancers; mental illness and a wide range of minor health problems including migraine, headache, stomach ulcers, hay fever, skin rashes, insomnia, panic attacks, feeling anxious or depressed, irritability, poor concentration, impotence and menstruation problems
- premature death

- forced early retirement
- sickness absence
- high labour-turnover
- poor performance and productivity
- ineffective managerial or supervisory style
- unsatisfactory employee relations
- low levels of motivation
- job and life dissatisfaction
- poor safety performance
- accident vulnerability (at work and/or while driving)
- delayed recovery from illness or accident
- poor health behaviours – e.g. alcohol, drug abuse problems/lack of exercise/dietary problems
- marital/relationship problems
- lack of self-esteem
- increased insurance premiums
- stress litigation.

The workplace – change and more change!

'Change is here to stay' is an old adage whose truth permeates all our lives (Cooper *et al.*, 1988) and it would seem that exposure to change can result in adverse and costly outcomes. In this section we will consider the nature of workplace changes and some of the strategies available to minimise the costs of exposure to change which is inevitable. These will be considered under the following headings: changes in the nature of the job; the job role; relationships at work; the concept of 'career'; and the 'organisational structure and climate'.

Changes in the job itself

A key change for many of us in 'volume of workload' and 'demand' means having too much rather than too little work to do (although both are potential stressors). Aided by rapid technological development, many organisations have 'streamlined, downsized, or rightsized' in order to continue to meet increasing competition from home and foreign markets. Therefore, fewer people are doing more and more work. Heavy volume of work, time pressures, difficult and demanding deadline pressures and lack of, or uneven distribution of, resources are some of the acknowledged pressures. A recent Institute of Management Services survey found that 47 per cent of a sample of 1,100 managers reported that their workload had increased greatly in the past year (1995–1996); nearly six in ten respondents claimed that they always work in excess of their official working week, while one in seven always work at weekends. Only half of the respondents looked forward to going to work, and 65 per cent felt that their professional and personal lives were not in balance. Unreasonable deadlines and office politics were identified as key sources of pressure.

Rapid development in technology was originally thought to be stressful because it was associated with deskilling of jobs. These fears seem to be unfounded (Lindstrom, 1991), and it would appear that the method of introduction of the new technology is more important as a potential source of pressure (Daus, 1991). Continual development of computer systems leads to constant changes, requiring retraining. The 'leaner' workforce experiences the pressures associated with trying to release colleagues for training in addition to time off needed for holidays and during periods of sickness absence, while trying to maintain performance and production demands. The workforce is required to become multi-skilled so that maximum use is made of the costly investment in new technology (plant and equipment), and the use of new technology often results in pushing responsibility for decision-making lower down the organisation.

Computer-led tasks are likely to be monitored, for example, in terms of number of key strokes, error rate or task completion rate, causing the worker to experience increased control and lack of discretion in the job. Lack of autonomy and discretion in a high-demand work condition defines a job as a 'strain' job, according to Karasek (1979), and is associated with reduced psychological and physical well-being. Thus, increasing the level of job control one has would be one way of minimising the impact of a high level of demand in a job.

In addition to the pressures of adapting to the changes associated with increased volume of work, the introduction of new technology and lack of job control, hours of working and patterns of employment are also altering. The global restructuring of production and the resulting shift of focus from manufacturing industries to the services sector and recessionary climates have caused the loss of jobs for many, and job insecurity for those who have remained in employment. Advent of the 'contingent worker' has led to changes in the nature of employment and the associated strains and pressures for individuals directly and indirectly employed (i.e. the differences that exist between 'core' and 'periphery' workers). Organisations now 'contract-out' many services, and staff are given fixed-term contracts. Jean Hartley (1995) provides a comprehensive account of this, but the following offers a flavour of the changes.

- Part-time employees made up 15.5 per cent of the workforce in 1971; this increased to 26 per cent by 1991 and is estimated at 32.3 per cent by the year 2001 (TUC, 1994). Many organisations now refer to these employees as 'key-time' workers rather than 'part-time', in order to signify the importance of this form of employment to their businesses (Hartley, 1995).

- In 1994 there were 157,000 temporary employees in the UK, and this represented a 10 per cent increase over the previous year.
- Women are more likely to occupy part-time jobs than men, and ultimately are more likely to be employed than men if the trend towards part-time working continues.
- Legislative changes have decreased the influence and scope of trade unions and membership has declined; since women are less likely than men to join trade unions, this decline will continue if action is not taken to rectify the problem.

Contingent employment, suggests Christensen (1995), is a cost framework in which the worker is perceived as a commodity rather than an investment. As such it creates a divided workforce where one group is perceived as a cost while the other, 'core' worker group is seen as 'the investment'.

Such changes described as intrinsic to the job will combine with existing pressures, which means that many of us will remain in a high state of arousal for much of the working day without any natural release of the normally protective 'fight' or 'flight' responses against exposure to stress. Since many also have sedentary jobs which require us to expend very little physical energy, and those in full-time employment are working very long hours, it is unlikely they will engage in physical activities outside work because they have no time, feel burned-out, psychologically exhausted, and/or depressed or anxious because they are not coping. Therefore, the problem becomes compounded and the vicious downward spiral of stress begins.

The changing nature of the job role

Changes to role structure are common as companies continually re-invent themselves, and change is often stressful because we try to resist it. The impact of change can cause role ambiguity (lack of clarity about the task) or role conflict (for example, coping with the conflicting demands of quantity versus quality, or safety versus quantity). Role ambiguity has been associated with tension and fatigue, leaving the job, high levels of anxiety, physical and psychological strain, and absenteeism (Breaugh, 1981). Role conflict has been associated with absenteeism (Breaugh, 1981), job dissatisfaction, abnormal blood chemistry and elevated blood pressure (Ivancevich and Matteson, 1980).

A situation is exacerbated if the workforce perceives lack of managerial and supervisory support in the workplace. Significant downsizing of the workforce also affects management-grade personnel and so they are experiencing high workload demands, long hours of working, and the realisation that they might not be available when needed. Feeling unable to fulfil others' expectations of one's job role is now a commonly cited source of pressure experienced by personnel with responsibility for other people. Since it is commercially essential to run a business continuously, it is also necessary to work a shift system. Managerial staff still mainly work day-shifts, with only a skeleton staff available for night-shift duties, and this compounds the problems associated with lack of availability of one's supervisor/manager when needed.

Change in the nature of our relationships at work

A recessionary climate which is characterised by job insecurity and the use of contingent workers is likely to lead to divisiveness, rivalry, unhealthy competition for jobs and poor interpersonal relations at work. These are defined as having low trust and low supportiveness. Mistrust is positively related to high role ambiguity, inadequate interpersonal communication between individuals, and psychological strain in the form of low job satisfaction and decreased well-being (French and Caplan, 1972). Indeed, levels of violence in the workplace appear to be on the increase, thus creating another dimension to stress in the workplace, and this includes violence between staff, towards staff and to the managers and supervisors themselves. As Sartre said, 'hell is other people', and having to live and work with others can be one of the most stressful aspects of life. However, computerisation has taken over many jobs, and individuals often work in relative isolation for many hours at a time, with limited opportunities to socialise. Communication channels are limited and opportunities to develop strong social support networks are denied if the employee is restricted to a workstation base. Although CCTV and video-phone links for isolated employees may help to overcome such problems, a fear that 'Big Brother is watching' pervades, and is sustained in, a climate of insecurity and distrust.

Changes in the concept of 'career'

In addition to the pressures associated with starting, developing and maintaining a career, a mismatch in expectations, feeling undervalued and frustration in attaining a sense of achievement are common 'career' stressors. As the pyramid shape of the organisational structure typically becomes flatter and many job levels are removed, this 'flatter' structure provides fewer opportunities for career progression. This, and job insecurity which results from downsizing and the increased use of a contingent workforce, are potent sources of stress, leading to the observation that perhaps the nature of 'career' is undergoing a radical transformation. Disruptive behaviours, poor morale and poor quality of interpersonal relationships are associated with the stress of perceived disparity between actual status within the organisation and expectations, while threat of job loss is a potent source of stress linked to several serious health problems, including ulcers, colitis, alopecia and increased muscular and emotional complaints. Since contemporary employment relations are in transition, the demise of loyalty and the need for employees to take care of themselves are viewed as a sign of the times (Hirsch, 1989), thus it might be said that the nature of 'career' is changing.

Changes to the organisational structure and climate

Threats to freedom, autonomy and decision-making imposed by the organisational structure and climate, and the way the organisation treats its people, are potential sources of stress. Lack of participation in decision-making processes, lack of effective consultation and communication, and unjustified restrictions on behaviour have been referred to above, and are associated with negative psychological mood, escapist drinking and heavy smoking. Buck (1972) found that managers and workers who felt 'most under pressure' reported that their supervisor always 'ruled with an iron hand' and rarely allowed participation in decision-making or trying out new ideas. Increased control and opportunity to participate has benefits in terms of improved performance, lower staff turnover, and improved levels of mental and physical well-being and accident reduction (Sutherland and Cooper, 1991). However, employees will often resist the offer to adopt a more participative style of working, since they suspect that the proposed changes in working practices to improve productivity and quality etc. are required to gain better competitive advantage. Levine (1990) suggests that employee support for a participative work climate is more likely when the industrial relations system within the organisation is characterised by:

1 the presence of some form of profit/gain-sharing
2 job security
3 ways in place for the development of group cohesiveness
4 guaranteed individual rights.

Many organisations are trying to 'empower' their employees, but encounter problems because they fail to realise that a workforce acclimatised to a dependency culture, where they are simply told what to do (and are not expected to solve problems and/or make decisions for themselves), cannot move easily or quickly to a condition of mutual dependence (i.e. control shared by mutual agreement) or to interdependence, which is characterised by flexibility, interchange of activities, joint decision-making and sharing of control, and which is vital for the success of the empowerment process. If management tries to give up control too quickly, or employees try to escape from being controlled when the authority figure will not relinquish control, the workforce will become counter-dependent (Cox and Makin, 1994). In these situations there will be a 'fight-back', with acts of rebellion such as overtime bans, wildcat strikes, and works-to-rule. A need to overcome a state of learned helplessness, low levels of confidence and lack of esteem is important in successfully changing to an empowerment model of working (McGrath, 1994). This explains why the changes proposed as 'stress reducers' sometimes become more damaging than the original source of strain.

Diagnosing occupational stress

Understanding the nature of stress and thinking positively and proactively about stress management, rather than taking a defensive, self-blaming stance, means that we must accept that each of us, at various times during our life, may be vulnerable and will need to know how to manage actively (Cooper amd Straw, 1998) and positively a potentially stressful situation without resorting to maladaptive ways of coping (e.g. excessive alcohol and nicotine; drug dependence; lack of exercise; comfort eating, particularly sugars and fatty foods which have low/poor nutritional value), which render us **less fit** to cope with the demand. Indeed, members of an organisation are unlikely to come forward and deal with potential problems within an organisation if there is a fear of being identified and labelled as a 'non-coper'. Therefore, conducting a risk assessment is the first step towards successful stress management. A confidential stress audit will:

1 measure sources of stress, stress outcomes (i.e. performance indicators), individual moderators of the stress response and biographical/job demographics of the workforce
2 identify predictors of performance and well-being
3 identify staff attitudes to options for the management of stress (Sutherland, 1993).

A variety of techniques and measures are available for this, including interviewing, focus group sessions, questionnaire administration (including standardised measures such as the Occupational Stress Indicator), and medical examinations. Analysis of the results could identify differences, for example, between departments, job grades, gender, age, length of service, location, etc. Audit benefits include the following.

1 It is a proactive rather than a reactive approach to managing stress at work.
2 It can identify organisational and individual strengths and weaknesses (similarly to an appraisal/training-development needs analysis – and so helps us to target scarce resources).
3 It can identify the level of stress management required (primary, secondary, tertiary); this includes guidance in the planning of Organisational Development (OD) strategies.

4 It provides a baseline measure from which to evaluate subsequent interventions.
5 It makes stress a respectable topic for discussion in the workplace.

Options for the management of stress in the workplace

The contemporary view is that stress management should be tackled at more than one level if it is to be successful, and both the organisation and the individual should actively be involved (Burke, 1993). It is suggested that we actively manage stress by:

1 eliminating or minimising the stressor at source
2 helping the individual cope with a source of stress that cannot be changed.

Both 'preventive' and 'curative' stress management options are available. We must decide whether we take the problem away, or manage the situation more effectively (i.e. primary level interventions – stressor-directed); do we need to help improve response to a source of pressure which can not be removed or minimised (i.e. secondary level interventions – response-directed); or do we already have problems associated with exposure to stress and a need to treat victims of mismanaged stress (i.e. tertiary level interventions – symptom-directed)? (Cooper, Linkkonen and Cartwright, 1996).

Primary level interventions – preventive stress management strategies

This type of intervention is directed at the source of stress. Job and task re-design, and changes to the macro environment are part of this approach to stress management. Job and/or task re-design includes more job rotation and job share opportunities so that no-one is exposed to a high-stress job for any substantial length of time. Career/job development opportunities are improved when levels of responsibility are increased through horizontal job enlargement or by 'vertical loading'. Providing more flexible work patterns can also eliminate the stress problems of travelling to work and coping with child-care/dependant demands. Provision of crèche facilities would be placed in this category as a stress management strategy.

As already noted, lack of worker control is acknowledged as a potent source of pressure, and increased perceptions of control seem to be important for job satisfaction, health and well-being. A variety of strategies exist to improve perception of worker control and decision-making, including building and developing semi-autonomous work groups, quality circles/groups and/or health circles. Ultimately they aim to re-design any system practice which induces stress at work and create a better balance between demands and level of worker control. Other sources of pressure can also be eliminated or reduced by making changes in the macro environment. This would include an audit of ergonomic factors in the workplace (e.g. the layout (open plan offices), noise, heat, cold, ventilation levels). Also, examination of the appropriateness of the organisational structure, climate and culture, safety climate and management/supervisory style (e.g. identification of 'stress carriers'!), might usefully identify sources of stress which could be more positively managed. For example, a supervisor with a 'type A behaviour' style can create a stressful milieu for their subordinates, and an abrasive manager will create an inordinately stressful work environment for staff. (Type A behaviour predisposes one to a stress-related illness; it is shown by someone who is rushed, impatient, time-conscious, overly competitive and aggressive in his or her relationships.)

Secondary level interventions – preventive stress management strategies

This type of programme is 'response-directed' since it improves the response to a source of pressure in order to avoid a negative outcome; it thus operates at the level of 'the individual employee' or 'work group' and includes a variety of skills training options, usually following an assessment-focused stress management programme (i.e. a training needs analysis). For example:

- assertiveness training – being able to negotiate or ultimately say 'no' without becoming aggressive or non-assertive is a useful stress management technique
- improved coping skills – understanding adaptive coping rather than using the maladaptive styles of coping which ultimately render us less fit to cope with pressure
- interpersonal/social skills training
- leadership skills training
- time management skills
- cognitive reappraisal of the situation, which aims to improve the balance between perception of 'demand' and our ability to cope with it (e.g. by examining faulty thinking)
- relaxation training/biofeedback
- type A coronary prone behaviour management.

Other options in this category aim to keep the individual fit to cope with the pressures of work and living, and many organisations already have this type of wellness programme in place. For example:

- exercise/keep fit programmes

- healthy lifestyle management
- dietary advice
- smoking/alcohol cessation programmes
- advisory clinics on drugs.

Tertiary level interventions – curative stress management strategies

At a tertiary level it is possible to provide stress management options for individuals who are suffering from the effects of exposure to strain and pressure. Many of the strategies described above take time to implement, and so it will be necessary to have in place some form of programme to catch the people who 'fall through the net' and become victims of exposure to stress. These include:

- provision of an employee assistance programme (EAP), i.e. access to a confidential telephone/counselling service for employees – this can also play a 'preventive' part in a stress management programme if the EAP company provides anonymous feedback about the types of stress involved, so that the organisation can implement a 'preventive' programme where possible
- internal/external psychological counselling services
- opportunities for a career sabbatical (can also be a 'preventive' strategy if implemented before the individual becomes a casualty of exposure to stress)
- development social support networks – social support can play a significant part in enhancing the level of employee well-being, particularly social support from a boss; self-managed work teams, action groups, etc. play a part in developing social support networks, particularly for employees who work in relative isolation in a computer-led work environment.

Conclusions

Some degree of pressure is an inevitable part of living in a constantly changing work environment. While this can be a spur to improved performance and motivation to respond, it can also be damaging in its consequences if it is mismanaged. Change is inevitable but distress is not, because many options are available for the management of stress. While the concept of 'healthy individual – healthy organisation' is not enshrined in legislation, common law and the general duties under Section 2 of the Health and Safety at Work Act 1974 require employers to ensure the health, safety and welfare at work of their employees as far as reasonably practicable. Thus, in recent years the practice of introducing stress management interventions has steadily grown. However, most of these have been aimed at the individual, either by treating people who have become victims of exposure to stressful conditions, with the aim of getting them fit to deal with stress, or by helping them to cope more effectively with the demands and strains of organisational life. Research evidence indicates that these strategies alone will not be totally effective, and organisations must begin to identify negative occupational stress and work to remove or minimise these problems. A stress audit (also recognised as a psychological risk assessment) is required to identify sources of stress and inform action. Recent audits show a clear role for those involved in the design and planning of productive work environments. In this way they can contribute to a proactive organisational approach to stress management which aims to prevent problems and the need to rely on costly, often ineffective curative strategies that are currently used to keep people healthy, happy and productive at work. Indeed, effective stress management will put you in control of change, which will be an inevitable part of working and living in the twenty-first century.

References

Breaugh, J.A. (1981) Predicting absenteeism from prior absenteeism and work attitudes. *J. Appl. Psychol.*, 36, 1–18.

Buck, V.E. (1972) *Working Under Pressure*. New York: Crane, Russak.

Burke, R.J. (1993) Organizational-level interventions to reduce occupational stress. *Work and Stress*, 7, 77–87.

Christensen, K. (1995) *Contingent Work Arrangements in Family-sensitive Corporations. Boston, MA: The Work and Family Institute, Boston University.*

Cooper, C.L., Cooper, R.D. and Eaker, L.H. (1988) *Living with Stress*. Harmondsworth: Penguin.

Cooper, C.L., Linkkonen, P. and Cartwright, S. (1996) *Stress Prevention in the Workforce*. Dublin: European Foundation of Living and Working Conditions.

Cox, C.J. and Makin, P.J. (1994) Overcoming dependence with contingency contracting. *Leadership Org. Dev. J.*, 15(5), 21–26.

Cooper, C.L. and Straw, A. (1998) *Successful Stress Management*, London: Hodder and Stoughton.

Daus, D. (1991) Technology and the organization of work, in A. Howard (ed.), *The Changing Nature of Work*. San Francisco, CA: Jossey Bass.

French, J.P.R. and Caplan, R.D. (1972) Organizational stress and individual strain, in A.J. Marrow (ed.) *The Failure of Success*, 30–66. New York: AMA Committee.

Hartley, J. (1995) Challenge and change in employment relations: Issues for psychology, trade unions and managers, in Lois E. Tetrick and Julian Barling (eds), *Changing Employment Relations: Behavioural and Social Perspectives*, 3–30. Washington, DC: American Psychological Association.

Hirsch, P. (1989) *Pack your Own Parachute*. Reading, MA: Addison Wesley.

Ivancevich, J.M. and Matteson, M.T. (1980) *Stress and Work*. Glenview, IL: Scott Foresman.

Karasek, R.A. (1979) Job demands, job decision latitude and mental strain. Implications for job redesign. *Admin. Sci. Q.*, **24**, 285–306.

Levine, D.I. (1990) Participation, productivity and the firm's environment. *California Man. Rev.*, **32**(4) 86–100.

Lindstrom, K. (1991) Well-being and computer-mediated work of various occupational groups in banking and insurance. *Int. J. Comput. Interaction*, **3**(4) 339–361.

McGrath, R., Jr (1994) Organizationally induced helplessness: The antithesis of empowerment. *Qual. Prog.*, **27**(4), 89–92.

Selye, H. (1956) *The Stress of Life*. New York: McGraw Hill.

Sutherland, V.J. (1993) The use of a stress audit. *Leadership Org. Dev.*, **14**(1) 22–28.

Sutherland, V.J. and Cooper, C.L. (1990) *Understanding Stress: A Psychological Perspective for Health Professionals*. London: Chapman and Hall.

Sutherland, V.J. and Cooper, C.L. (1991) *Stress and Accidents in the Offshore Oil and Gas Industry*. Houston, TX: Gulf Publishing.

Sutherland, V.J. and Cooper, C.L. (1994) Stress in the work environment, in *Human Stress and the Environment*, J. Rose (ed.) 131–160. Amsterdam: Gordon and Breach Science Publishers.

Part 2

The economic case for productivity

Chapter 8

The economics of enhanced environmental services in buildings

David H. Mudarri

The theme of this chapter is that the great disparity between the economic loss society suffers from poor indoor environmental quality and the cost necessary to improve it is caused in large measure by an imbalance in the marketplace. Private entities that want improved environments have been unable to translate this desire into an overt expression of market demand that would justify the expenditure and risk that the improvements require. Public policies which facilitate the easy expression of market demand by occupants and which ease the response to that demand by building owners and developers are recommended. These policies include the establishment of protocols of good building practices; a rational integration of energy and indoor environmental policies; and guidance and software packages for building owners and others that assist in calculating bottom-line impacts of indoor environmental quality projects.

Magnitude of economic loss associated with poor indoor environments

While economic loss from poor indoor environments has not been rigorously studied, there is good evidence that the economic loss to industrialised nations is substantial. The United States Environmental Protection Agency (USEPA, 1989) estimated a total annual cost of indoor air pollution in the USA of approximately $6 billion due to cancer and heart disease,[1] and approximately $60 billion due to reductions in productivity. The productivity loss was derived from self-reported survey data and represents an average productivity loss of approximately 3 per cent for all white-collar workers, or approximately twice that figure for just those workers reporting some loss.

Raw *et al.* (1990) looked at self-reported productivity loss in buildings in the United Kingdom. From their data it appears that about half of the office workers in the UK experience fewer than three indoor air quality symptoms, with little or no reported productivity loss, but that the remaining workers report productivity losses of about 7 per cent on average, or just over 3 per cent when considering all workers.

These estimates are based on self-assessments of productivity losses, and therefore must be interpreted with caution. But other evidence, including experimental evidence, supports the overall conclusion that the productivity losses are large (Wyon, 1996; Kroner *et al.*, 1992).

Mudarri (1994) estimated the costs and benefits of a proposed national ban on smoking in public buildings. The portion of the estimate of benefits over costs associated with the reduced impact of second-hand smoke on non-smokers was in the range of $30–$60 billion per year,[2] while the portion associated with reduced building housekeeping and maintenance was in the $4–$8 billion range.

Fisk and Rosenfeld (1997) estimated that annual savings and productivity gains from building improvements in the USA could be $6–$9 billion from reduced respiratory diseases, $1–$4 billion from reduced asthma, $10–$20 billion from reduced sick building syndrome, and $12 to $125 billion for improved worker performance that is unrelated to health. Example calculations also suggest that the value of the benefits from improvements might be between 18 and 47 times the cost of those improvements. And Dorgan Associates (1993) estimated that a productivity gain worth $55 billion annually could be achieved with an $88.6 billion initial investment (payback of 1.6 years) with an annual cost of $4.8 billion to sustain the improvements.

While these estimates differ in their detail, and while it is admittedly difficult to quantify national costs and benefits, all these studies are consistent in their implication that even modest improvements in the quality of environments inside buildings would result in an extraordinarily large social benefit. So it is worth while discussing why there is a problem, and what can be done about it.

Problems with the market for indoor environmental quality

With such apparent large net economic gains to be derived from improved building practices, why doesn't the private sector simply jump at the opportunity? Perhaps from the standpoint of the building owner or company executive, such changes would not be as profitable as the economic analysis above suggests. Or perhaps they could be profitable but would require some changes in old patterns of behaviour.

Indoor environmental quality is seldom a meaningful consideration in the most fundamental building market transactions – the sale and purchase of design and construction services, the sale and purchase of the building, the leasing of rental space, or the operation and maintenance of buildings. In part, this is because the market entity which benefits from improved indoor environmental quality (e.g. the tenant/occupant) is not the entity capable of providing it, or that must pay for it (e.g. the building owner). While it is true that this creates a barrier to having the latter respond to the former's desires, it should not prevent a response. After all, competitive markets are supposed to ensure that the desires of those that purchase or rent space is satisfied by those supplying space, as long as they are willing to pay for it. The fundamental problem here is that those who want and are willing to pay for improved indoor environmental quality (e.g. building tenants) are not asking for it during the market transaction (e.g. rental of space), so that those capable of providing improvements (e.g. building owners) are not motivated to do so.[3] The **desire** for improved environments has not been translated into a **market demand**, even though there is ample evidence that the desire exists.

Tables 8.1 and 8.2 suggest that the desire for improved indoor environments is widespread. Two independent surveys of major building tenants place indoor air quality and the related issues of thermal comfort and HVAC performance high on the list of major tenant complaints. Despite this, when it comes to seeking available space, these issues are rarely raised. Tenants look for, and owners market their space according to its location, its appearance, parking, and other items that are visible and tangible. Such measures as continuous and adequate outdoor air flow, HVAC operational parameters, the housekeeping and maintenance programme, or other factors that directly affect indoor environmental quality, but are not visible or tangible, rarely become major considerations in seeking space or in negotiating the rental agreement. If such demand is not expressed in the purchase and sale for space, it will not trickle up into either the design and construction, or the sale and purchase of buildings.

Table 8.1 BOMAI survey: office tenant moves and changes (BOMAI, 1988)

Worst problem	*% of responses*
HVAC and indoor air quality	30
Elevators	12
Building design	7
Loading docks	6
All other	45

Table 8.2 IFMA survey: top five complaints of corporate tenants (IFMA, 1991)

1. Too hot
2. Too cold
3. Storage and filing space
4. Indoor air quality (tie)
5. Janitorial service (tie)

Thus, the cost-conscious developer or building owner has every motivation to **avoid** expenditures that might improve the indoor environment. Commissioning new building construction to ensure that building systems are designed, integrated, and installed correctly is an expense; writing operation and maintenance manuals to instruct the building engineer how to operate and maintain the systems is an expense; and actually performing adequate maintenance once the building is constructed is an expense. For items such as these, for which there is no discernible market demand, the operating phrase is often 'make sure you meet code, but do no more'. These expenses are avoided. It is not surprising, therefore, that many indoor environmental problems are traced to faulty design, faulty construction, or faulty operating and maintenance procedures.[4]

The lack of expressed market demand for improved indoor environmental quality creates an enormous chasm between those that would benefit and those responsible for providing it. For example, in addition to inadequate maintenance, attempts to reduce energy costs can also create indoor air quality problems if they unduly reduce outdoor ventilation rates, or if they unduly relax the ability of the ventilation system to control temperature and humidity to suit occupant needs. But the savings resulting from reduced maintenance or from energy conservation strategies are insignificant when compared to the potential health, comfort, and productivity losses of the occupants. Consider, for example, the relative value of building costs (Table 8.3). Labour costs are clearly the most significant cost item in most commercial buildings. A 30 per cent saving in HVAC energy and/or in HVAC maintenance translates into something of the order of $0.25–$0.35 per square foot. However, the 3 per cent loss in productivity associated with poor environmental quality would correspond to approximately $4.50–$6.00 per square foot, with unmeasured additional impacts of discomfort and poor health. From a holistic view this makes no sense – society is sacrificing values of over $4.50–$6.00 to gain values of $0.25–$0.35.

Tenants and occupants ought not to tolerate such gross suffering for such

Table 8.3 Some typical office building expenses

Item	*$/square foot*
Rental income	10.00–20.00
Operating expenses	
HVAC energy	0.75–1.00
HVAC maintenance	0.30–0.50
Cleaning	0.75–1.00
Grounds	0.30–0.60
Leasing	0.20–1.75
Tenant expenses	
Employees	150.00–200.00

minuscule savings, but tolerate it they must unless they know **specifically** what changes to ask for. And building owners who want to advertise the environmental quality of their building need to know **specifically** how to achieve it and get credit in the marketplace for doing so.

Public policy remedies

Translating the notion of good indoor environmental quality into specific generally accepted protocols would go a long way to enable the expression of market demand. Combined with specific provisions to reduce costs and risks, the market for indoor environmental quality could very well become healthy, active and robust.

Indoor environmental quality protocols

Indoor environmental quality does not come in a can or a box, and you cannot order it by the ton or by the gallon. It has no clear definition or metric. Rather, it is wrapped into the fabric of the design, construction, operation, and maintenance of the building in hundreds of different ways. To make it a marketable product, it first needs to be packaged in an identifiable way, as, for example, through the development of specific protocols or standards of practice for the design, construction, operation and maintenance of buildings. Fortunately, much has been learned over the past 15 years and there has been significant progress in developing elements of these protocols in the form of guidelines and standards for indoor air quality, ventilation, and green buildings and products. While this process has been slow and is still in its infancy, there are signs that some groups are packaging elements of these protocols into specific design or operation and maintenance services with the idea of marketing these services to building owners. The gamble is that the 'unexpressed market demand' for good indoor environmental quality will be recognised by some building owners who will use the certification associated with these services to gain a market advantage. **As a matter of public policy, encouraging, supporting, and participating in the accelerated development of these protocols is surely an important thing to do.**

Rationalising indoor environmental policies with energy policies in buildings

Since ventilation is so critical to indoor air quality, and since the energy cost associated with ventilation systems is an important building operating cost, a set of procedures which integrates the needs for energy cost reduction with the needs for good indoor environmental quality ought to be adopted and codified. The attitudes which pit indoor environmental quality interests against interests in energy efficiency are wasteful and unnecessary. This is why, for example, the US Environmental Protection Agency's comment on the proposed revision to ASHRAE Standard 62 encouraged ASHRAE to focus more on improving energy recovery technologies to reduce the cost of outdoor air rather than attempting to limit outdoor air in buildings unreasonably.

In addition, the professional literature is beginning to explore objectively issues of compatibility and conflict between indoor environmental quality and energy efficiency (Coad, 1996; Eto and Meyer, 1988; Eto, 1990; Lizardos, 1994; Mudarri *et al.*, 1996a, 1996b; Mutammara and Hittle, 1990; Reddy *et al.*, 1996; Rengarajan *et al.*, 1996; Shirley and Rengarajan, 1996; Steele and Brown, 1990). This is a necessary first step in the formation of a rational integrated policy. Based on current knowledge, an integrated energy and indoor air quality protocol might include the items identified in Table 8.4.

To evaluate the potential savings from energy measures which compromise indoor air quality (see Table 8.4) relative to other energy-saving options, an office building in a moderate climate is being modelled using the DOE-2 energy simulation program. This modelling is an extension of previous work (Mudarri *et al.*, 1996a, 1996b). Preliminary results suggest that the energy-saving actions which compromise indoor air quality are not significant when compared to the very significant savings attributable to basic load reductions and equipment efficiencies. This gives impetus to the hope that an integrated energy/indoor air quality protocol can be widely accepted. It is appropriate, therefore, that the US Department of Energy and the US Environmental Protection Agency have begun a project to develop such a protocol. It is being developed in conjunction with a wide variety of other stakeholders as an indoor environmental quality appendix to the International Performance and Verification Protocol (IPMVP) – a protocol for the measurement and verification of building energy improvements.[5]

Reducing the cost burden to building owners

Two notable developments in the energy field have been the concepts of '*demand side management*' and '*performance contracting*'. Demand side management reflects the utility industry's awareness that it is more cost-effective in the long run to answer the growing demand for energy by promoting energy efficiency than by building additional power plant capacity. This response has resulted in the utility industry offering building owners and others financial incentives or discounts in purchasing energy-efficient equipment. Performance contracting provides the mechanism by which energy savings from energy-efficient investment can be used to pay for its financing. In such an arrangement, an energy service company (ESCO) may contract with the building owner to evaluate energy savings opportunities, to implement them, and to arrange for financing. The energy savings are

Table 8.4 Proposed elements of an integrated building environment and energy policy

Measures to improve energy with little impact on indoor air quality (IAQ): should be pursued on the basis of energy potential alone

- shell efficiencies
- internal load reductions (e.g. lighting, computers)
- air distribution upgrades (fans, fan motors and drives)
- central plant upgrades (chillers and boilers)
- energy recovery systems

Measures to improve IAQ with little energy impact: should be pursued as a matter of good building practice where practical

- choice of materials, furnishings, cleaning products
- timing and method of major pollutant source applications (e.g. painting, pesticides)
- filter changing
- trash storage and collection
- pressurisation controls for garage entrances, loading dock entrances, print rooms
- smoking restrictions

Measures to improve energy and IAQ: should be pursued as a matter of good building practice where practical

- economiser HVAC control
- night precooling/night-time flush
- preventive maintenance
- cleaning and disinfecting HVAC equipment
- calibration of thermostats and controls, reduced duct leakage
- exhausting pollution/heat-generating equipment (e.g. copying rooms)

Energy efficiency measures that may compromise IAQ: should be avoided or carefully adjusted to ensure a healthy, comfortable, and productive environment

- decreasing outdoor air below IAQ standards
- VAV systems using a fixed outdoor air damper
- aggressive equipment downsizing unless based on year-round outdoor air and thermal control
- late start-up and premature shutdown of HVAC
- overly wide temperature/humidity control deadband to save energy

then directed towards financing the project, paying the ESCO, and reducing the operating costs for the owner. The exact way in which the energy savings are proportioned will vary. One objective of the IPMVP is to facilitate these arrangements by standardising measurement and verification procedures.

These arrangements offer an opportunity to institutionalise indoor environmental quality into market transactions involving building construction or retrofit. Many indoor air quality improvement projects require that the HVAC system be upgraded. In itself, this is a substantial expense and a great disincentive to cost-conscious building owners. However, new HVAC equipment offers opportunities for energy upgrades. Thus, the opportunities to reduce costs and risks by packaging these retrofits into an overall energy upgrade programme while using performance contracting and utility incentives to reduce costs can go a long way to removing this disincentive and potentially creating a positive incentive for long-term financial gain.

A large number of indoor air quality problems could be avoided if building owners would simply maintain their HVAC system in order to ensure that the system operates according to its design intent (USEPA/NIOSH, 1991). This may also result in significant energy savings. In a study of seven existing buildings in Minnesota, easily achieved energy saving opportunities consistent with the operating intent of the building owner were estimated to account for 8–21 per cent of the buildings' utility cost with an average payback of only 0.67 years (Herzog and LaVine, 1992). Considering just those aspects associated with the HVAC system, savings would be of the order of 2–20 per cent with an average of 10 per cent savings in energy costs.[6] This is why 'HVAC tune-up' is recognised as an important element of both an energy conservation and an indoor air quality programmme for commercial buildings (USEPA/NIOSH, 1991; USEPA, 1995).

Housekeeping is an important operating expense, perhaps equivalent to the HVAC energy budget. Not only is good housekeeping important to maintaining good indoor environmental quality (USEPA/NIOSH, 1991; Franke *et al.*, 1997), but a reasonable pollutant source management programme, including smoking restrictions, medium efficiency air filtration, and simple strategies for managing outdoor sources of contaminants (e.g. from the loading dock and parking garage) can reduce housekeeping costs. Savings in housekeeping from smoking restrictions alone have been estimated to be of the order of 10 per cent (Mudarri, 1994).

These three ways which reduce the cost of indoor environmental services – energy savings through HVAC equipment retrofit, energy savings through good maintenance, and savings in housekeeping expenses – makes indoor environmental quality an issue with potentially little or no cost burden. **If this cost containment strategy were to be combined with the ability of building owners to market and advertise the environmental quality of their space, there is a good chance that the building owner's net revenue potential would be enhanced.**

Bottom line analysis for building owners and managers

Let's look into the future for a moment, and consider the situation of a typical 200,000 square foot office building located in a temperate climate in the United States. Suppose the building is reasonably well run, but the owner has been hearing rumblings of dissatisfaction from one or two tenants about air quality and thermal comfort. The owner believes that his building is about average in terms of how well it is managed, and he has always had the impression that there isn't much more he can do about improving the indoor environment within his current budget. Then he picks up an article in the trade press that describes a newly internationally recognised[7] energy and indoor environmental protocol, and a new software package that shows building owners how to improve the energy efficiency and indoor air quality of a building at modest to negligible cost. He notes that the same software is available to building tenants. He is interested in this concept, but is also concerned that potential tenants might become aggressive in negotiating for space if such resources are available to them.

He obtains a copy. The software convinces him that the concept is viable for his building, so he hires a new kind of company which may one day come to be known as energy and environment service companies (EESCOs). This EESCO recommends a series of changes including a building tune-up, lighting upgrades, increases in outdoor air ventilation, upgrades to chillers and boilers, improved housekeeping to control outdoor pollutant sources, a non-smoking policy, and a rigorous HVAC maintenance programme. The EESCO recommends that this effort be achieved over a three-year period. The expense analysis of the software programme, based on the EESCO's recommendations, is reflected in Fig. 8.1. It reflects some initial expenses to remediate previously deferred maintenance, to tune-up the HVAC system, revise maintenance procedures, reorganise records, train personnel, and generally start the programme. He will have some capital expenses for new HVAC equipment and controls, which provide for basically a 4–5 year pay-back. The startup expenses and initial capital outlays are distributed over a three-year period. The capital outlays are financed over a ten-year period.

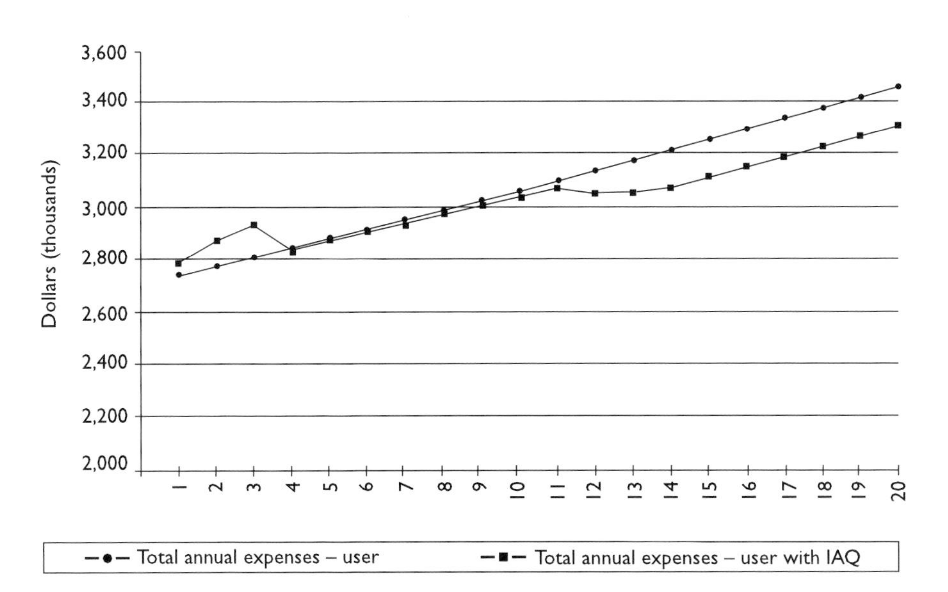

Figure 8.1 Annual expenses projections with and without energy/IAQ upgrades

However, the annual expenses associated with this programme are essentially paid for by the reduced operating costs. After about 10–12 years, the expenses are effectively reduced.

The owner now considers his revenues and leasehold expenses. Since the owner has been hearing rumblings of dissatisfaction from tenants concerning indoor environmental conditions, he considers the possibility that over the long haul, an improved indoor environment may convince some tenants to renew their lease rather than seek alternative space. This could be very advantageous to him. With the same rental rate, retaining a tenant will reduce leasehold expenses associated with buildout and brokerage fees. It will also allow him to avoid the delay in occupancy associated with a new tenant, as well as the need to offer free rent to attract new tenants. He quickly sketches out the difference between renewing the lease of an existing tenant with this new programme and having to seek a new tenant when the lease expires under the old programme (Fig. 8.2). This possibility is attractive and suggests some long-term gains even if just a small portion of his tenants reverse a decision to leave and decide to renew their lease instead.

Then he considers the potential impact on his 20-year pro forma budget. He loses about four months in occupancy every time a space turns over. In addition, on average, he is offering two months free rent for new tenants. He tries to recoup some of this expense with an increased rental rate for new tenants, but in a tight market, there is a definite trade-off between months of

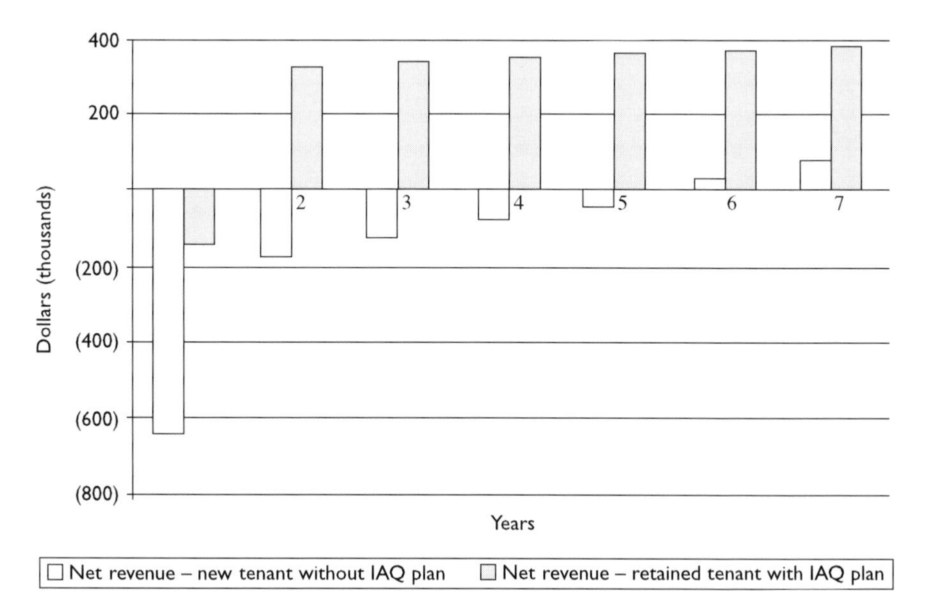

Figure 8.2 Net revenue comparison for a single lease with and without energy/IAQ upgrades

free rent, the time it takes to attract a new tenant, and the rental rate. He anticipates that he will use the upgraded features of the building as a marketing and negotiating tool. The software package shows him how his effective rents (rental revenue less leasehold expenses) will change with changes in occupancy and rental rates (Table 8.5). He estimates that if he undertakes the recommended improvements, his long-run occupancy rate might improve from its current 97 per cent to 98 per cent as he is better able to retain tenants. He postulates that, with this strategy, he might be able to increase his average rent modestly from $19.00 to $19.25 per square foot.

With these inputs, he uses his software package to run a 20 year pro forma budget (Fig. 8.3). The building owner is impressed with the results, and implements the package. He is modestly successful and decides to do the same with other buildings he owns. He does this for 20 years and, as he ages and matures, he reflects on how all his competitors are now doing the same. It's a matter of survival. Indoor environmental quality has become a

Table 8.5 Changes in effective revenues (i.e. revenues net of leasehold expenses) after improvements (before improvements: rents = $19/square foot; occupancy = 97%)

Annual rental rate (per square foot)	*Occupancy rate (%)*		
	97.0	*97.5*	*98.0*
$19.00	0	4	7
$19.50	7	7	10
$20.00	6	10	13

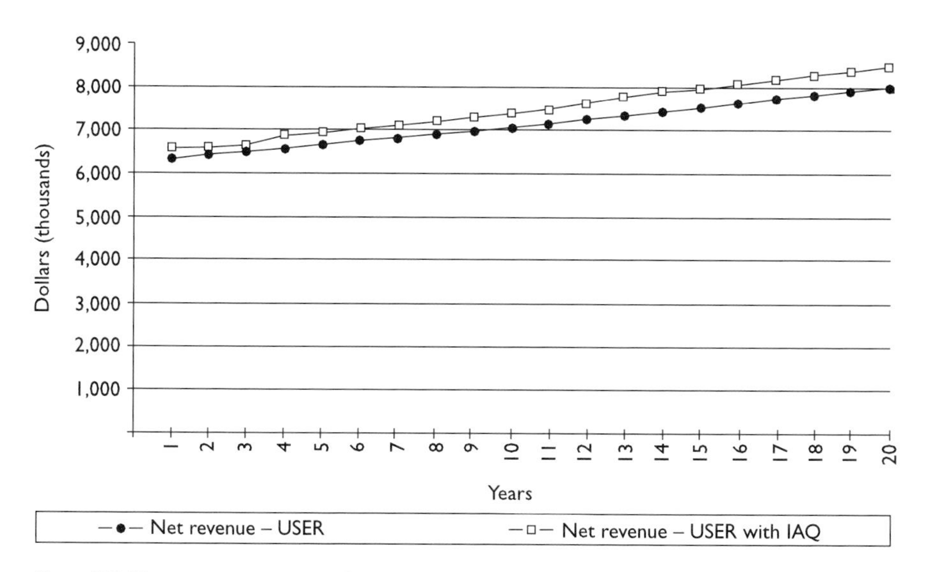

Figure 8.3 Twenty-year net annual summary with and without energy/IAQ upgrades

commonly marketed item, with very specific and well-defined parameters, known as much to occupants and tenants as to building managers, architects and builders. The term 'building performance' no longer just refers to energy performance and operating costs. It also refers to how well the building services the occupants in the space with comfortable and healthy conditions that maximise their performance and productivity. He reflects on how his world has changed from what it used to be, and he is delighted in the change.

Disclaimer

The opinions expressed in this paper are those of the author, and do not necessarily represent the policy of the US Environmental Protection Agency.

References

BOMAI (1988) *Office Tenant Moves and Changes: Why Tenants Move, What They Want, Where They Go.* Building Owners and Managers Association International, Washington, DC.

Dorgan Associates (1993) *Productivity and Indoor Environmental Quality Study.* Report prepared for National Energy Management Institute, Alexandria, VA.

Eto, J. (1990) The HVAC costs of increased fresh air ventilation rates in office buildings, part 2. *Proc. Indoor Air 90: 5th International Conf. on Indoor Air Quality and Environments*, Toronto.

Eto, J. and Meyer, C. (1988) The HVAC costs of fresh air ventilation in office buildings, *ASHRAE Trans.*, **94**(2), 331–345.

Fisk, W.J. and Rosenfeld, A.J. (1997) Estimates of improved productivity and health from better indoor environments. *Indoor Air: Int. J. Indoor Air Quality and Climate*, 7(3).

Franke, D.L., Cole, E.C., Leese, K.E., Foarde, K.K. and Berry, M.A. (1997) *Indoor Air*, 7.

Herzog, P. and LaVine, L. (1992) Identification and quantification of the impact of improper operation of mid-size Minnesota office buildings on energy use: a seven building case study. *Proc. 1992 Summer Study on Energy Efficiency in Buildings*, American Council for an Energy Efficient Economy, Washington, DC.

IFMA (1991) Results of IFMA Corporate Facilities Monitor. *IFMA News*, International Facility Management Association, Oct.

Kroner, W., Stark-Martin, J.A. and Willemain, T. (1992) *Rensselaer's West Bend Mutual Study: Using Advanced Office Technology to Increase Productivity.* Renssalaer Polytechnic Institute, Troy, NY.

Lizardos, E. (1994) IAQ and energy efficiency: guidelines for achieving both. *Consulting-Specifying Eng.*, Sept., 41.

Mudarri, D. (1994) *The Cost and Benefits of Smoking Restrictions: An Assessment of the Smoke-Free Environment Act of 1993 (H.R. 3434).* Report prepared by the Indoor Air Division, United States Environmental Protection Agency, Washington, DC.

Mudarri, D., Hall, J. and Werling, C. (1996a) Energy costs and IAQ performance of ventilation systems and controls. In *IAQ 96. Paths to Better Building Environments*. Conference of the American Society of Heating, Refrigerating, and Air Conditioning Engineers, Atlanta, GA.

Mudarri, D., Hall, J., Werling, E. and Meisegeier, D. (1996b) Impacts of increased outdoor air flow rates on annual HVAC energy costs. Paper presented at the *1996 Conference of the American Council for an Energy-Efficient Economy*, Asilomar, CA.

Mutammara, A. and Hittle, D. (1990) Energy effects of various control strategies for variable air-volume systems. *ASHRAE Trans.*, **96**(1).

Raw, G.J., Roys, M.S. and Leaman, A. (1990) Further findings from the office environment survey: productivity. *Indoor Air 90: The Fifth International Conference on Indoor Air Quality and Climate*, vol. 1, 231.

Reddy, T., Liu, M. and Claridge, D. (1996) Synergism between energy use and indoor air quality in terminal reheat variable air volume systems. Paper presented at the *1996 Conference of the American Council for an Energy-Efficient Economy*, Asilomar, CA.

Rengarajan, K., Shirley, D.B. and Raustad, R.A. (1996) Cost-effective HVAC technologies to meet ASHRAE Standard 62–1989 in hot and humid climates. *ASHRAE Trans.*, **102**(1), 39–49.

Shirey, D. and Rengarajan, K. (1996) Impacts of ASHRAE Standard 62–1989 on small Florida offices. *ASHRAE Trans.*, **102**(1).

Steele, T. and Brown, M. (1990) *Energy and Cost Implications of ASHRAE Standard 62–1989*. Bonnyville Power Administration, May.

USDOE (1993) *Commercial Building Characteristics 1991*. Energy Information Administration, Washington, DC.

USEPA (1989) *Report to Congress on Indoor Air Quality; Volume II: Assessment and Control of Indoor Air Pollution*. EPA/400/1–89/001C, United States Environmental Protection Agency, Washington, DC.

USEPA (1995) *Energy Star Buildings Manual: A Guide for Implementing the Energy Star Buildings Program*. EPA 430-B-95–007. United States Environmental Protection Agency, Washington, DC.

USEPA/NIOSH (1991) *Building Air Quality: A Guide for Building Owners and Facility Managers*. EPA/400/1–91/023. United States Environmental Protection Agency and United States National Institute for Occupational Safety and Health, Washington, DC.

Wyon, D. (1996) Indoor Environmental Effects on Productivity. Keynote address at *IAQ 96. Paths to Better Building Environments*. Conference of the American Society of Heating, Refrigerating, and Air Conditioning Engineers.

Notes

1 No attempt was made to assess the economic value of premature death due to indoor pollution, but only the direct medical cost and lost earnings from the major illnesses which could lead to premature death. The true economic value of society's loss is thus considerably underestimated as a result.

2 This estimate was based on a valuation of each premature death avoided using a willingness to pay measure of $4.8 million.

3 Similarly, in owner-occupied buildings, the managers interested in the health and productivity of the employees need to ask for improved environmental quality from the managers responsible for building operations.
4 In the USA, in only about half of the commercial buildings is the HVAC system subject to a regular maintenance programme (USDOE, 1993).
5 Information about the IPMVP protocol and the indoor environmental quality appendix is available at http://www.ipmvp.org/committee.
6 In effect, this would ordinarily be enough to pay for over half of the cost of a good HVAC preventive maintenance programme.
7 Such a software package is being developed by the USEPA. For information contact mudarri.david@epamail.gov.

Chapter 9

Assessment of link between productivity and indoor air quality

Charles E. Dorgan and Chad B. Dorgan

An assessment of the health status of all commercial buildings in the United States was made based on nineteen research reports and related information on indoor air quality projects, and medical and epidemiology evaluations of existing reports and papers. This information was used to determine the cost to bring all buildings up to ASHRAE Standard 62-1989 and other accepted indoor air quality (IAQ) practice. The productivity benefits were determined: both employee-related productivity benefits and general customer or client benefits. An evaluation of annual cost to sustain the IAQ measures in buildings was included in the study. The study determined the 'healthy building' status of existing buildings by commercial type and five US climatic regions. The economic productivity return for bringing all buildings up to our defined ASHRAE 62-1989 level is recovered in about half a year. However, for a specific building this can range from weeks to several years. Many buildings in the current inventory are very close to ASHRAE 62-1989, and these require only a small investment to bring them up to this level.

Introduction

Due to the large portion of time, up to 95 per cent (ASHRAE, 1993), an average individual spends indoors, the quality of the indoor environment has a significant effect on the health and productivity of employees and other occupants and visitors to commercial buildings. The indoor environmental quality (IEQ) is composed of factors such as space, temperature, humidity, noise, lighting, interior design and layout, building envelope, and structural systems. A subset of the IEQ is indoor air quality (IAQ). The factors that define IAQ are temperature, humidity, room air motion and contaminants.

When IAQ is not satisfactory, both employee and other building occupants' health can be impacted, resulting in work degradation and dissatisfaction by customers. This results in a reduction in productivity. With poor IAQ, employees can suffer significant health and productivity degradation. Productivity is a measure of the quality/quantity of accomplishments actually completed by an employee compared to what could be accomplished under ideal circumstances. Due to the wide variation in tasks in commercial buildings, productivity can be measured in many ways. It can be a measured value, such as sales, profits, the number of errors per hour or actual time at work. It can be a subjective value, such as a personal evaluation of productivity, benchmarked satisfaction of either employees or customers, better students' or supervisors' annual evaluations.

For the productivity link to IAQ research project that is reported in this chapter, all commercial building stock that was not either industrial or residential as defined in Energy Information Agency (EIA) reports for the USA was included. This includes such buildings as government and institutional buildings, retail buildings, offices, schools, hospitals and other non-industrial buildings.

IAQ is only one of the components that affect productivity in commercial buildings and organisations. Others factors, including management styles, education, training, experience, salary, business stress, competition and workload, must be accounted for in any productivity benchmarking study to obtain valid results.

This chapter is a summary of two studies completed for and funded by a national contractors' association on the health costs and productivity benefits of improved IAQ. The original study (Dorgan Associates, 1993) documented the general health costs and productivity benefits of improved IAQ. The second study (Dorgan *et al.*, 1995) expanded the scope to include reduction of specific illnesses, which could result in medical cost reductions, by improving the IAQ. The impact of non-employee productivity benefits was also determined, based on other prior reports and papers and from non-published work completed by Dorgan Associates for a number of clients, and by EPRI in some of its energy research projects.

Objectives

The studies were conducted with the following objectives:

- develop a classification system for known and recognised IAQ degradation in commercial buildings for the United States
- quantify the health cost benefits of good IAQ
- quantify the productivity benefits of good IAQ
- determine the cost to implement known measures and solutions for typical IAQ degradation.

Assumptions

Due to the relative lack of uniform methods for linking productivity to IAQ and of prior information on a recognised format for presenting productivity, several assumptions were made and it was decided to report conservative results. The information from the existing reports could justify productivity benefits 3–6 times those reported in the research and survey. Likewise, some published reports and papers suggest that the link is small or random. However, these are minor and some seem to be basically faulted and based on opinion. Similarly, the reports that conclude benefits of 20 per cent to 40 per cent probably have misinterpreted the data. Our results are further dependent and based upon the following.

- Building and employee distribution data from the Energy Information Agency (EIA) are accurate and representative of the whole population of commercial buildings.
- High benefit estimates reported in other studies and papers that did not provide detailed research procedures were discounted for this study. For this report, a conservative percentage of these values was used. This was typically a third or sixth of the implied and reported benefits, based on the research team judgement.
- Only the direct benefits of improving the IAQ were considered. Secondary benefits, such as fewer inter-office problems due to reduced stress caused by improved IAQ, are difficult to estimate and were not included.
- Experienced professionals would be involved in determining the proper renovations to be implemented to improve the IAQ in buildings and that, once installed, the improvements would be properly maintained to avoid future IAQ problems (includes training of operation and maintenance personnel). This is clearly a key deterrent to full implementation of the benefits. However, it does not impact a specific building owner who has qualified professionals available to implement upgrading of all the existing building to our 'ASHRAE 62-1989 level'.

Results

The productivity and health benefits determined by the results in this study affect a large proportion of those who work, visit, do business or are a customer in the commercial buildings in the United States. The benefits are monetary (profits or income), and relate to quality (education and art), health (fewer low-level health issues) and satisfaction (reduced complaints). These results for direct employee health related productivity benefits are summarised in Table 9.1. The total productivity benefits for commercial buildings are summarised in Table 9.2. Both quantifications are based on all buildings meeting ASHRAE Standards 62-1989 and 55-1992 and related IAQ aspects of these standards that may be considered recommended, not mandatory. These include humidification, filtration and maintenance.

The data in Table 9.2 are based on information gathered, but not reported in the original reports (Dorgan Associates, 1993; Dorgan *et al.*, 1995), in

Table 9.1 Summary of worker productivity benefits

Inventory	
Number of commercial buildings in United States	4,149,000
Total space	58.1 billion ft^2 (5.4 million m^2)
Number of workers	68.9 million
Productivity and health benefits	\$54.7 billion/yr
Annual total productivity benefits	\$8 billion/yr
Annual reduced health cost	\$62.7 billion/yr
Annual total productivity and health benefits	\$910/worker/yr
Annual employee-related benefits, total	\$1.08/ft^2 per yr (\$11.61/m^2 per yr)
Cost to implement	
Implement all identified IAQ improvements	\$87.9 billion
Average cost per area	\$1.51/ft^2 (\$16.28/m^2)
Average cost per worker	\$1,276
Initial average economic simple payback	1.4 years
Annual cost to sustain all improvements	\$4.8 billion/yr
Net 20-year present value of benefits less cost (interest = 3%)	
For all improvements	\$774 billion
Per area for all improvements	\$13.31/ft^2 (\$143.33/m^2)
Per worker for all improvements	\$11,227/worker

that the client was only interested in employee health-related productivity and health benefits linked to IAQ in commercial buildings (non-industrial).

Definitions

During the literature review for this research project it became apparent that there were multiple definitions for almost every key term used in the IAQ field. Therefore, consistent definitions were developed based on guidance from published information from ASHRAE and the United States Environmental Protection Agency (USEPA). Three of the key definitions are for IAQ, sick building syndrome (SBS) and building-related illness (BRI).

Indoor air quality (IAQ)

ASHRAE (1991) defines indoor air quality as:

> attributes of the respirable atmosphere (climate) inside a building including gaseous composition, humidity, temperature, and contaminants.

Table 9.2 Summary of total productivity benefits

Productivity and health benefits	
Annual total productivity benefits	\$54.7 billion/yr
Annual reduced health costs	\$8 billion/yr
Annual total productivity and health benefits	\$62.7 billion/yr
Including annual sales benefits	\$211.2 billion/yr
Annual employee-related benefits, total	\$3,065/worker/yr \$3.64/ft² per yr (\$39.11/m² per yr)
Cost to implement	
Implement all identified IAQ improvements	\$120 billion
Average cost per area	\$2.07/ft² (\$22.22/m²)
Average cost per worker	\$1,742
Initial average economic simple payback	0.56 years
Annual cost to sustain all improvements	\$6.6 billion/yr
Net 20-year present value of benefits less cost (interest = 3%)	
For all improvements	\$2,924 billion
Per square metre for all improvements	\$50.33/ft² (\$541.48/m²)
Per worker for all improvements	\$42,438/worker

Air contaminants can be classified (ASHRAE, 1993) as:

> particulate or gaseous, organic or inorganic, visible or invisible, toxic or harmless, submicroscopic, microscopic or macroscopic, stable or unstable.

Sick building syndrome (SBS)

The ASHRAE (1987) Position Paper on indoor air quality defines SBS as follows:

> The term 'sick building' is used to describe a building in which a significant number (more than 20 per cent) of building occupants report illness perceived as being building-related. This phenomenon, also known as 'sick building syndrome', is characterised by a range of symptoms including, but not limited to, eye, nose, and throat irritation, dryness of mucous membranes and skin, nose bleeds, skin rash, mental fatigue, headache, cough, hoarseness, wheezing, nausea, and dizziness. Within a given building, there will usually be some commonality among the symptoms manifested as well as temporal association between occupancy in the building and appearance of symptoms. Note: some experts place the percentage at 10.

Building-related illness (BRI)

BRI is defined as a specific recognisable disease entity that can be clearly related to chemical, infectious or allergic agents in buildings. The cause of the illness is determined by clinically lab testing patients and by identifying the source in the building. Hypersensitivity pneumonitis, humidifier fever, occupational asthma and *Legionella* infection are often included as BRI. Indicators of BRI (USEPA, 1991) are that:

- building occupants complain of symptoms such as cough, chest tightness, fever, chills, sinus congestion, headaches and muscle aches
- the symptoms can be clinically defined and have clearly identifiable causes
- complainants may require prolonged recovery times after leaving the building.

Research methodology

The research methodology used for these studies focused on the compilation of previous research and reports, both published and unpublished, dealing with the link between IAQ and productivity. Over 500 reports were reviewed. These include published reports from the USA and Europe, international proceedings and personal correspondence. These are listed in the full National Energy Management Institution (NEMI) Report of 1993 and 1995 (unpublished). From this review, the key reports were sorted and their data summarised. A description of the specific methodology used for individual topics is included below. These topics are:

- wellness categories (baseline)
- building and employee inventory
- health and medical effects
- health cost benefits
- productivity benefits, including economics and future preparedness
- recommended improvements
- other benefits.

Wellness categories

Each of the reports reviewed appeared to classify building wellness for commercial buildings differently. The majority of the reports used two general categories. Typically these were sick and healthy, but variations such as buildings without known problems and problem buildings (Woods, 1989) did exist. Other categories of building wellness (Burge *et al.*, 1987) include:

- buildings with high and low rates of IAQ-related complaints
- SBS buildings as a percentage of the total
- occupant response above and below a given level, including SBS symptoms
- healthy building characteristics, such as humidity, temperature, outside air
- number of SBS-type symptoms reported
- other, or combinations of, methods of reporting the data.

In order to estimate the national implication of unsatisfactory and poor IAQ in terms of affected employees' health and productivity, a detailed classification of buildings in terms of wellness level was necessary. For this report, the various categories found in the different reports were compared and integrated to obtain consistent and definable categories. The results from each of the reports were then put into the new categories, and an average distribution of buildings by wellness category was determined and assigned. The sub-categories included public and private buildings.

Building and employee (worker) inventory

Building categories based on building use were obtained from the US Energy Information Administration (1989). The number of buildings, floor space and the number of workers (employees or occupants) in each building category were obtained from this source. For a given building type, the number of employees allocated to each wellness category was based on the valid published reports and wellness categories developed.

Health and medical effects

The primary concern with unsatisfactory and poor IAQ is the degradation of the employees' health and related illnesses. Degradation of health not only affects the personality and happiness of an employee, but also increases medical costs for employees and employers. A review of research articles from medical, engineering and legal databases and from unpublished sources was conducted to evaluate the health and medical effects of unsatisfactory and poor IAQ. Three diseases were reviewed from an epidemiological viewpoint in the second study (Dorgan *et al.*, 1995):

- hypersensitivity pneumonitis (HP)
- occupational asthma (OA)
- sick building syndrome (SBS).

Health cost benefits

From the results of the epidemiological investigations, health cost benefits were determined for acute respiratory disease (ARD), building-related illnesses (BRI), and mild IAQ and SBS symptom-related illnesses. The cost data were from several sources, including:

- Brundage *et al.* (1988), Dixon (1985) and Woods (1989) for ARD
- researchers' experience and judgement
- conversations with operation and maintenance people on illness prevalence in buildings related to poor IAQ
- conversations with medical researchers on clinic visits and hospitalisations related to poor IAQ.

Productivity

Published and unpublished research projects and reports dealing with productivity in relation to the IAQ were reviewed. The definition of productivity and the methodology used in the reports were clearly analysed to ensure that only conservative values of productivity were used in this study. The results of these previous studies and reports were combined with the researchers' experience to determine the percentage productivity increase for each building wellness category. With the number of workers in each wellness category, and the annual salary and fringe benefits of the worker known, the total productivity cost benefits achievable by upgrading a specific wellness category building to a healthy level was computed.

Recommended improvements

Recommended improvements were developed from the researchers' experience on remediation of IAQ problems in commercial buildings. The ultimate goal of implementing the improvements was that all commercial buildings in the United States would meet or exceed the relevant ASHRAE Standards (ASHRAE 1989, 1992).

Other benefits

During the course of the study, other non-health/productivity benefits were identified. These included benefits to individuals, businesses and communities, and avoidance of litigation. The benefits for individuals, businesses and communities are due to jobs generated to implement the IAQ improvement projects, and were estimated using average wage rates and the cost of improvements.

Avoidance of litigation was a major benefit, often overlooked. While there are no guarantees, case studies were developed showing what happens when the IAQ degrades. This can be viewed as a risk management economic benefit, in that it is involved on a percentage of all buildings. It appears to have more benefits than catastrophic losses related to structure, fire and security. There is another productivity benefit that was not included in this study. This is the non-productive cost related to IAQ complaints and problems. There is some information on costs related to fixing a problem or remediation of a complaint. However, these do not include the cost of time that should be allocated to the complainant, others in the area, administrators, managers and building service unit overhead. If we look at a complaint in a hotel room, it usually is ignored on first reporting, and may consume many hours of hotel staff time and loss of productive output by the hotel guest, loss of time at the business meeting as they repeat their complaint, and the time to record and react to solving the IAQ dissatisfaction. This cost is several times the direct cost to correct the IAQ degradation or marginal performance. These are benefits that could substantially increase the benefits related to the link between productivity and IAQ.

Results of research

Using the research methodology described above, results were obtained for:

- building wellness categories
- building inventory
- health and medical effects of poor or degraded IAQ
- health cost benefits
- productivity benefits
- recommended improvements.

Building wellness categories

Five building wellness categories were developed in order to classify fully the existing commercial building stock in the United States. These are:

- healthy
- generally healthy
- unhealthy, source unknown
- unhealthy, source known
- SBS/BRI.

The rationale for selecting these five categories was based on how other publications reported information on buildings. Although a number of publications use 'sick', 'problem building', 'satisfactory' and other terms as well as healthy and unhealthy, it was determined that the above categories reflected the most positive terminology. They also reflected the health relationship of productivity to IAQ.

For each category, characteristics were developed in order to define clearly what was right and what was wrong with a facility, and to make it easier to determine how to improve IAQ in unhealthy buildings.

Healthy

- *Always* meets ASHRAE Standards 62-1989 and 55-1992 during occupied periods.
- Eighty per cent or more of the occupants do not express dissatisfaction with the indoor air quality.
- Building systems are well maintained.
- Building health management exists.

Generally healthy

- Meets ASHRAE Standards 62-1989 and 55-1992 during *most* occupied periods.
- The factors leading to indoor conditions that do not comply with the ASHRAE standards for some of the occupied hours are:
  - relative humidity falls below recommended minimum 30 per cent in winter
  - relative humidity temporarily rises above recommended maximum 60 per cent during summer days
  - outdoor air ventilation rate temporarily drops below recommended minimum due to occupant density being greater than design value
  - lack of maintenance leads to periodic IAQ degradation
  - the HVAC system does not operate during low occupancy.

Unhealthy, source unknown

- Would *fail* to meet ASHRAE Standards 62-1989 and 55-1992 during *most* occupied periods, if evaluated.
- More than 20 per cent of the building occupants would consistently express dissatisfaction with the indoor air quality.
- Less than 20 per cent of the occupants would complain of SBS symptoms.
- Components of HVAC systems and controls that are sources of IAQ problems have not been identified, by the building management or operating staff.
- SBS symptoms and illnesses shown by occupants would not be related specifically to the building, if evaluated.

- This category included information from several reports that determined during investigations or surveys that the building had the above unhealthy characteristics, but that the building owners or operators did not recognise that a problem existed.

Unhealthy, source known

- *Fails* to meet ASHRAE Standards 62-1989 and 55-1992 during *most* occupied periods.
- More than 20 per cent of the building occupants consistently express dissatisfaction with the indoor air.
- Less than 20 per cent of the occupants show SBS symptoms.
- Subsystems causing poor IAQ have been identified. However, the source of the IAQ and SBS problems cannot be linked to a specific HVAC component.
- Occasional high levels of IAQ-related complaints or symptoms.

SBS/BRI

- More than 20 per cent of the building occupants complain of SBS symptoms.
- One or more cases of BRI have been documented.
- Occupants report daily symptoms of IAQ-related illness while in the building.

Building and worker inventory

The distribution of specific buildings types by wellness category is given in Table 9.3. The specific sources for Table 9.3 (Woods, 1989; Woods *et al.*, 1987) include:

- Woods – 50 to 70 per cent of non-industrial buildings in the USA are healthy buildings (this also includes generally healthy buildings)
- Burge – 9 per cent SBS/BRI buildings
- Putnam – 20 per cent SBS/BRI and unhealthy buildings, source known
- Robertson – 15 to 30 per cent SBS/BRI and unhealthy buildings, source known
- Woods – 20 to 30 per cent SBS/BRI and unhealthy buildings, source known.

Therefore, 60 per cent of all commercial building types were considered to be healthy or generally healthy, 10 per cent each for SBS/BRI and unhealthy buildings, source known, and 20 per cent for unhealthy, source unknown buildings. The rationale for distribution of office buildings, education,

Table 9.3 Distribution of all and specific building types by wellness categories (%)

Building wellness category	*National building wellness*	*Office buildings*		*Educational*	*Mercantile and service*	*Lodging (humid climate)* *	*Food service*
		Government	*Others*				
Healthy	20	10	30	10	30	12	15
Generally healthy	40	20	40	20	35	18	35
Unhealthy, source unknown	20	35	15	35	20	40	30
Unhealthy, source known	10	20	10	20	10	15	12
SBS/BRI	10	15	5	15	5	15	8

Note
* Humidity is an IAQ problem concern in all lodging buildings. Humid climates typically worsen the IAQ problems.

mercantile and service, lodging and food service was primarily researchers' experience and judgement and other reports. This is detailed in the full report.

The building inventory data for a number of the building categories are given in Table 9.4. Data for the remaining building sectors are included in the full report.

Health and medical effects of poor IAQ

The results of an epidemiological review of the available research articles on health and medical effects are summarised below.

- **Hypersensitivity pneumonitis (HP).** Although most studies did not specifically investigate the effect of IAQ on the outbreaks of respiratory diseases, it was determined that hypersensitivity pneumonitis in office buildings is usually associated with the microbial contamination of ventilation air of HVAC systems. Culprits identified included open cold-water-spray humidification and cooling systems.
- **Occupational asthma (OA).** No statistically significant evidence was found to relate the number of asthma cases to indoor air quality in office building environments. While a prevalence of asthma was found, apparently related to the poor indoor air environment in the office building, the sampling size made the study inconclusive (Hoffman *et al.*, 1993).
- **Sick building syndrome (SBS).** Based on the research results analysed, it is certain that SBS is linked to IAQ problems related to HVAC systems. Sufficient data show a link between SBS symptoms and the indoor air quality in a building. This conclusion was supported by a statistically significant odds ratio. An odds ratio is the ratio of the risk of disease in exposed individuals to the risk of the disease in unexposed individuals.

The odds ratios obtained by Jaakkola and Miettinen (1995) can be used to measure the strength of association between exposures (mechanical ventilation and natural ventilation) for SBS symptoms. Their odds ratios ranged from 1.31 to 2.32 for SBS symptoms. This meant that the individuals exposed to mechanically ventilated system are 1.31 to 2.32 times more likely to develop SBS symptoms than those exposed to naturally ventilated systems. All the OR values cited in the Jaakkola and Miettinen study were statistically significantly larger than 1.0. This is also true of the adjusted relative risk, 1.51, obtained by Brundage *et al.* (1988) for acute respiratory diseases (ARD). The ARD incidence rate obtained by Brundage *et al.* in mechanically ventilated army barracks was 0.67 per 100 trainee-weeks, as compared to 0.46 per 100 trainee-weeks in old barracks with natural ventilation.

Table 9.4 Building inventory data

Building category	*Total number of buildings (thousands)*	*Total floor space, ft² (m²), millions**	*Number of workers (thousands)*					
			Healthy	*Generally healthy*	*Unhealthy, source unknown*	*Unhealthy, source known*	*SBS/BRI*	*Total†*
Food service	241	1,167 (108)	291	680	583	233	155	1,942
Health care	80	2,054 (191)	845	1,690	845	423	423	4,226
Mercantile and service	1,278	12,365 (1,149)	3,724	4,345	2,483	1,241	621	12,414
Office	679	11,802 (1,096)	7,223	10,001	5,278	3,334	1,945	27,718
All other ‡	1,871	30,652	3,679	7,280	5,834	3,034	2,673	22,500
Total used	4,149	58,040	15,762	23,996	15,023	8,265	5,817	68,863

Notes

* US Energy Information Administration, *Commercial Buildings Energy Consumption Survey: Commercial Buildings Characteristics 1989*, Table 28.

† US Energy Information Administration, *Commercial Buildings Energy Consumption Survey: Commercial Buildings Characteristics 1989*, Table 19.

‡ This building category includes assembly, education, food sales, public order and safety, skilled nursing, warehouse and other (hangar, crematorium and public restrooms/showers and telephone exchange).

Odds ratios do not need to be significantly greater or less than 1.0 to show a causal relationship between the disease of interest and the exposure if a significant general population is exposed tc the disease. Odds ratios derived from epidemiology studies in many areas, such as formaldehyde, are often low. However, the impact of low odd ratios on the public health is immense due to the large population that is affected. Therefore, for IAQ where at least 10 per cent of the general population are exposed to SBS symptoms, the impact is sufficiently significant to warrant action. Some of the best evidence of the impact of IAQ is the comparison of mechanically ventilated buildings with naturally ventilated buildings. Within these studies, the best mechanically conditioned buildings performed better than naturally ventilated buildings. Since humidity is an issue in most US buildings, and new buildings are constructed energy-efficient, almost all US construction can achieve adequate indoor air quality only with a mechanical system. A review of existing epidemiological studies indicates a strong need for adequate mechanical systems in buildings to achieve acceptable IAQ. This is required to prevent degradation of occupant health and to provide a control environment that will increase functional productivity.

Health cost benefits

Health cost benefits were determined for ARD, miscellaneous diseases and SBS problems. The total benefits to health costs due to improving the IAQ were estimated to be $8 billion per year.

- **ARD-related health cost reduction.** The ARD cost reduction was calculated using three separate methods in order to obtain a more accurate estimate. The average medical cost benefit related to ARD as a result of improving the IAQ in commercial buildings was estimated to be $1.2 billion per year.
- **Other BRI-related cost reduction.** The incidence rate for all other miscellaneous diseases, which include humidifier fever, legionnaires' disease, occupational asthma, and lower respiratory diseases from mould and bacteria, will likely be greater than that for only ARD. However, to maintain a conservative approach, the research team estimated that the incidence rate was two-thirds of the ARD rate. Therefore, the health costs attributable to the combination of these other illnesses were estimated to be $800 million per year.
- **Mild IAQ-related cost reduction.** There are health costs associated with general IAQ illnesses, including SBS symptoms of rashes, eye irritation, nausea, headaches and mild coughs. There were no statistically reliable data available on the frequencies of hospitalisation and clinic visits due to these illnesses. However, based on the review of buildings over a 20-year period, and on conversations with operation and maintenance

personnel and with medical researchers, the medical costs due to mild IAQ-related hospitalisation and clinic visits were estimated to be $6.0 billion per year.

Productivity benefits

Research studies document a productivity loss of 2 to 100 per cent in SBS buildings. The 100 per cent loss resulted from the complete shutdown of a 22-storey office building in greater Washington, DC while the experts tried to identify the indoor air pollutants (Hansen, 1991). A majority of the studies indicate an average productivity loss of 10 per cent due to poor IAQ. Therefore, by improving the IAQ, a conservative benefit of 6 per cent could readily be achieved. The percentage increase in productivity for the remaining building wellness categories is as follows:

- unhealthy, problem known – 3.5 per cent
- unhealthy, problem unknown – 3.5 per cent
- generally healthy – 1.5 per cent.

The development of these estimates is provided in the full report.

The annual productivity benefits that can be obtained by upgrading the buildings of various wellness categories to a healthy level were determined for each building type. The total annual and per worker productivity benefits related to workers only in commercial buildings are presented in Table 9.5. This is the information that was used to develop the summary data in Table 9.1. Additional data were developed for cost and benefits per unit area, and are reported in Table 9.1. Table 9.2 includes the additional productivity benefits related to economics of doing business. This includes profits, increased business, economic benefits to society (such as better education, more use of museums, faster recovery in hospitals, satisfaction with government). Buildings with floor space less than 10,000 ft^2 were considered small buildings. Buildings with floor space between 10,000 and 25,000 ft^2 were considered medium buildings, and with floor space greater than 25,000 ft^2 large buildings.

Recommended improvements

A description and cost for the recommended improvements is detailed in the following sections.

Description of recommended improvements

The recommended measures required to improve the HVAC systems of existing buildings with poor IAQ are listed under two categories:

Table 9.5 Annual productivity benefits

Building type	*Wellness category*	*Total number of workers*		*Productivity increase per worker ($)*		*Benefits ($ million)*		
		Small	*Medium/large*	*Small*	*Medium/large*	*Small*	*Medium/large*	*Total*
Totals	1. Healthy	3,896	12,059	0	0	0	0	0
	2. Generally healthy	5,871	18,174	435	435	3,048	9,434	12,482
	3. Unhealthy, problem unknown	3,679	11,390	1,016	1,016	4,596	14,228	18,824
	4. Unhealthy, problem known	2,036	6,304	1,016	1,016	2,613	8,089	10,702
	5. SBS/BRI	1,413	4,374	1,742	1,742	3,117	9,647	12,764
	Total	16,895	52,300	4,209	4,209	13,373	41,398	54,772

- meet or exceed the requirements of ASHRAE Standard 62-1989
  - change the rate of outdoor air per person to 15 cfm or more
  - monitor outdoor air quantity to meet ventilation requirements
  - install local exhaust
  - increase ventilation effectiveness
  - maximise economiser cycle
  - relocate air vents
  - change the air filtration method
  - reduce unwanted infiltration and/or exfiltration
- improve space control to meet the health needs of Standard 62-1989 and meet or exceed the generally accepted requirements of ASHRAE Standard 55-1992
  - improve space temperature control
  - improve control or provide positive control of humidity (dehumidification)
  - install humidification, self-contained steam humidifiers.

Costs to implement IAQ improvements

The estimated cost of implementing each improvement in a building was estimated as a fraction of the new construction cost of a mechanical system for the building. The labour and material costs per building area were obtained. This number was then multiplied by the area inventory of the building category and the percentage of buildings affected by each recommended HVAC measure to yield the cost of improvement. Costs were computed for five climatic zones of the United States. The total costs of improvements for small and medium/large buildings were estimated to be $7.6 billion and $80.4 billion respectively. The grand total one-time cost was $87.9 billion.

Maintenance programme

Some of the improvements identified required an increase in the preventative maintenance programme. This included record-keeping of occupant complaints and training of maintenance and operation personnel. The cost for this increased maintenance was $4.8 billion per year.

Conclusions and future research requirements

A number of non-industrial buildings in the US stock have degraded indoor air quality. The percentage of buildings in each of the five levels of IAQ degradation (i.e. wellness categories) was determined based on previous research findings, researchers' experience and expert judgement, conversations with operation and maintenance personnel, and with medical personnel. A one-time upgrade cost of $87.9 billion, an annual operating cost of $4.8 billion and a total annual benefit of $62.7 billion resulting in a simple pay-back period of 1.4 years was estimated as the productivity and health benefits for employees in all commercial (non-industrial) buildings. This also results in a net 20-year present value of $774 billion ($13.31/ft^2 or $11,227/worker). The productivity benefits related to IAQ are an employee health issue. The gains are related to a healthier indoor environment and healthier employees.

Therefore, to ensure the future viability of the United States on an international front, the investment must be made today to reap the benefits. Further, all new and major system modifications must have IAQ as a priority design intent requirement.

In order to achieve this goal, further research is required, including:

- research studies to investigate fully the causal relationship between the indoor air quality of commercial buildings and hypersensitive pneumonitis (HP) and occupational asthma (OA)
- a set of guidelines to perform productivity studies and benchmark existing buildings properly
- case studies to determine the actual health and productivity benefits that result from improved indoor air quality due to the implementation of HVAC improvement measures.

Although the evidence is strong for relating health and economic benefits, many require further scientific data. This issue is following the same historic development as the debate on health issues and smoking. The evident health and economic benefits are significant and need to be acted on now.

References

ASHRAE (1989) Standard 62-1989, *Ventilation for Acceptable Indoor Air Quality*. Atlanta, GA: American Society of Heating, Refrigerating and Air-Conditioning Engineers.

ASHRAE (1991) *ASHRAE Terminology of Heating, Ventilation, Air Conditioning, & Refrigeration*. Atlanta, GA: American Society of Heating, Refrigerating, and Air-Conditioning Engineers.

ASHRAE (1992) Standard 55-1992, *Thermal Environmental Conditions for Human Occupancy*. Atlanta, GA: American Society of Heating, Refrigerating and Air-Conditioning Engineers.

ASHRAE (1993) Air contaminants, *Handbook of Fundamentals*, I-P Edition, 11.8. Atlanta, GA: American Society of Heating, Refrigerating, and Air-Conditioning Engineers.

ASHRAE Environmental Health Committee (1987) *Indoor Air Quality Position Paper*. Atlanta, GA: American Society of Heating, Refrigeration, and Air-Conditioning Engineers.

Brundage, J.F., R.M. Scott, W.M. Lednar, D.W. Smith and R.N. Miller (1988) Building-associated risk of febrile acute respiratory diseases in army trainees. *J. Am. Med. Assoc.*, **259**(14) Apr, 2108–2112.

Burge, S., A. Hedge, S. Wilson, J.H. Bass and A. Robertson (1987) Sick building syndrome: a study of 4,373 office workers. *Br. Occupational Hygiene Soc.*, **31**(4A), 493–504. See Table 5, p. 502.

Dixon, R.E. (1985) Economic cost of respiratory tract infections in the United States. *Am. J. Med.*, **78** (suppl. 6B), 45–51.

Dorgan, C.E., C.B. Dorgan and M.S. Kanarek (1995) *Productivity Benefits Due to Improved Indoor Air Quality*. Alexandria, VA: National Energy Management Institute.

Dorgan Associates (1993) *Productivity and Indoor Environmental Quality Study*. Alexandria, VA: National Energy Management Institute.

Energy Information Administration, US Department of Energy (1991) *Commercial Buildings Characteristics 1989*.

Hansen, S.J. (1991) *Managing Indoor Air Quality*, 5, 37. Lilburn, GA: Fairmont Press.

Hoffman, R.E., R.C. Wood and K. Kreiss (1993) Building-related asthma in Denver office workers. *Am. J. Publ. Health*, **83**(1), Jan., 89–93.

Jaakkola, J.J.K. and P. Miettinen (1995) Type of ventilation system in office buildings and sick building syndrome. *Am. J. Epidemiol.*, **141**(8), 755–765.

Putnam, V.L., J.E. Woods and T.A. Bosman (1989) Objective measures and perceived responses of air quality in two hospitals. *Proceedings of IAQ '89: The Human Equation: Health and Comfort*, April 1989, 241–250. American Society of Heating, Refrigerating and Air-Conditioning Engineers, Atlanta, GA. See p. 246.

Robertson, A.S., P.S. Burge, A. Hedge, J. Simes, F.S. Gill, M. Finnegan, C.A. Pickering and G. Dalton (1985) Comparison of health problems related to work and environmental measurements in two office buildings with different ventilation systems. *Br. Med. J.*, **291**, 373–376. See Table 1, p. 374.

Skov, P., O. Valbjorn, B.V. Pedersen and the Danish Indoor Climate Group (1989) Influence of personal characteristics, job-related factors and psychosocial factors on the sick building syndrome. *Scand. J. Work Environ. Health*, **15**, 286–295. See Table 2.

US EPA (1991) *Indoor Air Quality Facts No.4 (revised), Sick Building Syndrome*. Washington, DC: US Environmental Protection Agency.

Woods, J.E. (1989) Cost avoidance and productivity in owning and operating buildings. *Occupational Medi.: State of the Art Rev.*, **4**(4) Oct.–Dec. 753–770.

Woods, J.E. (1991) An engineering approach to controlling indoor air quality. *Environ. Health Perspect.*, **95**, 15–21.

Woods, J.E., G.M. Drewry and P.R. Morey (1987) Office worker perceptions of indoor air quality effects on discomfort and performance. *Proceedings of Indoor Air '87: Fourth International Conference on Indoor Air Quality and Climate 2*, Berlin, Aug. 1987, 464–468.

Part 3

The nature of productivity

Chapter 10

Assessment and measurement of productivity

Derek Clements-Croome and Yamuna Kaluarachchi

Productivity depends on good concentration, technical competence, effective organisation and management, a responsive environment and a good sense of well-being. The economic assessment of environment in terms of both health (medical treatment, hospitalisation) and decreases in productivity (absenteeism) has received very little attention by researchers as yet. However, this assessment is absolutely necessary in order to assess the effectiveness of improved design and management protocols (Barbatano, 1994). Until now there have been no standard procedures to measure productivity, therefore it has been difficult to persuade clients to accept the concept of a relationship between economic productivity benefits and indoor environment. The challenge is to investigate productivity and develop a methodology to assess the link between indoor environment and productivity using scientific principles and the experiences of occupational psychologists.

The direct beneficiaries of productivity research are clients who commission buildings and employers who hope to achieve high productivity, as well as construction managers and contractors who want to achieve a high quality of production; planners; architects, designers and engineers who are involved in a building project from its genesis; and building and facilities management consultants who are involved in the day-to-day running and maintenance of the building and who wish to provide a high level of well-being to its occupants. Most importantly it will be beneficial to building users who wish to work in, and enjoy being in, healthy buildings. It is also relevant to academics and researchers in a variety of fields due to its multidisciplinary nature covering health, well-being and comfort.

Thorndike (1949) described four criteria for performance measures: **validity, reliability, freedom from bias**, and **practicality**. The inclusion of validity and reliability implies that the standards for performance measures are similar to those of the tests. Bias most frequently originates from rater bias: in this case, those who use the rating may systematically rate the performance of particular individuals either higher or lower across a number of dimensions than is justifiable from the rated performance. Finally, the concern for practicality as a measurement criterion is obvious, yet very little attention is given to the discussion of issues of practicality in the literature. There has been little disagreement abut the four criteria for performance measurement over the past forty years.

Ilgen and Schneider (1991) classify the methods of performance measurement into three categories: (i) **physiological**; (ii) **objective**; and (iii) **subjective**. The rationale for using **physiological methods** is based on the reasoning that physiological measures of activation or arousal are associated with increased activity in the nervous system, which is equated with an increase in the stress on the operator. Common physiological measurements include: (a) cardiovascular measures (heart rate, blood pressure); (b) respiratory system (respiration rate, oxygen consumption); (c) nervous system (brain activity, muscle tension, pupil size); (d) biochemistry (catecholamines). Three fundamental criticisms of physiological measures of workload have been raised by Meister (1986). First, he questions the validity of the measures. The evidence supporting a relationship between physiological and other workload indices is not strong, and the meaning of these relationships when they do occur is frequently difficult to interpret. Second, the measures themselves are highly sensitive to contaminating conditions. Third, he argues that the measures are intrusive and/or impractical. Physiological measures can restrict or interfere with the operator's task performance. For example, eye pupil dilation changes with concentration, but measuring it is intrusive. Also, restrictions imposed by the job (e.g. task demands, safety considerations) can often limit the number and kind of physiological measures used at one time.

The second of three classes of measures of mental workload has been labelled **objective measures** (O'Donnell and Eggemeier, 1986) or **task performance**. Task performance is frequently used to infer the amount of workload, both mental and physical. Task measures are typically divided into primary and secondary task measures. In primary task performance measurement, the task difficulty for a single task is manipulated and performance variations are assumed to reflect change in workload. In secondary or comparative task performance measurement, the person is first presented with a single task and then a second task is added, or performance is compared across two different tasks and changes in performance are recorded. Task-based measurement has advantages in that it has high face validity and is amenable to quantitative and empirical testing. Task measures present a number of challenges. However, first conclusions based on the task performance allude to the limited resource model, namely that individuals have a finite pool of resources which can be devoted to one task or distributed among tasks. If this model does not hold, then the conclusions from this method are not valid. This procedure has a great deal of utility in the laboratory where task performance and the introduction of new tasks can be highly controlled (Meister, 1986). It is often difficult to cross-calibrate (scale) diverse measures across tasks.

A final set of measures of workload comprises **subjective measures** (Moray *et al.*, 1979). Subjective measures of workload are applied to gain access to the subjects' perceptions of the level of load they are facing in task performance. Rating scales, questionnaires, and interviews are used to collect opinions about workload. While these methods may not have the empirical or quantitative appeal of physiological or objective measures, it is often argued that subjective measures are most appropriate since individuals are likely to work in accordance with their feeling regardless of what physiological or behavioural performance measures suggest (Moray *et al.*, 1979). An assumption of subjective assessment is that people are aware of, and can introspectively evaluate, changes in their workload, and that this assumption holds regarding general impressions of the difficulty of ongoing experiences. When comparing subjective measures with performance measures, high correlations have been found during early and middle stages of overload (Lysaught *et al.*, 1989). Usually higher subjective ratings of workload correspond to poorer performance, yet there is evidence that respondents rate workload more highly in a task that they perform better. The advantages of subjective measures are that they are easy to implement, non-intrusive, low-cost, valid, sensitive to workload variations, and they offer a wide range of techniques (O'Donnell and Eggemeier, 1986).

There is some evidence that air pollution, or the perception that it exists, can create stress among employees who believe that it poses a threat to their health. The stress may be particularly intense among people who believe they have no control over the pollution. It is further argued that psychosocial factors, such as labour–management relations and satisfaction or dissatisfaction with other factors in the work environment, can have a profound influence on the level of response of the occupants to their environment. Productivity measurement may be carried out by using physiological methods, as described above, and environmental psychological methods.

Early studies have shown that the emotional and behaviour effects due to environment can be assessed more more easily by questionnaires than by physiological measurements. Mehrabian and Russell (1974) state that there are three basic emotional responses (**pleasure**, **arousal**, and **dominance**) which combine and can be used to describe any emotional state. The effects of the physical environment on emotional and behavioural responses to work performance can be compared by considering the impacts of environment on pleasure, arousal and dominance. Environmental psychology has been concerned with two major topics: the emotional impact of physical stimuli and the effect of physical stimuli on a variety of behaviours such as work performance or social interaction.

Cooper and Robertson (1990) stated that if you were to look at stress levels in the organisation, you would find that people reporting a negative attitude to the indoor environment would also be the people expressing high job dissatisfaction or a low mental well-being. Our work shows that this can be true, but is not exclusively so. Generally speaking, occupational stress is regarded as a response to situations and circumstances that place special demands on an individual with negative results. Cooper and Robertson (1990) designed an occupational stress indicator to gather information on groups of individuals. The indicator has six questionnaires concerned with: how one feels about the job; how one assesses one's current state of health; the way one behaves generally; how one interprets events around one; sources of job pressure; and how one copes with general stress. The sources of stress are multiple, as are the effects. It is not just a function of being 'under pressure'. The sources may be work-related, but home life can also be implicated. The effects in terms of health may not just concern how one feels physically, but how one reacts and behaves in work and at home. The occupational stress indicator (OSI) attempts to measure: (i) the major sources of occupational pressure; (ii) the major consequences of occupational stress; and (iii) coping mechanisms and individual difference variables which may moderate the impact of stress (Cooper, 1988). Questionnaires and semi-structured interviews are used for subjective assessment, and a study of the medical records and the attendance of the staff members is also made. An environmental aspect has been built into this indicator covering temperature, ventilation, humidity, indoor air quality, lighting, noise, crowded work space, and is referred to as EPOSI.

The analytic hierarchy process

In Part 3 of the questionnaire data are collected and then analysed using the analytic hierarchy process (AHP) originated by Saaty (1972, 1988), as it provides a powerful tool that can be used to make decisions which involve several variables; it is an effective process to describe unquantitative systems quantitatively (Fig. 10.1). Schen and Lo (1997) use AHP to assign weightings for criteria in the prioritisation process of building maintenance. Here it is used to establish an empirical model under 'real' dynamic working conditions for assessing the impact of indoor environment on health and productivity, by analysing the priority factors derived using the occupational stress indicator (Cooper and Robertson, 1990).

Figure 10.1 represents a value tree where each of the attributes are compared in terms of importance using a pair-wise rating process. Ratings are entered on a comparative basis or derived from questionnaire data analysis. The AHP calculates eigenvectors which assign weighting factors (Salvendy, 1997).

AHP needs fine judgements using detailed questionnaires aided by semi-structured interviews to derive the levels or layers (or strata) and the elements within them. The theory of hierarchies is a way of structuring complex

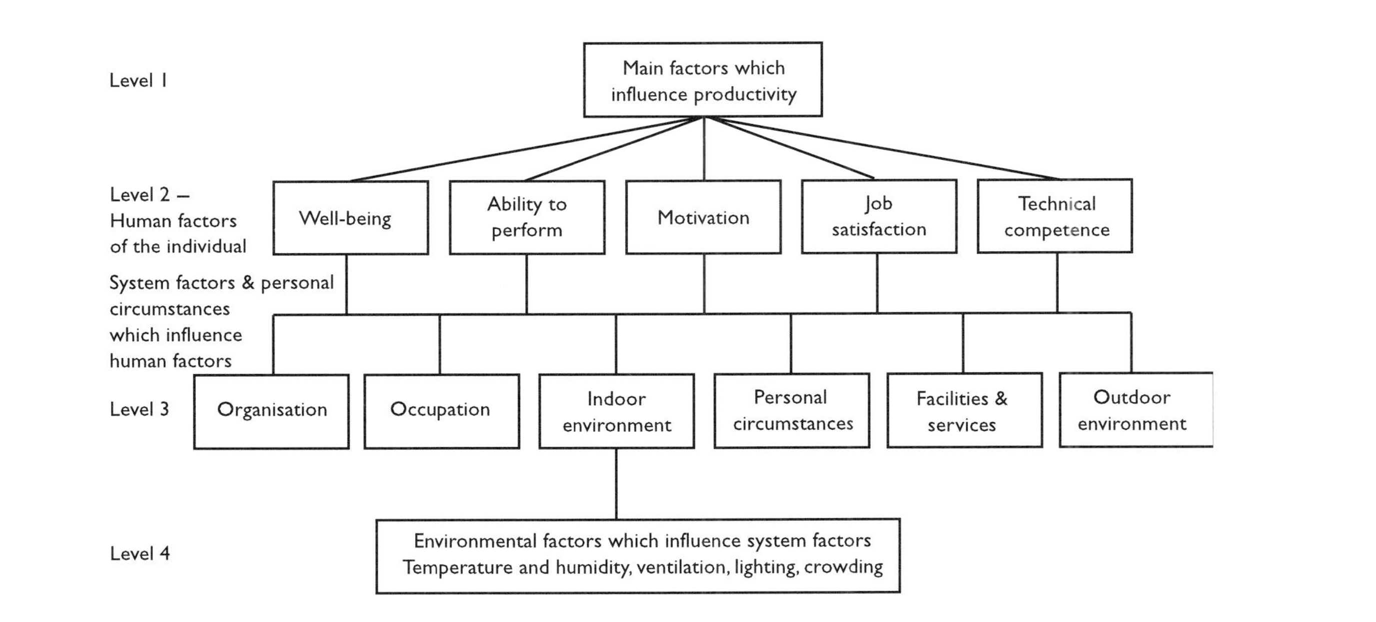

Figure 10.1 The analytic hierarchy process

multidimensional systems, and analysing the interactions of elements in each stratum of the hierarchy in terms of their impact on elements in the stratum immediately above. A five-stratum hierarchy which assesses the productivity of occupants in offices and the influence of indoor environment on productivity is proposed. It is divided into various factors according to their properties and the main objectives (Li *et al.*, 1996; Li, 1998). The top stratum is a single element defined as the productivity of occupants in offices. The elements of the strata have been evaluated using EPOSI. The second and third strata cover human and system factors (e.g. organisational factors, personal factors, environmental factors); each factor in the third stratum is evaluated in terms of all the elements in the second stratum. The fourth and fifth strata deal with health and environmental factors. AHP is an example of a transdisciplinary model in which coordination is achieved across a level (or discipline) and between levels (or disciplines).

Subjective measurements and design of questionnaire

A number of research studies have demonstrated that there is no simple relationship between single environmental factors and complex human behaviour. The analytic hierarchy process (AHP) method together with multi-regression and correlation analysis can be used to establish an empirical model of 'multi-sensory' well-being of occupants. AHP is an effective process to describe unquantitative systems and factors quantitatively (Saaty, 1972). The system should first be divided into various factors according to their properties and the main objective. These factors are then classified in successive hierarchies (levels or layers) through which the analytic hierarchical model is developed. For example, the importance of the factors in the lowest hierarchy can be related to the factors in the highest hierarchy on the basis of questionnaire and semi-structured interview surveys (Yao Runming *et al.*, 1992). On the basis of analysing factors such as the visual field, hearing, smell, warmth, dust, touch, freshness, and space, a relationship between 'multi-sensory' occupant well-being (indoor environment) and productivity can be developed.

The subjective assessment used questionnaires answered by occupants across various work grades and tasks (Cooper, 1988; Raw, 1994). The questionnaire was designed to elicit:

- background information about the organisation and the workplace
- how frequently the subject suffers from the environment or the job when at work
- the feelings of the subject about their current working situation
- the priority factors influencing health symptoms of occupants in offices

- the priority order among these factors which influence job satisfaction and productivity of occupants in offices.

The questionnaire contains a common core of questions. Section 1 begins with general **office physical environment** and the information regarding the occupant, including **type and status of their workplace, the effect the weather has on their well-being, health and productivity**. Section 2 concentrates on **how occupants feel about their working situations, how much they like the office and how much personal control they have** on a seven-point rating scale. Also of interest are the four self-report productivity items, consisting of amount of work accomplished, quality of work, feeling creative, and taking responsibility. Respondents were asked how much their office environment affected their productivity at work (they may increase or decrease their productivity because they are affected by current working environment) on a nine-point scale: decrease by 40 per cent or more, 30 per cent, 20 per cent, or 10 per cent; not at all; increase by 10 per cent, 20 per cent, 30 per cent, or 40 per cent or more.

Respondents were also asked to rate their level of productivity on a seven-point rating scale, from extremely dissatisfied to extremely satisfied, and then asked to rate on a five-point rating scale (no change, increase by 10 per cent, 20 per cent, 30 per cent, or 40 per cent or more) by what percentage it would increase if the related office environment problems were resolved.

In section 3, the factors of greatest interest include those concerned with office environmental conditions which act positively or negatively on the productivity of the occupants. Also of interest are the responses in terms of physical health conditions which give rise to sick building syndrome (SBS). The respondents were asked to rate on a seven-point scale how often they had suffered from SBS conditions in the year prior to the survey. SBS conditions are prevalent in the work environment, but disappear when people leave it.

The mental health conditions related to the job were:

- **job stress** (the stress experienced may cause physical symptoms or emotional and psychological difficulties)
- **job dissatisfaction** (dissatisfaction with the job itself, with organisational design and structure, with achievement value and growth, or with personal relationships)
- **an overall dissatisfaction with the indoor environment.**

Part 2 of the questionnaire was based on, and developed using, the analytic hierarchy process which adopts a pair-wise comparison scale in making judgements against various influencing factors. Five **human factors** of the individual were selected by using the occupational stress indicator and after reviewing other related research surveys and literature, which influence productivity; six **system factors** (including personal circumstances) which in turn influence the five human factors were compared with each other. The various strata shown in Fig. 10.1 can be described as follows:

- human factors of the individual which influence productivity
  - well-being (physical and mental health of an individual)
  - ability to perform (physical aptitude to carry out an assigned task)
  - motivation (mental drive and enthusiasm to carry out a task)
  - job satisfaction (the enjoyment attained by carrying out tasks in an occupation)
  - technical competence (the qualifications and the know-how to carry out a task)
- system factors which influence the human factors of an individual
  - indoor environment (the immediate surroundings of one's internal workspace (including light, sound, temperature, ventilation, indoor air quality and pollution))
  - occupation (the interest in the job, salary, bonuses, career prospects and job stress)
  - organisation (organisational structure, managerial role and the social environment)
  - personal circumstances (private life, psychological factors, sex, age, health and behaviour patterns)
  - facilities and services (communications, networks, hardware, software and facilities for occupants provided by organisations)
  - outdoor environment (weather conditions and outdoor views).

Environmental factors that influence the indoor environment are: temperature; stale or stuffy air; draught; dry or humid air; poor indoor air quality; noise; discomfort from sunlight; insufficient/excessive lighting; crowded workspace.

Health factors influenced by the environmental factors are:

- respiratory problems (dryness, hoarseness, dry/sore throat, changes in voice, wheezing)
- skin problems (soreness, itching, dry skin, rashes)
- nervous problems (headaches, nausea, drowsiness, tiredness, lethargy, reduced mental capacity, dizziness, forgetfulness, fatigue)
- nasal-related problems (itchy or teary eyes, runny nose, asthma-like symptoms among non-asthmatics)
- odour complaints (changes in odour, unpleasant odours or tastes).

Occupant and environmental survey of buildings

This research focuses on the relationship between productivity and the indoor environment in offices, and takes into account the fact that productivity depends on other factors by using an 'occupational stress indicator' (Cooper, 1988), which has been developed to include an environmental dimension (Li *et al.*, 1996; Clements-Croome and Li, 1995). The modified occupational stress indicator (EPOSI) has been used in the design of the questionnaire to gather information on the occupants in the surveyed buildings. This method of self-assessment provides valuable information on individuals, as well as collective responses. In this research a number of environmental surveys have been carried out to gather data on occupants' responses on productivity as well as office physical environmental criteria including air temperature, relative humidity, air velocity, quantity and quality of light, noise levels, radiation and electromagnetic fields around workspaces, indoor CO_2 concentrations, and thermal comfort indices.

Pilot studies were carried out at the University of Reading and a local consultancy office. This survey was limited to six non-air-conditioned buildings using about 170 subjects (Li Baizhan, 1996a). The study used the above methodology to establish:

- how factors such as workplace, environment, management style and job stress interrelate to affect job satisfaction and productivity in offices
- which are the dominant factors in given situations, by asking subjects to complete the questionnaires.

The research results showed that there was a strong relationship between job stress, or job dissatisfaction, and an overall unsatisfactory indoor environment. Based on the results of the questionnaire survey, an increase of about 10 per cent in office productivity could result from improving the indoor environment. With the experiences from the survey, the questionnaire was refined.

Detailed environmental surveys were carried out in the office headquarters of a building in London in August 1997, and one in Maidenhead in November 1997, in order to gather data and information on the main factors influencing productivity. About 250 occupants voluntarily participated in the questionnaire survey and provided information regarding the workplace, the way they feel about their work situations, environmental criteria, physical and mental health, and factors which they considered influenced their productivity. The buildings consisted of air-conditioned and non-air-conditioned spaces, and physical environmental data were measured and monitored for a two-week period.

The survey adopts an approach which gathers data on occupants' perceived personal productivity and relates these responses to those responses given regarding their comfort levels, the degree of control over their working environment and the physical symptoms in the workplace. The study does not measure the actual productivity but establishes the key factors affecting productivity. It does measure the relative productivity between one floor and another, between zones on the same floor and between individuals.

Case study 1: office building in London

In this office building about 120 occupants participated in the questionnaire survey during August 1997. The staff mainly belonged to the age group of 20–30 years and consisted of the ratios of male and female occupants given in Table 10.1.

The **type of office space** for the majority was **open plan**. The **type of work** varied from floor to floor and belonged to four categories (Table 10.2).

The **overall working hours per week** were 36–45; about 77 per cent of the group spent up to 35 hours in front of a **visual display unit** (VDU). In comparing the number of working hours reported by individuals, it was seen that the fourth floor occupants reported a high percentage of people working longer hours. This coincides with a high percentage of staff spending longer hours in front of visual display units such as computers or television screens (Table 10.3).

Regarding the **weather and their work performance**, more than 60 per cent responded that the weather has an effect on the individual work performance and also that they work better in the mornings compared to the afternoons. Planting and its use in the layout of the office interior was insignificant, but some occupants had attempted to personalise their workspace with a few plants.

Table 10.1 Proportions of male/female occupants, case study 1

Location	*Male*	*Female*
First floor	53.85%	46.15%
Third floor	67.74%	32.26%
Fourth floor	28.57%	71.42%

Table 10.2 Type of work, case study 1

Location	*Managerial*	*Professional*	*Clerical*	*Other*
First floor	7.69%	61.54%	7.69%	23.08%
Third floor	9.35%	61.29%	19.35%	—
Fourth floor	14.29%	28.57%	28.57%	28.57%

Table 10.3 Working hours, case study 1

Location	*Normal working hours (%)*		*Hours spent in front of a VDU (%)*	
	25–35 hours	*36–55 hours*	*25–35 hours*	*36–55 hours*
First floor	30.76	61.53	46.16	30.76
Third floor	32.26	70.96	13.79	24.14
Fourth floor	14.29	85.72	57.14	14.29

In the three floors surveyed, the responses to direct questions regarding dissatisfaction with 'the job' and the 'indoor environment' varied widely (Table 10.4).

The results for the first floor contradict those anticipated by Cooper (1990), whereas those for the third and fourth floors support his contention that people dissatisfied with their job are likely to be those dissatisfied with their environment. On the third floor a quarter of the occupants (24.14 per cent) reported that they were dissatisfied with their job, while the proportion of people who were dissatisfied with their work environment was about 15 per cent. On the fourth floor half of the occupants stated that they were dissatisfied with their job and 40 per cent reported a high level of dissatisfaction with their workplace. On these two floors the dissatisfaction regarding the job was 10 per cent higher than with the indoor environment.

The first-floor data for dissatisfaction illustrate something quite fundamental from all these aspects. While the third and fourth floors show that there is a clear relationship between people feeling dissatisfied about their job and the work environment, the first floor illustrates that even if people are wholly satisfied regarding their jobs, they can be quite unhappy with their work environment and this will affect their productivity. The third and fourth floors are air-conditioned while the first floor is a non-air-conditioned space which could provide extremely uncomfortably warm conditions on a hot summer day (when the survey was conducted).

In section 2 of the questionnaire the occupants were given the opportunity to rate on a seven-point scale their feelings regarding their work situations under four categories: feelings regarding the job, self-rated level of productivity, the layout of the office, and the decoration of the office. The responses correlate with the dissatisfaction they expressed in section 1 of the questionnaire (see Table 10.4) regarding their job and the indoor environment. In the first instance the question put to the occupants was 'Is there any particular day of the week when you feel dissatisfied with your job?'. In the second instance (see Table 10.5), they were asked to rate their satisfaction or dissatisfaction according to a seven-point rating scale.

Table 10.4 Dissatisfaction with job and indoor environment, case study 1

Location	*% of people dissatisfied with*	
	job	*indoor environment*
First floor	0.00	41.67
Third floor	24.14	14.81
Fourth floor	50.00	40.00

Table 10.5 Satisfaction/dissatisfaction with various factors, case study 1

Location	*Dissatisfied*	*Not sure*	*Satisfied*
	Feelings about the job (%)		
First floor	0.00	7.69	92.31
Third floor	17.24	17.24	65.52
Fourth floor	16.67	16.67	66.67
	Self-rated level of productivity (%)		
First floor	0.00	25.00	75.00
Third floor	9.67	9.67	80.65
Fourth floor	0.00	33.33	66.67
	Office layout (%)		
First floor	7.69	7.69	84.62
Third floor	45.16	6.45	48.39
Fourth floor	66.67	16.67	16.67
	Office decor (%)		
First floor	38.46	15.38	61.54
Third floor	25.81	25.81	48.39
Fourth floor	50.00	16.67	33.33

The data in Table 10.5 strengthen the responses in Table 10.4. In Table 10.4, the first-floor respondents report 0 per cent dissatisfaction regarding their jobs and in Table 10.5 they report the highest percentage of occupants satisfied with their jobs, which reinforces the consistency of the methodology and its capability for obtaining the same results for clarification, in a variety of ways, from individual responses.

For the **self-rated level of productivity** the fourth floor reports the lowest percentage of satisfied occupants (66 per cent) which could be related to the higher number of working hours, and the dissatisfaction about the job and the workplace environment. It was also observed that this floor had a majority of female occupants (71 per cent) involved in clerical (28 per cent) and other types of work (28 per cent), compared with the high percentage of professionals on the other two floors (61 per cent). This suggests that occupants with lower status jobs are more likely to feel a higher level of dissatisfaction. The third-floor occupants report the highest self-rated level of productivity in accordance with low satisfaction scores for the job and the environment.

Regarding the **layout** of the office, the occupants on the third and fourth floors reported a high level of dissatisfaction but the first-floor occupants reported a high level of satisfaction with the indoor environment. The third floor consisted of the largest open plan office which was shared by about 55 occupants. The dissatisfaction they showed could have been due to the crowding and the size of the individual spaces, which were smaller than for the other floors. For example, the fourth-floor open plan office was smaller and more private than the third floor.

Concerning the **decoration** of the office in terms of colour, texture and fittings, the majority of users from the third and first floors stated that they were satisfied, but the responses for the fourth floor illustrated that half of the occupants were dissatisfied. These responses echo some of the research findings described in Section 1.1 of the Ellis study (1994).

The fourth-floor respondents consistently express dissatisfaction regarding various aspects of their jobs as well as the office environment; in the section below it is seen that these respondents also show more physical and mental health symptoms.

The responses from all three floors are important, as they illustrate that at a strategic level:

- work and indoor environment are related
- the methodology provides consistent results to the same questions presented in a variety of brainstorming methods for clarification
- the methodology is sensitive enough to detect various differences in individual responses
- productivity depends on a broad range of issues.

The influence of indoor environment on well-being and productivity

Scientific studies indicate that productivity can increase by as much as 15 per cent when workers are satisfied with their environment. In addition, current case studies show that high financial returns can be realised from increased productivity within an organisation (Lomonaco and Miller, 1997). There are lots of tell-tale signs when productivity is suffering: absenteeism, sickness leave, people coming late and leaving early, extended tea and lunch breaks, increased risk of accidents, increased complaints, excessive socialisation, slow work output and increase in mistakes.

Factors such as poor lighting, both natural and artificial; poorly maintained or designed air-conditioning; and poor spatial layouts are all likely to affect performance at work. In a survey of 480 UK office occupiers, Ellis (1994) found that 96 per cent were convinced that the design of a building affects productivity. When asked an open-ended question to categorise the aspects of design that they felt would contribute, to this 43 per cent used words such as **attractive, good visual stimulus, colours and windows**; 41 per cent mentioned **good morale, 'feel-good' factor** and **contented happy staff**; 19 per cent said more **comfortable, relaxing, restful conditions** to work in; 16 per cent said **increases in motivation** and **productivity**; 15 per cent said **improved communications**; 3 per cent or less said **reduced stress**. The highest percentage (43 per cent) is regarding the layout and decoration of the office space which demonstrates the importance of this criterion and strengthens the findings presented in Table 10.5. Here the highest percentage of occupant dissatisfaction is regarding the layout and decoration of the office.

The results of the analysis for this survey show that productivity is affected by environmental conditions. Even though there were mixed responses to direct questions regarding whether the indoor environment affects physical and mental health, the responses for well-being and productivity show that the majority consider the office physical environment as a major aspect which affects these factors (Figs 10.2 and 10.3).

On the first floor, a high percentage of people responded that the office physical environment affected their physical and mental health, well-being and productivity, and this floor contained the highest percentage dissatisfied with their indoor environment. The responses for the fourth floor vary and even though there was a high degree of job dissatisfaction, 50 per cent of people thought that indoor environment affects health, well-being and productivity.

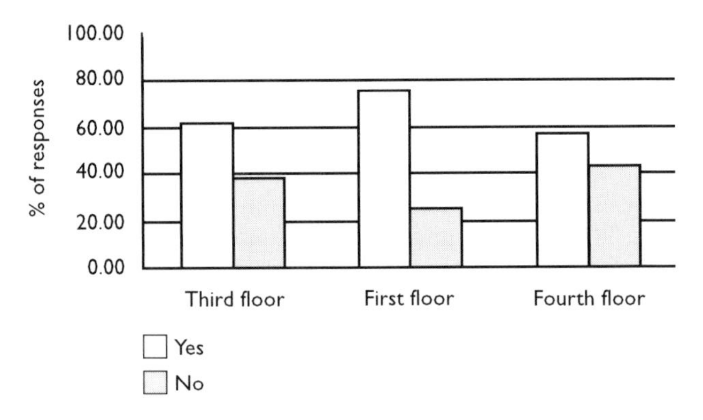

Figure 10.2 Does the physical office environment affect well-being?

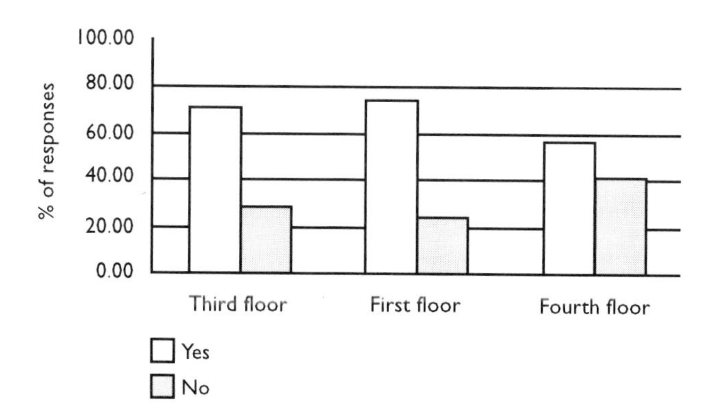

Figure 10.3 Does the physical office environment affect productivity?

For direct questions concerning the degree of influence the office environment has on the **quantity and quality of work** and on **being creative** and **taking responsibility**, the responses ranged from it has a little to a large effect on these factors. More than 33 per cent of the occupants from the first and fourth floors reported that the office environment has a large effect on being creative and the quantity of work that they can carry out; more than 50 per cent from the fourth floor felt the same regarding the quality of the work.

The amount of increase in productivity by improved environmental conditions was felt by about one-third of the occupants on the third and first floors to be as much as 30 per cent; a third of the occupants on the fourth floor felt that a 50 per cent increase in productivity was possible (Table 10.6).

It should be noted that the fourth-floor occupants reported a low level of productivity compared to the other floors and higher degree of dissatisfaction with their conditions. The data are interrelated and illustrate the basic relationship of productivity, satisfaction with job and workplace.

Personal control and its influence on productivity

A healthy building must satisfy individual user needs. As there are innumerable reasons for different users having different requirements at different times, this implies that the user must have a degree of control over what the building provides for them. The ratings from respondents regarding their typical office working conditions (comfort, temperature, ventilation, air quality, humidity, and satisfaction) show that there are diverse responses to the same environmental conditions. These differences between individuals can be physiological and/or psychological due to age, sex, clothing, activity, fatigue, thermal history, current state of health, allergy or hypersensitivity (Wyon, 1992).

Table 10.6 Self-projected possible improvements in productivity through improved conditions, case study 1

Location	*0%*	*10%*	*20%*	*30%*	*40%*	*50%*	*60%*	*70%*
First floor	16.67	0.00	8.33	*33.33*	8.33	16.67	8.33	8.33
Third floor	6.9	24.14	*31.03*	24.14	0.00	6.90	3.45	3.45
Fourth floor	16.67	16.67	16.67	16.67	0.00	*33.33*	0.00	0.00

The analysis indicated that, as found in other research, improving personal control over temperature, ventilation and lighting improves self-rated productivity. Perception of having personal control of one's own environment in the workplace leads to improved comfort judgements as well as psychological benefits and territorial needs. Psychological benefits could result in increased job satisfaction, increased satisfaction regarding the workplace and increased motivation, which in turn can result in an optimum level of productivity.

The survey results indicated that there was very little personal control in adjusting the temperature, ventilation and lighting levels around each individual working space. This is one of the main reasons why people were dissatisfied with their workplace, because personal control gives one the ability and the flexibility to change the environmental conditions as required. Figures 10.4 and 10.5 illustrate that the majority of the occupants felt there was little personal control at their workplaces.

Thermal conditions, air flow, lighting and background noise masking for privacy are all needs which are significantly different and vary over the course of a day. For example, older employees typically need more light (Lighting Research Institute, 1989), and female workers are known to prefer higher temperatures (Weiner, 1994). From personal comments provided by the respondents, a repeated complaint was that **the temperature, ventilation**

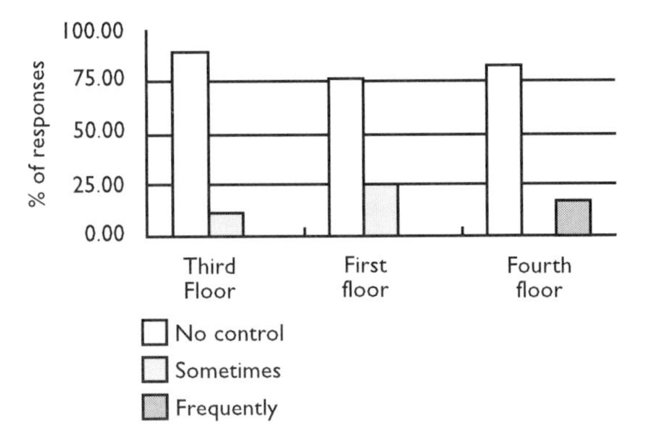

Figure 10.4 Personal control over temperature

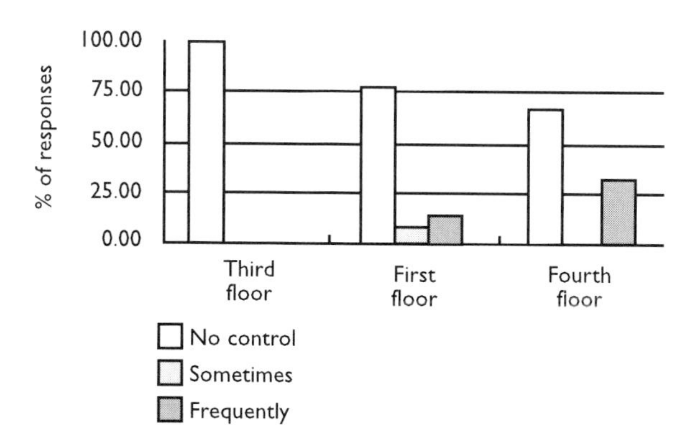

Figure 10.5 Personal control over lighting

and lighting provided in the workplace were the same throughout the day. This condition gets worse in the winter, as the conditions provided for the morning are uncomfortable in the afternoon. If **flexible personal control** methods can be provided, the occupants can adjust their environmental conditions to suit individual comfort levels and needs throughout the occupancy period.

Comfort and health in the workplace

Unless a person feels a positive sense of well-being they will not perform as effectively, and an optimum level of productivity will not be achieved. To enjoy a good sense of well-being a person should enjoy their workplace. The physical environment has a marked influence on the occupants' mood and comfort level. Temperature and humidity, ventilation, light, noise and crowding are the most important factors in determining a good working environment. While the occupants were responding to the questionnaire, a number of environmental criteria were monitored. These include temperature and humidity levels, air speed, light and sound levels, indoor CO_2 concentrations, radiation and electromagnetic levels.

In section 3 of the questionnaire, a list of adverse environmental conditions were presented to the participants and they were asked to rate their responses on a seven-point rating scale ranging from 'never' to 'always'. The percentage of occupants who 'frequently/very frequently' or 'always' endured these conditions are given in Table 10.7.

The first floor reported the highest number of adverse environmental conditions endured by occupants; the fourth floor reported the second highest percentages (Table 10.7). According to the analysis, from the percentage of occupants that frequently endured adverse environmental conditions, the most common conditions suffered were:

Table 10.7 Cited presence of adverse environmental conditions, case study 1

Negative environmental condition	*% of occupants*		
	First floor	*Third floor*	*Fourth floor*
Draught	0.00	6.45	0.00
Stale/stuffy air	69.23	35.48	57.14
Dry/humid air	69.23	25.81	42.86
Dusty air	15.38	6.45	28.57
High or low temperature; temperature changes	76.92	38.71	71.43
Little or too much daylight	15.38	19.35	28.57
Little or too much light	30.77	16.13	0.00
Flickering light	0.00	6.45	0.00
Annoying reflections	0.00	12.90	0.00
Vibrations, sounds or noise	46.15	22.58	14.29
Static electricity	0.00	3.23	0.00
Crowded workspace	7.69	38.71	85.71

1 high/ low temperature or temperature changes
2 stale/ stuffy air and dry or humid air
3 insufficient daylight and low-quality general lighting
4 crowded workspaces.

These findings are related to the health symptoms suffered by the occupants. Those who experience symptoms of ill-health are the least satisfied with the environmental conditions at their place of work. It is evident that the occupants believed that health symptoms can affect their productivity because it can result in absenteeism, leaving work early, feeling lethargic, depression or headaches.

The majority of participants that suffered health symptoms belonged to the fourth floor. This could be related to the high rate of dissatisfaction with the indoor environment and their jobs which the occupants reported, and also the low level of self-rated productivity. The first floor reported the second highest number of people suffering from health symptoms. The most common physical symptom felt by the majority was '**do not want to get up in the morning**' or '**feeling lethargic**' at workplace. **Inability to sleep** and **nasal-related problems** were the second most common conditions felt by the occupants.

Table 10.8 illustrates clearly the facts which have been discussed throughout the analysis. The highest dissatisfaction regarding office environment is expressed by people on the first floor and is reinforced by the occupants' attitude to the office environment and its effect on physical and mental health, well-being and productivity. It was also observed that it was not the

Table 10.8 Levels of stress and dissatisfaction, case study 1

Location	*The percentage of people who frequently suffered*		
	job stress	*job dissatisfaction*	*overall unsatisfactory indoor environment*
Third floor	6.67	23.33	15.38
First floor	7.69	7.69	30.77
Fourth floor	14.29	14.29	28.57

layout or the decoration that they were dissatisfied about, but the overall design and the performance of the office space.

The highest level of job stress and a high level of job dissatisfaction were evident on the fourth floor, which was reflected in the low level of self-rated productivity. The majority of occupants who suffered health symptoms were from this floor, which again correlates to the high level of job stress illustrated.

The third floor reports a low level of job stress and dissatisfaction with the indoor environment, but a high level of job dissatisfaction.

In summarising all these factors it was clearly seen that the responses for the first floor were consistent in their dissatisfaction towards the indoor environment while maintaining a quite high level of satisfaction regarding the job and related stress and health symptoms. The fourth floor reported a high level of dissatisfaction regarding the job and the indoor environment, but the compiled information shows that the dissatisfaction is quite high in factors related to the work organisation rather than the indoor environment.

Priority factor analysis according to the analytic hierarchy process model

Using the responses from the questionnaires, pair-wise comparison of judgements was compiled. Each wing on each floor surveyed was looked at independently and then averaged to arrive at results which concern all floors. From the results it was clear that **well-being** is the main human factor which influences productivity of occupants in offices. The other two human factors which had a very significant influence on productivity were **ability to perform** and **motivation**. These are well illustrated in the responses for several wings and most floors. **Technical competence** appears to play a less important role in influencing productivity (Figs 10.6 and 10.7).

Figures 10.8 and 10.9 show that the system factors that influence the human factors of the individual are interrelated; the responses show that the **indoor environment** is given varying degrees of importance. **Personal, occupational** and **organisational factors** have a considerable influence but it was

Figure 10.6 Main human factors influencing productivity: all floors

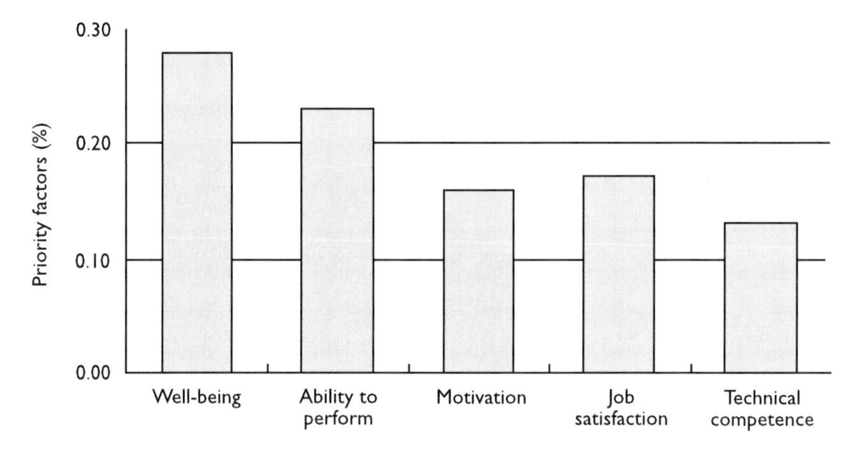

Figure 10.7 Main human factors influencing productivity: fourth floor

clearly observed that in many office wings surveyed, factors which relate directly to a person's physical and mental health and feelings about an individual's work situation, such as **well-being, ability to perform** and **motivation**, are significantly influenced by the indoor environment.

The research findings are summarised as follows.

- The responses for the third and fourth floors support Cooper and Robertson's (1990) views that people dissatisfied with their jobs are likely to be dissatisfied with the indoor environment (Tables 10.4 and 10.5).
- The responses for the first floor contradict Cooper and Robertson's (1990) views and show that even if people are wholly satisfied with their

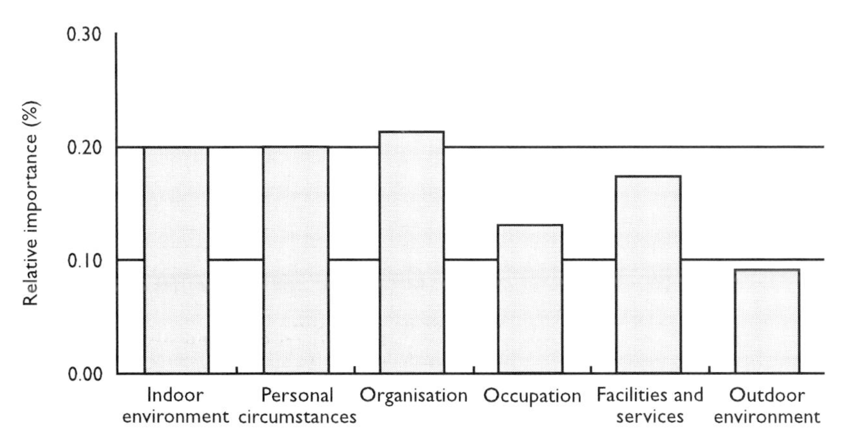

Figure 10.8 Relative importance of system factors with respect to ability to perform: fourth floor

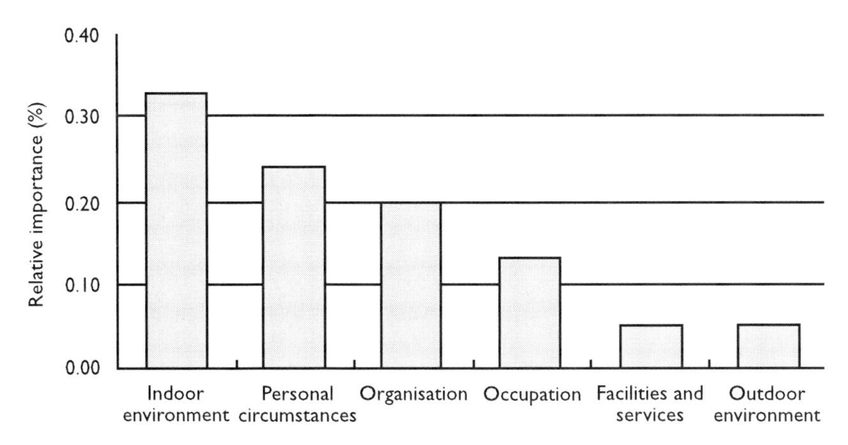

Figure 10.9 Relative importance of system factors with respect to motivation: third floor

jobs, they can be quite unhappy about the work environment (Table 10.3).

- The first-floor occupants expressed satisfaction with the layout and the decoration, but dissatisfaction about the overall design and the performance of the workspaces. The third floor behaved contrarily and expressed satisfaction regarding the overall office but dissatisfaction regarding the layout, which could relate to crowding and small individual spaces (Table 10.5).
- When dissatisfaction regarding the job and the indoor environment is high, a low level of self-rated productivity is evident. The lowest

percentage of satisfied occupants belonged to the fourth floor (Table 10.5).

- The fourth floor consisted of a majority of female occupants involved in clerical and other types of work compared to the high percentage of male professionals on the other two floors. This indicates that the occupants with lower status jobs tend to feel a higher dissatisfaction (Table 10.5).
- A high percentage of people from the first floor responded that the office physical environment affected their physical and mental health, and this floor contained the highest percentage dissatisfied with their indoor environment).
- A very high percentage of occupants from all floors considered the office physical environment as a major factor in enhancing well-being and productivity (Figs 10.2 and 10.3).
- Occupants from the first and fourth floors reported that the office environment has a large effect on being creative and the quantity of work they carry out; occupants from the fourth floor felt the same regarding the quality of work.
- The amount of increase in productivity by improved environmental conditions was perceived to be as much as 30–50 per cent (Table 10.6).
- A repeated complaint was that the temperature, ventilation and lighting provided in the workplace were the same throughout the day.
- There was very little personal control in adjusting the temperature, ventilation and lighting levels around each individual working space (Figs 10.4 and 10.5).
- Flexible personal control methods were desired by the occupants to adjust their environmental conditions to suit individual comfort levels and needs.
- The adverse environmental conditions suffered by the majority were:

  1 high/low temperature or temperature changes
  2 stale/stuffy air and dry or humid air
  3 insufficient daylight and low-quality general lighting
  4 crowded workspaces (Table 10.7).

- The fourth floor reported the highest number of occupants suffering from health symptoms; the first floor reported the second highest.
- The most common physical symptom felt by the majority was 'do not want to get up in the morning' or 'feeling lethargic' in the workplace. Inability to sleep and nasal-related problems were the second most common conditions.
- The highest level of job stress and a high level of job dissatisfaction were on the fourth floor, which was reflected in the low level of self-rated productivity (Tables 10.5 and 10.8).
- The third floor reported a low level of job stress and dissatisfaction with

indoor environment, but showed a high level of job dissatisfaction (Table 10.8).

- Well-being is the main factor which influences productivity of occupants in offices. The other two factors which had a very significant influence on productivity were ability to perform and motivation. Technical importance appears to play a less important role in influencing productivity (Figs 10.6 and 10.7).

Factors which relate directly to a person's physical and mental health and feelings about an individual's work situation, such as **well-being**, **ability to perform** and **motivation**, are greatly influenced by the indoor environment (Figs 10.8 and 10.9).

Case study 2: office building in Maidenhead

In this office about 100 occupants participated in the questionnaire survey during November/December 1997. It was conducted in a two-week period and environmental data were recorded and monitored for this time period. Three zones, including the right and left wings of the first floor and the left wing of the ground floor, were taken into consideration. Planting in the layout of the office interior was minimal except for the occasional potted plant. Similarly to case study 1, the staff mainly belonged to the age group of 20–30 years but included a spread in age groups. The first floor, had a majority of female occupants while the ground floor had a majority of male occupants (Table 10.9).

Again, the main type of office space was 'open plan'; the type of work is categorised in Table 10.10.

Table 10.9 Proportions of male/female occupants, case study 2

Location	*Male*	*Female*
First floor, left wing	38.46%	61.52%
First floor, right wing	44.44%	55.56%
Ground floor	56.00%	44.00%

Table 10.10 Type of work, case study 2

Location	*Managerial*	*Professional*	*Clerical*	*Other*
First floor, left wing	21.62%	37.84%	29.73%	10.81%
First floor, right wing	33.33%	16.67%	44.44%	5.56%
Ground floor	20.83%	54.17%	20.83%	4.17%

The ground floor housed a large number of professionals and managers, while the first floor had a good distribution of all types of work groups.

As for case study 1, the occupants had **overall working hours per week** in the range 36–45 and 77 per cent of the group spend up to 35 hours in front of a **visual display unit** (VDU). A higher proportion of the occupants in this office reported longer working hours than the occupants in the London office. Of the three zones, the ground floor reports the highest number of working hours (Table 10.11), and the **type of work** illustrates that this is mainly occupied by professionals and managers.

Regarding the **weather and its effect on work performance**, contrary to the results obtained by the London study, more than 66 per cent responded that the weather has no effect on the individual work performance. The results for the first floor illustrate that the occupants work better in the mornings than in the afternoons, while the results for the ground floor indicate that the occupants perform similarly whether it is morning or afternoon.

Results regarding dissatisfaction with 'the job' and the 'indoor environment' varied considerably from those obtained in the London office (Table 10.12).

The results in all three cases support the contention of Cooper and Robertson (1990) that people dissatisfied with their job are likely to be dissatisfied with their indoor environment. On the first floor, left wing, more than 25 per cent state that they are dissatisfied with their job and a similar percentage express dissatisfaction with the indoor environment. The other

Table 10.11 Working hours, case study 2

Location	*Normal working hours (%)*		*Hours spent in front of a VDU (%)*	
	25–35 hours	*36–55 hours*	*25–35 hours*	*36–55 hours*
First floor, left wing	28.21	71.80	71.80	28.2
First floor, right wing	22.23	77.78	77.78	22.20
Ground floor	19.23	80.77	76.92	23.08

Table 10.12 Dissatisfaction with job and indoor environment, case study 2

Location	*% of people dissatisfied with*	
	the job	*the indoor environment*
First floor, left wing	28.21	25.64
First floor, right wing	16.67	22.22
Ground floor	13.04	17.39

two zones illustrate low percentages of dissatisfaction for the job and indoor environment; the ground floor reports lowest dissatisfaction. The ground floor consisted of a majority of male professionals compared to the majority of female occupants involved with clerical and other types of work on the first floor (similar to the London study), indicating that people with low status jobs tend to feel a higher level of dissatisfaction regarding their job and indoor environment. In these two cases (first floor, right wing and ground floor) the dissatisfaction regarding the indoor environment is much higher than with the job; this is prominently illustrated in the first floor, right wing data.

The responses for section 2 of the questionnaire correlate with the dissatisfaction expressed regarding the job and the indoor environment (Table 10.13).

In accordance with the data illustrated previously (Table 10.12), the first floor, left wing respondents report the lowest level of satisfaction for all criteria while the ground floor respondents report the highest. For the self-rated level of **productivity** all three zones report a high percentage of more than 82 per cent of satisfied occupants, which again relates to the high percentage of occupants being satisfied about the job and the

Table 10.13 Satisfaction/dissatisfaction with various factors, case study 2

Location	*Dissatisfied*	*Not sure*	*Satisfied*
	Feelings about the job (%)		
First floor, left wing	7.69	17.95	74.36
First floor, right wing	11.11	11.11	77.78
Ground floor	7.4	7.4	83.4
	Self-rated level of productivity (%)		
First floor, left wing	7.69	10.26	82.05
First floor, right wing	0.00	16.67	83.33
Ground floor	8.00	4.00	88.00
	Office layout (%)		
First floor, left wing	25.63	15.38	58.97
First floor, right wing	22.22	0.00	77.78
Ground floor	33.33	12.50	62.49
	Office decor (%)		
First floor, left wing	64.12	25.64	56.41
First floor, right wing	16.67	22.22	61.11
Ground floor	32.00	12.00	56.00

workplace. The ground floor occupants report the highest level of self-rated productivity (88 per cent), which strengthens the previous findings.

The level of satisfaction with the **layout** and the **decoration** of the office is lower. The first floor, left wing again reports the lowest level of satisfaction, while the ground floor reports the highest. The first floor, right wing seems to be the happiest regarding the physical environment. The first floor is mainly open plan office, compared to the more cellular offices on the ground floor. The low level of satisfaction on the first floor could be due to the crowding, the size of the individual spaces, giving lack of privacy. Even though the ground floor shows a high level of satisfaction regarding the job and productivity, it shows a lower satisfaction rate regarding the indoor environment.

Individual comments provided on the questionnaire indicate that the occupants were generally unhappy about the lighting system and considered it to be of poor quality. This was reflected in all three zones; the company was in the process of replacing the entire lighting system due to increasing complaints about its quality and about headaches believed to be due to the lighting system.

The influence of indoor environment on well-being and productivity

To questions regarding whether the office physical environment has an effect on physical and mental health, well-being and productivity, the responses varied widely from those found by Ellis (1994). On the first floor the majority considered that the office physical environment has an effect on the physical and mental health of the individuals, while the majority of occupants on the ground floor considered that this was not so. Even though these responses varied from the previous study, the responses regarding whether the indoor environment affects well-being and productivity were similar and showed that the majority considered that the office physical environment has an effect (Figs 10.10 and 10.11). This fact is not as significant as in the London study. This methodology enables priority factors to be established and detects strong and weak aspects which contribute to the productivity.

A high percentage of people on the first floor considered that the office physical environment affected their physical and mental health, well-being and productivity, and this floor contained the highest percentage dissatisfied with their indoor environment. The responses for the ground floor varied, and the percentage of people who thought that the indoor environment affected health, well-being and productivity was lower at about 50 per cent.

Responding to direct questions concerning the degree of influence the office environment has on the **quantity** and **quality of work** and on **being creative** and **taking responsibility**, the answers ranged from its having a **little**

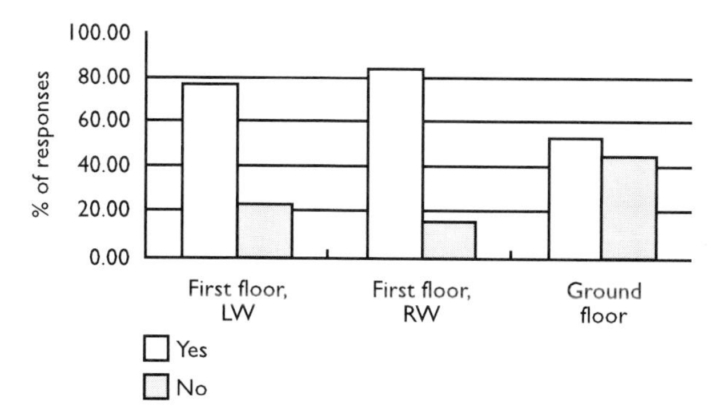

Figure 10.10 Does the physical office environment affect well-being?

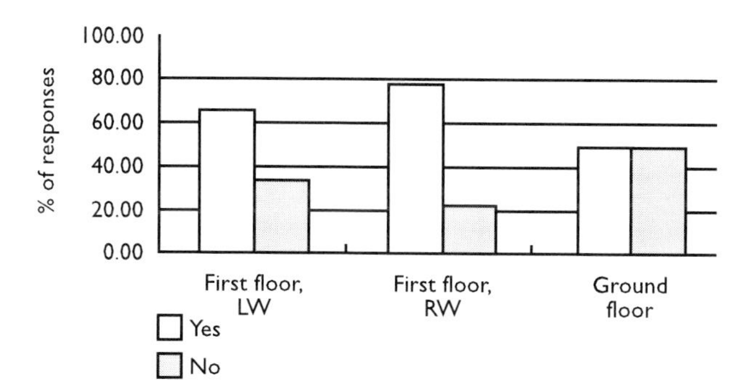

Figure 10.11 Does the physical office environment affect productivity?

to a **large** effect on these factors (similarly to the London office). More than 20 per cent of the occupants from the first floor reported that the office environment has a **large** effect on the quantity and quality of work that they can carry out; more than 20 per cent from the first floor, right wing reported that the office environment has a **very large** effect on being creative.

The amount of possible increase in productivity by improved environmental conditions was felt by the occupants to be 20–30 per cent; the highest number of occupants who felt this was on the ground floor, where job satisfaction was high and the satisfaction related to the indoor environment was lower (Table 10.14).

Personal control and its influence on productivity

The survey results indicated that there was very little personal control in adjusting the temperature, ventilation and lighting levels around each individual working space. Figures 10.12 and 10.13 illustrate that the majority of

Table 10.14 Self-projected possible improvements in productivity through improved conditions, case study 2

Location	*0%*	*10%*	*20%*	*30%*	*40%*	*50%*	*60%*	*70%*
First floor, left wing	14.29	48.57	20.00	5.71	0.00	5.71	0.00	5.71
First floor, right wing	17.65	35.29	23.53	5.88	5.88	5.88	0.00	5.88
Ground floor	21.74	9.45	26.07	8.70	0.00	0.00	0.00	0.00

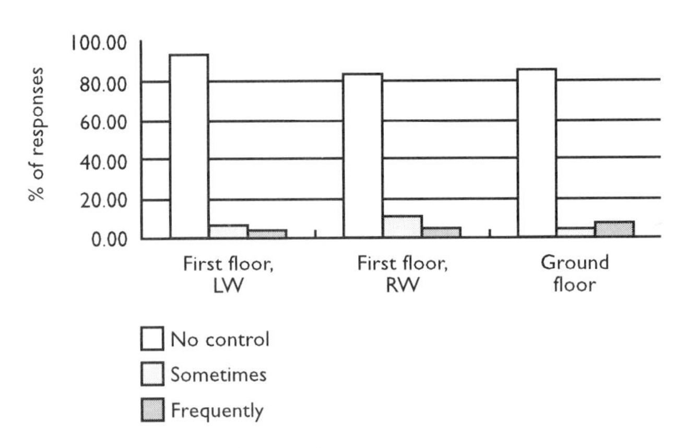

Figure 10.12 Personal control over temperature

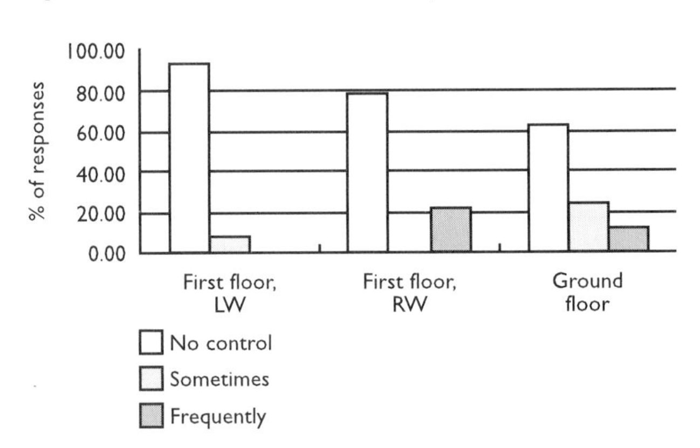

Figure 10.13 Personal control over lighting

the occupants felt there is little personal control at their workplaces (the same conclusion as in the London office survey).

A complaint voiced by most occupants was that **they have no access to a window** and that **they are deprived of a view to the outside which resulted in depressing conditions.** Even though the windows provide light and ventilation it seems that there is more of a psychological need to have access to a window and consciously know what is happening outside. The majority of occupants stated that it would be ideal if they could have views to a green area, but they would settle just to have an outside view.

Comfort and health in the workplace

Table 10.15 shows that first floor, left wing reported the highest number of adverse environmental conditions endured by occupants, and the ground floor reported the lowest percentages for many but not all aspects. From the percentages of occupants that frequently endured adverse environmental conditions, the most common conditions suffered by a majority were:

1 high/low temperature or temperature changes
2 insufficient daylight and low-quality general lighting
3 stale/stuffy air and dry/humid air
4 noise problems.

These findings are similar to the adverse conditions reported by occupants in the London office survey.

The findings in the environmental survey correlate with the health symptoms suffered by the occupants; it is clearly seen from the data for the first floor, left wing that the highest number of health symptoms occur there (Table 10.16). A common and repeated complaint was that illnesses caught by an individual noticeably circulate fast and spread to almost everybody

Table 10.15 Cited presence of adverse environmental conditions, case study 2

Negative environmental condition	*% of occupants*		
	First floor, left wing	*First floor, right wing*	*Ground floor*
Draught	14.29	16.67	0.00
Stale/stuffy air	42.86	27.78	37.50
Dry/humid air	50.00	22.22	37.50
Dusty air	9.76	11.11	12.00
High or low temperature; temperature changes	58.14	50.00	39.13
Little or too much daylight	45.24	38.89	30.43
Little or too much light	28.57	33.33	13.04
Flickering light	4.88	0.00	0.00
Annoying reflections	17.07	11.11	12.50
Vibrations, sounds or noise	26.83	22.22	20.83
Static electricity	9.76	11.11	4.17
Crowded workspace	7.32	15.79	8.33

Table 10.16 Occurrence of symptoms, case study 2

Symptom	*% of occupants*		
	First floor, left wing	*First floor, right wing*	*Ground floor*
Inability to sleep	19.05	18.75	12.00
Indigestion or sickness	7.50	0.00	8.33
Shortness of breath/dizziness	2.50	18.75	0.00
Changes in appetite	10.26	25.00	0.00
Do not want to get up in the morning or lethargy	40.00	31.25	24.00
Sweating/heart pounding	10.00	31.25	8.33
Shivering/muscle, joint pain	7.32	22.22	4.17
Nervous problems	30.00	11.76	16.67
Nasal-related problems	25.00	0.00	16.67
Sensory irritations in mouth	19.51	17.65	8.00
Skin problems	23.68	5.88	12.50
Odour or taste problems	2.63	17.65	0.00

surrounding that particular area within days or weeks. This mainly referred to colds and viral flus; there have been instances where a number of people have been off sick with the same symptoms.

The occupants on the first floor, left wing complained about the highest number of health symptoms. This is related to the low level of self-rated productivity and the dissatisfaction regarding the job and the indoor environment. The first floor, right wing reported the second highest number of people suffering from health symptoms; this is in accordance with the other high levels of satisfaction shown for the ground floor, where the lowest number of health symptoms were reported. The most common physical symptom felt by the majority was **'do not want to get up in the morning'** or **'feeling lethargic'** in the workplace (similarly to the London office). **Nervous problems** were the second most common conditions felt, followed by **nasal-related problems** and **sweating/heart pounding.**

Even though the occupants report a high level of productivity and a low overall dissatisfaction with a variety of factors (Table 10.17), the levels of job stress and job dissatisfaction are much higher than in the London office building. Contrarily, the percentage that felt they had an overall unsatisfactory indoor environment was much lower than in the London study. On the first floor, left wing 25 per cent reported dissatisfaction with their indoor environment and the job. The ground floor report the lowest dissatisfaction with the indoor environment, and experienced the lowest level of job stress.

In summary, the ground floor clearly housed a group of professionals who were fairly satisfied with their job and less satisfied with their indoor

Table 10.17 Levels of stress and dissatisfaction, case study 2

Location	*Percentage of people who frequently suffered*		
	job stress	*job dissatisfaction*	*overall unsatisfactory indoor environment*
First floor, left wing	12.82	25.64	25.00
First floor, right wing	17.65	23.53	17.65
Ground floor	12.50	25.00	12.50

environment, but did not depend on any external factors to provide the motivation and inspiration they needed. They enjoyed a good sense of well-being and this was reflected in all aspects of the responses and the environmental and health conditions that they experienced. Contrarily, on the first floor, left wing occupants reported the highest level of dissatisfaction regarding their jobs and workplaces. This work group consisted of a majority of female workers involved in a variety of job types and showed the maximum number of environmental and health symptoms. The first floor, right wing produced data which lay between these two areas.

Priority factor analysis according to the analytic hierarchy process model

Results from the AHP analysis illustrate similar trends as for the London office. Well-being and other factors relating to mental and physical health play an important role in influencing productivity of occupants in offices. The results for the three zones varied considerably and do not conform to such a consistent pattern as they did in the London study. The other factors that had a very significant influence on productivity, in order of importance, were **ability to perform, motivation** and **job satisfaction**. Similarly to the previous study, **technical competence** appears to play a less important role in influencing productivity (Figs 10.14 and 10.15).

The **system factors** that influence the **'human factors of the individual'** are interrelated and the responses show a varying level of importance. It should be noted that in the London office, the **indoor environment** was given considerable importance as a system factor which influences productivity, compared to the results achieved in this office. **Personal, occupational, organisational factors** seemed to have a much higher importance in influencing productivity. Results for the first floor, left wing show that with respect to well-being, **indoor environment** is given quite a high priority (Figs 10.16 and 10.17).

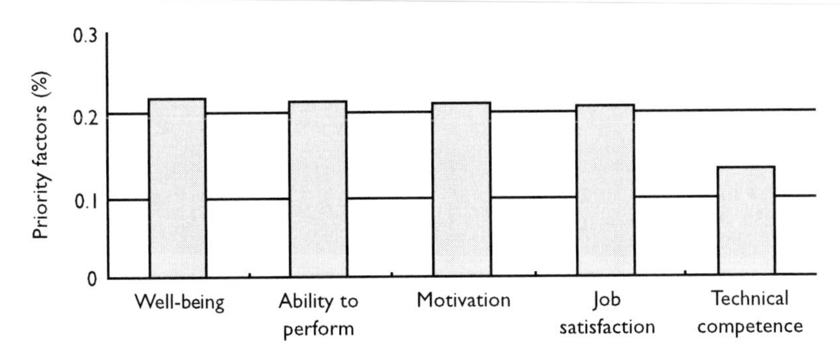

Figure 10.14 Main human factors influencing productivity: all floors

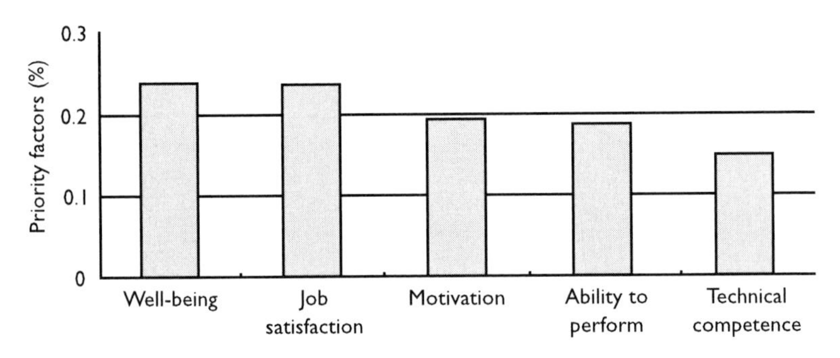

Figure 10.15 Main human factors influencing productivity: ground floor, left wing

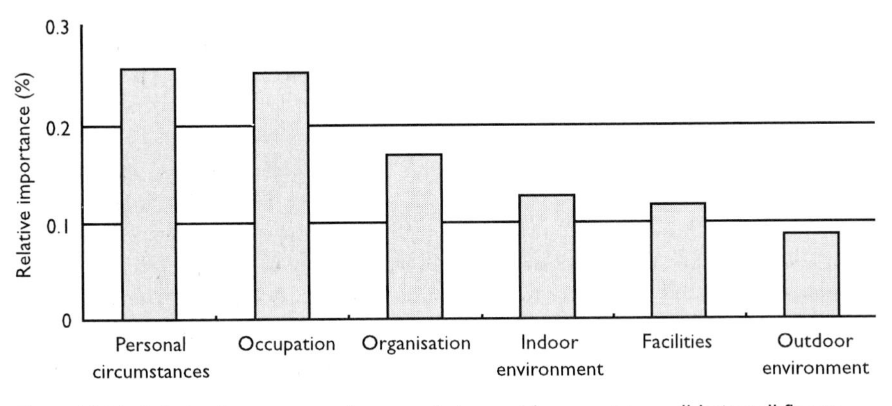

Figure 10.16 Relative importance of system factors with respect to well-being: all floors

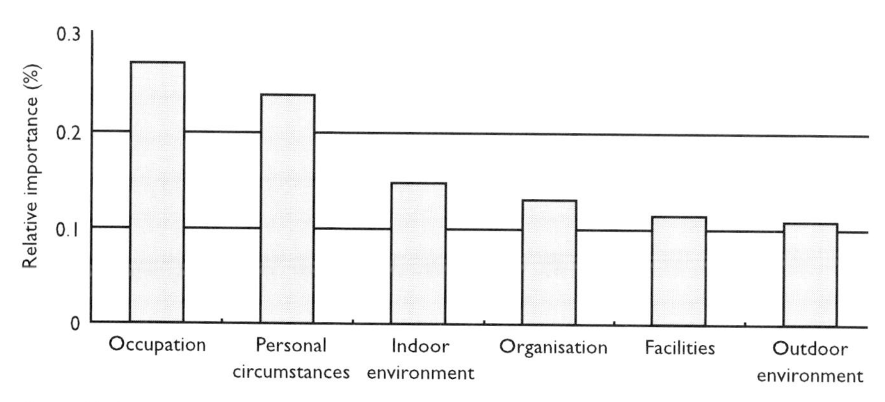

Figure 10.17 Relative importance of system factors with respect to well-being: first floor, left wing

Summary of the research findings

- A higher proportion of the occupants in this building reported longer working hours than in the London building, and the dissatisfaction regarding the indoor environment was much less than in the London office (Tables 10.11 and 10.12).
- It was noted that occupants with higher job status expressed a lesser degree of dissatisfaction regarding the job and the indoor environment. This was evident in both buildings (introductory information and Table 10.12).
- The dissatisfaction levels expressed were higher for indoor environment than for the jobs (Table 10.12).
- The self-rated productivity levels were much higher than in the London building (Table 10.13).
- The amount of increase in productivity by improved conditions as expressed by occupants was much less than in the London study. This could again relate to the high levels of self-rated productivity (Tables 10.13 and 10.14).
- There was very little personal control in adjusting temperature, lighting and ventilation according to individual needs. Frequent complaints were attributed to the poor-quality lighting system and the resulting headaches and other health symptoms. Inaccessibility of windows and outside views was disliked.
- The adverse environmental conditions suffered were similar to those in the London building, and could be summarised as follows (Table 10.15):

  1 high/low temperature or temperature changes
  2 insufficient daylight and low-quality general lighting

3 stale/stuffy air and dry/humid air
4 noise problems.

- The occupants on the first floor reported the highest number of health symptoms. They also felt a low level of self-rated productivity and a higher degree of dissatisfaction regarding the job and the indoor environment (Table 10.16).
- The most common physical symptom felt by the majority was '**do not want to get up in the morning**' or '**feeling lethargic**' at the workplace. **Nervous problems** were the second most common condition felt, followed by **nasal-related problems** and **sweating/heart pounding**. A common complaint was that illnesses spread rapidly among staff (similar to the London building).
- Even though the staff reported a high level of productivity and a low level of dissatisfaction regarding the indoor environment, they expressed a high level of job stress and job dissatisfaction.
- **Well-being** seems to be a major factor in enhancing productivity. The other factors which had a very significant influence on productivity, in order of importance, were **ability to perform, motivation** and **job satisfaction**. As in the London study, **technical importance** appears to play a less important role in influencing productivity.
- The **indoor environment** as a system factor which influences productivity, was given considerable importance in the London office compared to the results achieved in this study, where **personal, occupational, organisational factors** seemed to have much higher importance in influencing productivity. However, results for the first floor, left wing show that with respect to well-being, **indoor environment** is given quite a high priority.

Conclusions

A good working environment will help to provide the user with a good sense of well-being, inspiration and comfort. The main advantage of good environments is in terms of reduced upgrading investment, reduced sickness absence, an optimum level of productivity and improved comfort levels. Individuals respond very differently to their environments and the research supports a correlation between worker productivity, well-being and environmental comfort. The results illustrated that the occupants who report a high level of dissatisfaction about their job are usually the people who suffer more work- and office environment-related illnesses which affect their well-being, but not always so. There is a connection between dissatisfied staff and low productivity; and a good sense of well-being is very important as it can lead to substantial productivity gain. Well-being expresses overall satisfaction. If the environment is particularly bad, people will be dissatisfied irrespective of job satisfaction.

Comfort should be viewed in the context of well-being and hence links the quality of the indoor environment with employee productivity. **Well-being** is not concerned only with personal health. It is how you feel about yourself and your family as well as your surroundings. **Ability** and **motivation** are factors which require a stable and stimulated mind besides the drive to perform. Clearly these factors can be influenced by indoor environment.

Sensory pleasure indicates the sign of a stimulus; it is transient, and motivates behaviour. It has been shown experimentally that the wisdom of the body leads organisms to seek pleasure and avoid displeasure, and thus achieve behaviours which are beneficial to the subject's mind and body (Cabanac, 1971, 1997). Pleasure indicates a positive stimulus and serves both to reward behaviour and to provide the motivation for eliciting behaviour that optimises psycho-physiological processes. A sensitive and qualitative working environment which provides occupants with these stimuli increases the likelihood that motivation and performance will be at a high level.

There is a need to develop an empirical model to enable greater understanding of multi-sensory well-being of occupants under realistic dynamic working conditions, and to develop a correlation between productivity, well-being, and multi-sensory occupant comfort. Most work concentrates on thermal comfort for **groups** of people, but this is unrealistic as a basis of environmental design because productivity depends on the accumulated efforts of **individuals.** This alternative holistic approach will enable standards to be evolved which are realistic and recognise the combined value of low energy, health, comfort and productivity in various situations. It recognises the responses of the senses as a whole and does not focus exclusively on thermal factors.

The individual responses illustrated that the office physical environment has a direct influence on the health, well-being and productivity of occupants, and a properly designed working place with good-quality lighting, comfort conditions and a stimulating environment can invigorate the minds and inspire the occupants to excel in their performance.

Acknowledgements

The authors would like to acknowledge the financial support and assistance given by Clearvision International Ltd, Optimum Work Place Ltd, Canary Wharf Management Ltd and Jardinierie Interiors Ltd to carry out this research project, and would also like to thank the staff at Richard Ellis Ltd for participating in the questionnaire survey; Dr Jagit Singh of Oscar Faber Ltd for help with biological sampling, Professor Saaty of Pittsburgh University for assessing our work with AHP and Li Baizhan for his doctoral work on AHP.

References

Barbatano, L. (1994) *Home Indoor Pollution: A Contribution for an Economic Assessment, Proceedings of Healthy Air, '94 Conference*, Milan pp. 395–402.

Cabanac, M. (1971) Physiological role of pleasure. *Science*, **173**, 1103–1107.

Cabanac, M. (1997) Pleasure and joy and their role in human life, *Workplace Comfort Forum*, London.

Chalmers, D.J. (1997) The puzzle of conscious experience. *Scientific American* Special Issue, 'Mysteries of the Mind', Jan. 30–37.

Clements-Croome, D.J. (1997) *Naturally Ventilated Buildings*. London: E & FN Spon.

Clements-Croome, D.J. and Li, B. (1995) Impact of indoor environment on productivity. *Workplace Comfort Forum*, Royal Institute of British Architects, London.

Cohen, J. and Cohen, P. (1983) *Applied Multiple Regression/Correlation Analysis for the Behavioural Sciences*, Hillsdale, NJ and London: Lawrence Erlbaum Associates Publishers.

Collins, J.G. (1989) Health characteristics by occupation and industry: United States 1983–1985. *Vital Health and Statistics* **10**(170), Hyattsville, MD: National Center for Health Statistics.

Cooper, C.L. (1988) *Occupational Stress Indicator Management Guide*. Windsor: NFER Nelson.

Cooper, C.L. (1995) Personal communication, University of Manchester Institute of Science and Technology.

Cooper, C.L. and Robertson, I.T. (1990) *International Review of Industrial and Organizational Psychology*, vol. 9. Wiley.

Croome, D.J. and Roberts, B.M. (1981) *Airconditioning and Ventilation of Buildings*. Oxford: Pergamon.

Ellis, R. (1994) *Tomorrow's Workplace*, a Survey for Richard Ellis by The Harris Research Centre.

Donnini, G., Van Hiep Nguyen and Moline, J. (1994) Office thermal environments and occupant perception of comfort. *La Riforma Medica*, **109**, Suppl. 1 (2), 257–263.

Holcomb, L.C. and Pedelty, J.F. (1994) Comparison of employee upper respiratory absenteeism costs with costs associated with improved ventilation. *ASHRAE Trans.*, **100**(2), 914–920.

Ilgen, D.R. and Schneider, J. (1991) Performance measurement: a multi-discipline view. *International Review of Industrial and Organizational Psychology*, vol. 6, Cooper, C.L. and Robertson (eds), 71–108, Wiley.

Kroner, W.M. and Stark-Martin, J.A. (1992) Environmentally responsive work stations and office worker productivity. *Workshop on Environment and Productivity*, June (contact Department of Architecture, Rensselaer University, Troy, NY).

Leaman, A. (1994) Dissatisfaction and office productivity. *The Facilities Management Association of Australia, Annual Conference*, Sydney.

Leaman, A. (1997) Probe 10, bench marks for better buildings, *Build. Serv. J.*, May.

Lehto, M. (1997) Decision making, in *Handbook of Human Factors and Ergonomics* (ed. Salvendy, G.), section 37.3.4.2. Wiley Interscience.

Li, B., Croome, D.J. and Yao, R. (1996) Analytic hierarchy process model for assessing priority indoor environmental factors influencing well-being of occupants in the offices. *7th International Conference on Indoor Air Quality and Climate*, 21–26 July, Nagoya, Japan.

Li, B. (1998) *Assessing the Influence of Indoor Environment on Productivity in Offices*, PhD thesis, University of Reading.

Li Baizhan (1996a) *Assessing the Influence of Indoor Environment on Health, Well-being & Productivity*, Questionnaire For Environment Survey of Offices, The University of Reading, UK.

Li Baizhan (1996b) *Assessment of the Influence of Indoor Environment on Job Stress and Productivity of Occupants in Offices*, The University of Reading, UK.

Lomonaco, C. and Miller, D. (1997) Comfort and control in the work place. *ASHRAE J.*, **39**(9).

Lysaught, R.J. *et al.* (1989) *Operator Workload: Comprehensive Review and Evaluation of Operator Workload Methodologies* (Analytic Tech. Rep. 2075–3). Alexandria, VA: US Army Research Institute for Behavioural and Social Sciences.

Mehrabian, A. and Russell, J. (1974) *An Approach to Environmental Psychology*. Cambridge, MA: MIT Press.

Meister, D. (1986) *Human Factors: Testing and Evaluation*. Amsterdam: Elsevier.

Moray, N. *et al.* (1979) Final report of the Experimental Psychology Group. *Mental Workload: Its Theory and Measurement*, ed. Moray, N., 101–114. New York: Plenum.

O'Donnell, R.D. and Eggemeier, F.T. (1986) Workload assessment methodology. *Handbook of Perception and Human Performance: Cognitive Processes and Human Performance*, ed. Boff, K.R. *et al.* New York: Wiley.

Oppenheim, A.N. (1992) *Questionnaire Design, Interviewing and Attitude Measurement*. London: Pinter.

Oseland, N.A. (1996) Impact of the indoor environment on comfort and productivity. Workplace Comfort Forum, London.

Pepler, R.D. and Warner, C.G. (1968) Temperature and learning; an experimental study. *ASHRAE Trans.*, **74**, Part II.

Raw, G.J., Roys, M.S. and Leaman, A. (1990) Further findings from the office environment survey: *Productivity, Indoor Air '90. 5th International Conference on Indoor Air Quality and Climate*, **1**, 231–236.

Saaty, T.L. (1972) *The Analytic Hierarchy Process*. New York: McGraw-Hill.

Saaty, T.L. (1988) *Multicriteria Decision Making: The Analytic Hierarchy Process*. Pittsburgh, PA.

Schweisheimer, W. (1962) Does air conditioning increase productivity? *Heating Vent. Eng.*, **35**(419), 669.

Schen, Q. and Lo, K.K. (1997) A prioritisation method for planned maintenance. *Proceedings of Building Surveying Conference*, Hong Kong Institute of Surveyors, Hong Kong, 1–10.

Sutermeister, R.A. (1976) *People and Productivity*. New York: McGraw-Hill.

Thorndike, R.L. (1949) *Personnel Testing*. New York: Wiley.

Weiner, P. (1994) *Architectural Record*, May.

Wilson, S. and Hedge, A. (1987) *The Office Environment Survey*. London: Building Use Studies Ltd.

Wyon, D.P. (1992) Healthy buildings, barriers to productivity. *Build. Serv. J.*, June.
Yao Runming, Liu Antian and Li Baizhan (1992) Application of the AHP method to the planning of a solar house in China. *Proceedings of Thermal Performance of the Exterior Envelopes of Buildings V*, Clearwater Beach, FL, 602–606.

Chapter 11

Productivity in buildings: the 'killer' variables

Adrian Leaman and Bill Bordass

Introduction

This chapter deals with the somewhat vexing question of human productivity in the workplace. It sets out to answer: 'What features of workplaces under the control of designers and managers significantly influence human productivity?'. The main theme is how individual occupants are affected. We are seeking building or organisational features which most readily improve or hinder human productivity. The findings can then be used in the brief-making, design and management processes.

Observations are mainly based on surveys carried out since 1985 in the UK by Building Use Studies and William Bordass Associates, together with new and spin-off projects from the Building Research Establishment and UK Department of Environment among others. Some of this work has been published before, but the bulk of data collected remains to be analysed and reported on in greater detail.

There is also a substantial wider literature, much of it from the US, reviewed by Lorsch and Abdou (1994a, 1994b, 1994c) and Oseland (1996). Quite a lot is known about how well people respond to different conditions of temperature, humidity, lighting, ventilation and noise, for example, and regulations for building design are based on many of the findings (although with a considerable time lag).

Most of these studies come from military, industrial and commercial sources. Their findings can be contradictory (although there is a reasonable consensus on key points) and sometimes they can be hard to make sense of when productivity is linked to the indoor environment. For instance, Pepler and Warner (1968) found that young people worked best (and were thus more productive) for short periods when they were uncomfortably cold. Periods of relatively uncomfortable arousal can thus be important. It is unlikely that people will continue to perform well in conditions of prolonged discomfort. De Dear *et al.* (1993) showed that large numbers of office staff considered their working environments to be thermally unacceptable despite measured conditions falling within industry-standard comfort envelopes, so perceived and measured conditions can be different. De Dear *et al.* (1993) also demonstrated that 23.5°C is the temperature which people in offices prefer, but even with this there is a sizeable minority of about 35 per cent who wanted it to be warmer or cooler, so minority needs cannot be ignored.

Human productivity in workplaces is fraught with difficulty because:

- studies of individual occupants often miss the wider context of differences between buildings, and their operational and managerial circumstances
- buildings and their occupying organisations are rarely even similar to each other from case to case, which complicates comparisons
- methodological and interpretational problems result from the above and lead to begged questions about assumptions and spurious detail, so it can be difficult to filter out the most important points from case to case
- people behave differently in groups.

On balance, we prefer the approach of 'real-world research' (Robson, 1993) which deals with human activities in their real contexts. This involves smaller surveys (10–15 buildings is ideal), controlling for context through a thorough understanding of prevailing circumstances, detailed occupant, technical and energy surveys, based on Building Use Studies occupant questionnaires and EARM (Energy Assessment and Reporting Methodology) techniques, intensive peer group review and quality control, and reporting based on the 'exception proving the rule' (that is, finding the most benign energy case and the happiest occupants and understanding thoroughly why this happens). Laboratory experiments and statistical tests based on experimental designs often try to isolate cause and effects, but in so doing can over-simplify or interfere with the behaviour of the study group. A celebrated example is that people who have greater control over their indoor environment are more tolerant of wider ranges of temperature. Theory based on laboratory tests in climate chambers seemingly understates this effect, which is context-dependent (Bunn, 1993).

Buildings are complex systems made up of physical and human elements and their many associations, interactions, interfaces and feedbacks. Elements and interactions create '**emergent**' properties (those like aesthetic qualities which are greater than the sum of their parts and consequently harder to pin down). Because of interdependencies, it is often fruitless to try to separate out different variables and treat them as 'independent', as many statistical methods require. Characteristics such as depth of space from wall to wall, open-plan space and air-conditioning all depend on each other to a greater or lesser extent. As a result, they produce cat's cradles of statistical interdependencies which can be impossible to disentangle causally. Feedback loops also add to the complex dynamics.

As well as physical complexity, designers, managers and occupiers also have different preferences, priorities and personal agendas. Changing circumstances add to the richness and dynamism of contexts, but also make them less yielding to conventional analysis. Although it is tempting to seek out general theories governing behaviour in buildings (and enshrine principles in design guidance and building legislation), in practical terms constraints imposed by contexts frequently turn out to be more important for the designer, manager or occupier.

Contexts – locational, social, economic, technical and environmental – are always subtly changing, and can sometimes have surprisingly direct effects on local situations. For example, it can be baffling why two schools – identical in design and intake profiles and on adjacent sites – should differ markedly in vandalism damage rates. This can often be explained by environmental strategies – especially with respect to the speed of repair of damage and the extent to which particular individuals take responsibility for ('own') the problem of repair (CIRIA, 1994).

This example might tempt a designer to say that management is the cause, not design. However, as recent post-occupancy surveys show, buildings whose management strategy has been developed from the outset of the design, with a clear understanding of client requirements and management capacity, are more likely to perform better. Hence management and design factors, like many others in buildings, depend on each other and cannot be meaningfully isolated in real situations (see Leaman and Bordass (1997), many of whose hypotheses are being verified in the Probe studies (Bordass *et al.*, 1995b, 1996, 1997; Standeven *et al.*, 1995, 1996a, 1996b; Asbridge and Cohen, 1996; Standeven and Cohen, 1996; Cohen *et al.*, 1996a, 1996b, 1997; Leaman, 1997; Leaman *et al.*, 1997; Bordass and Leaman, 1997)).

Designers and managers constantly strive to create conditions which bring out the best in people and add value to investments and services. Occupants will also usually want to achieve reasonable conditions for themselves. To complicate matters further, designers, managers and occupiers can all behave perversely as well. For example, in a study of the top floor of an office building we asked the occupants 'Why do the lights always seem to be on when you don't need them, especially as the switches are easy to reach?'. The reply came: 'We only do it to annoy the manager. He is obsessive about switching the lights off so we switch them back on again when he goes out just to annoy him.' More seriously, different working groups in the same building can use the lighting to reinforce their team identity.

Too often, though, elements unwittingly interact and conspire to create unforeseen and unwanted chronic problems. In the context of technology and its side-effects, Tenner (1996) has memorably labelled these **'revenge effects'**. In buildings, technical elements often work reasonably well in isolation or in theory, but when included as part of a wider system of operation induce inefficiencies which ultimately affect the ability of people to perform their work properly. These wider aspects of the design, use management and operation of buildings concern us most here.

Terminology

By '**productivity**' we mean the ability of people to enhance their work output through increases in the quantity and/or quality of the product or service they deliver. Work output is impossible to measure meaningfully for all building occupants. How do you compare, for instance, the productivity of telephonists in a call centre with their managers? Our answer is to use scales of perceived productivity, rather than measure productivity directly. The question on productivity which has been incorporated into most Building Use Studies questionnaires since the pioneering *Office Environment Survey* (Wilson and Hedge, 1987) is shown in Fig. 11.1.

On balance, advantages with perceived productivity scales outweigh disadvantages. Advantages include:

- a single question covers the topic so it can be incorporated in surveys with wider objectives (although we find that building managers are still wary of the question and sometimes forbid us to use it)
- the question is common to all respondents so that fair comparisons can be made between most of them
- it can be incorporated in questionnaires across different building types (although strict comparability between types may need to be treated circumspectly, for example between teachers and administrative staff in a school where working conditions are slightly different between the two)
- large samples may be surveyed relatively cheaply
- benchmarks of averages or medians may be used to assess how occupants' perceptions in individual buildings score against the complete dataset
- data analysis and verification are easier across large samples in many different buildings.

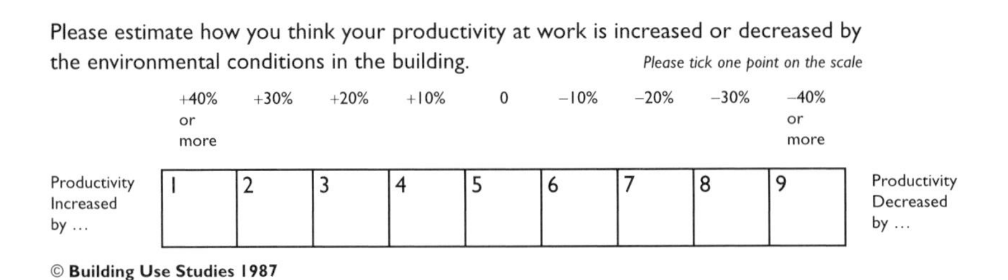

Please estimate how you think your productivity at work is increased or decreased by the environmental conditions in the building. *Please tick one point on the scale*

	+40% or more	+30%	+20%	+10%	0	−10%	−20%	−30%	−40% or more	
Productivity Increased by ...	1	2	3	4	5	6	7	8	9	Productivity Decreased by ...

Figure 11.1 Perceived productivity question used in Building Use Studies surveys

Disadvantages are:

- the nagging doubt that perceived productivity as measured may not associate well with the actual productivity of the occupants (although many agree on the key point that perceived and actual productivity are strongly associated (see the review of sources in Oseland, 1996, 1997))
- the need for occupants to judge their own reference point when answering the question (they sometimes want to know 'Productivity with respect to what?')
- the possible effects of context and other ruling factors at the time of the survey, for example, rumours of possible redundancies.

Objectives

The main objectives of the productivity parts of post-occupancy and diagnostic studies are to:

- give designers and managers indications of the main factors within their control that might influence human productivity at work
- help prepare design and management strategies which measurably aim to improve performance for organisations and individuals without compromising individuals' needs and introducing unhelpful side-effects.

Our emphasis is always on appropriate design and management strategies – as expressed in the design briefing process and troubleshooting studies – and the major risk factors affecting productivity. We are not attempting a 'theory' of productivity at work, nor a detailed analysis of cause and effect or costs and payoffs.

'Killer' variables

A 'killer' variable – to use hyperbole from the language of computing – has a critical influence on the overall behaviour of a system. With the present state of knowledge we can guesstimate that losses (or gains) of up to 15 per cent of turnover in a typical office organisation might be attributable to the design, management and use of the indoor environment. Fifteen per cent gains/losses as a ballpark figure crops up in the work of Brill (1986) and Vischer (1989), for example. Lorsch and Abdou (1994a, 1994b, 1994c) talk about the productivity of 20 per cent of office workers in the USA being raised simply by improved indoor air quality. Data from the Probe studies show perceived differences of up to 25 per cent between comfortable and uncomfortable staff, with uncomfortable staff showing consistently lower productivity (as common sense predicts – unless the arousal mechanism is more important) (Leaman, 1997). The difference gets narrower as overall satisfaction with the building improves.

Whatever the actual figures (no-one knows, of course), there is consensus that indoor environment factors improve output, as well as a lot of evidence to show associations with a cluster of related factors such as perceived health, comfort, and satisfaction. There are also data to show that some of the management, design and use characteristics which improve perceptions of individual welfare also contribute towards better energy efficiency, thereby closing the loop on a potential '**virtuous**' circle (Bordass *et al.*, 1995a; Oseland, 1997).

This said, there are not many grounds for optimism, because the vast majority of occupied buildings do not exhibit such self-reinforcing qualities and many are unmanageably complex (Bordass *et al.*, 1995b). From the perspective adopted here – that of strategic guidance for building designers, managers and occupiers – it is sufficient to know simply that there are positive and negative relationships between indoor environmental factors and human productivity. The question then becomes: Which are the most important?

Important factors – the 'killer' variables – have been arranged here into four '**clusters**'. Each represents a group of features which have more connections among themselves than with others. There are also connections between the four: as we have implied, there is no such thing as an independent variable in a building! Their relative importance also depends on prevailing circumstances – the stage of the design process, for example. We have not prioritised here, because contexts and interconnectedness alter in different situations, changing priorities as well.

The clusters are:

- Personal control
- Responsiveness
- Building depth
- Workgroups.

Personal control

Research work in the 1980s into what was then called '**sick building syndrome**' ('**building-related sickness**' is the preferred term now) confirmed to a new generation of researchers what was already well known to an older one – that people's perception of control over their environment affects their comfort and satisfaction. Work on thermal comfort, notably that of Humphreys and McIntyre in the 1970s (Humphreys, 1976, 1992; McIntyre, 1980), has shown that the range of temperatures that building occupants reported as 'comfortable' was wider in field studies than in controlled conditions in the laboratory. People were more tolerant of conditions the more control opportunities – switches, blinds and opening windows, for instance – were available to them.

Similar results on the relationships between perceived control and sickness symptoms have been reported by, for example, Raw *et al.* (1993). More recently, in studies on heart disease in civil servants, higher incidence of heart attacks seem to be related to people's perception of control over their work (Marmot, 1997). There are many other such examples, including the renewed interest in the 1990s in adaptive comfort (Baker, 1996).

Table 11.1 shows results for office workers in 11 UK buildings examined by Building Use Studies in 1996–97. Self-assessed productivity is significantly associated with perceptions of control in 7 out of 11 buildings. Perceptions of control are measured by the average of five variables for perceived control over heating, cooling, lighting, ventilation and noise. Figure 11.2 shows that the relationship in Table 11.1 probably gets weaker as buildings get better – as overall satisfaction gets better there is less need for discomfort alleviation. However, buildings which are designed to *provide* comfort but

Table 11.1 Relationships between control and perceived productivity for office workers in eleven UK buildings surveyed in 1996–97

Building	*Type*	*Average overall percentile*	*Spearman's rho (corrected for ties) between mean control and productivity*	*p value*	*Significant association between perceived productivity and mean control?*
A	AC	52	0.12	0.4133	
B	AC	43	0.17	0.0043	Yes
C	NV	81	0.08	0.4469	
D	NV	12	0.34	0.0348	Yes
E	NV	66	0.30	0.1546	
F	AC	67	0.31	0.0053	Yes
G	MM	91	0.24	0.0425	Yes
H	ANV	43	0.49	0.0002	Yes
I	ANV	22	0.35	0.0033	Yes
J	NV	54	0.16	0.0031	Yes
K	NV	74	0.07	0.6356	

Interpretation
Buildings: 11 studied by Building Use Studies in 1996–97 for which productivity data are available.
Type: AC = air-conditioned; NV = conventional naturally ventilated; MM = mixed mode; ANV = advanced natural ventilation. Average overall percentile: average from percentile score for seven variables from BUS dataset (see Bordass *et al.* (1996) for further details).
Spearman's rho: Correlation between scores for individual occupants between the mean of five perceived control variables (mean control – heating, cooling, lighting, ventilation, noise) and perceived productivity (see also Fig. 11.1).
p *value*: *p* values less than 0.05 indicate a significant association.

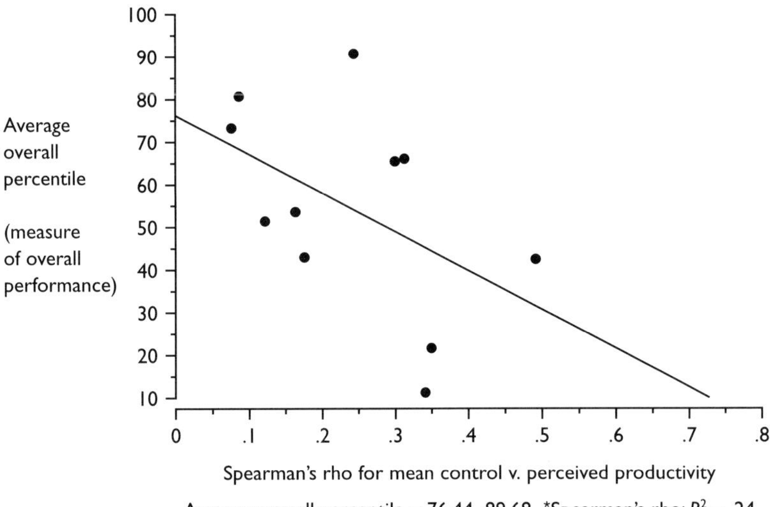

Interpretation: The 'average overall percentile' is a measure utilising seven summary variables from the Building Use Studies dataset of 50 buildings. The average percentile score (built from individual percentiles for each of the seven variables) shows how a particular building scores relative to all others. A percentile score of 50 is in the middle of the range. The best buildings – those with higher percentiles – tend to have lower correlation coefficients. The association is verging on significance ($p = 0.06$ for rho), and quite strong (rho = –0.58). A larger sample of buildings will help test this more thoroughly.

Figure 11.2 Relationship between perceived control and productivity decline as buildings perform better

do not deliver it (through technical, management or usability problems) tend to come out badly.

Table 11.2 shows that of the five perceived control variables – heating, cooling, lighting, ventilation and noise – the last, noise, is most strongly associated with perceived productivity, but the relationship is quite weak. Even so, perceived control over lighting is the only one that is not significant.

Tables 11.1 and 11.2 and Fig. 11.2 tell a stark statistical story about personal control and productivity. Building users in their personal comments on questionnaires are much more forthcoming. In study after study, people say that lack of environmental control is their single most important concern, followed by lack of control over noise. Taking one typical comment from many in the same vein from a building study carried out in 1996: 'Noise has the most disturbing effect on my work. Other factors such as heat and light are not so disrupting.' Many people, some almost instinctively, oppose the idea of open-plan working because they immediately suspect that

Table 11.2 Associations between perceived control and productivity for eleven study buildings and five perceived control variables

	Spearman's rho (corrected for ties)	*p value*	*Significant association?*
Heating	0.10	0.0001	Yes
Cooling	0.08	0.0001	Yes
Lighting	0.03	0.2513	
Ventilation	0.06	0.0001	Yes
Noise	0.12	0.0001	Yes

Interpretation: Given that perceptions of mean control are related to productivity, which of the five variables making up the mean control statistic are most important? This table shows that noise produces the strongest association with productivity, significant but relatively weak. Heating is the next strongest. Control over lighting is not significant. The order of these variables tends to confirm earlier work by Building Use Studies which showed that lighting, which is the easiest to change in a building, also is the least effective in its impact! [Editor's note: if lighting variables other than level could be changed (e.g., a vector-scalar ratio, daylight factor, colour, temperature) the result might be different.]

they will lose control and privacy and it will become more noisy. This might not necessarily actually happen in practice, but people suspect that it will.

In spite of the wealth of research and occupier evidence that high perceptions of personal control bring benefits such as better productivity and improved health, designers, developers, and sometimes even clients seem remarkably reluctant to act on it. There are many reasons for this, including the absence of thorough cost-in-use analysis in the calculation of future payoffs (and the problem of who actually receives the benefit), but four are prominent.

1 Environmental control operates at the interface between a building's physical and technical systems and its human occupants, or, less visibly, automatically and often under the supervision of computer-controlled building management systems. Perhaps seduced by the promise of technology rather than its delivered performance, designers assign more functions to automatic control than are usually warranted and, knowingly or not, make the interfaces obscure. They then often do not seem to make clear to the client the management implications of the technology, and whether these are acceptable to them. Simpler and more robust systems are required, with greater opportunities for users to intervene – especially for opportunities to override existing settings, better feedback on what is supposed to be happening and whether or not the system is actually working (Leaman and Bordass, 1997).
2 Building design is split into architectural and building services tasks, often with suprisingly little integration between them. Poor attention to detail in building controls is a common symptom of an incomplete design and specification process and gaps between areas of professional

responsibility. As well as a lack of recognition of the problems here, there is also an absence of tools for specification and briefing, and a lack of suitable standard componentry and systems. Manufacturers find it difficult to invest in suitable new or modified products to meet such requirements, owing to a diffuse market and a lack of well-articulated demand. Those who have tried have found success elusive. For example, the promising environmentally advanced Colt window system has recently been taken off the market as a complete package (Lloyd-Jones, 1994).

3 Designers do not fully appreciate the important difference between comfort provision and discomfort alleviation. For example, the ability to alter workstation position – a seemingly trivial feature – can be crucial to office users' comfort. By making tiny changes to their immediate environment to avoid the worst effects of (say) glare from the winter sun, or down draughts, occupants can turn intolerable conditions into marginally tolerable ones without management intervention. Most control adjustments will be at margins of discomfort, triggered by something experienced as uncomfortable, rather than in anticipation. The absence of this capability to fine-tune, especially in space-planned offices with fixed furniture systems and little or no user control, can make the difference between tolerable comfort and dissatisfaction.
4 Sadly, few building occupiers are motivated enough to gain control of systems which are troublesome.

Responsiveness

To many people, the relationship between better personal control and human performance is common sense; so too is the cluster of variables related to responsiveness. Many of the buildings which work well in post-occupancy studies appear to have the capability to meet people's needs very rapidly either in anticipation or as they arise. This applies to personal control, but it also works at other levels: the ability to reconfigure furniture, for example, or adaptability of spaces to accommodate change, or speed of response to complaints by the facilities management department.

The importance of responsiveness first became blindingly clear to us in a study in 1992 which included One Bridewell Street, Bristol (Energy Efficiency Office, 1991; Building Use Studies Ltd, 1992/93) revisited by Joanna Eley in 1996. This building is noteworthy because, although air-conditioned, it uses little more energy than a good-practice, naturally-ventilated, open-plan office. In addition, occupant satisfaction is unusually high. Was this just coincidence or is something more profound at work?

At One Bridewell Street high occupant satisfaction seemed to be related to the speed with which the facilities management department dealt with complaints of discomfort: the response was exceptionally fast, and occupants were told exactly what the outcome was. The facilities manager also learned to anticipate common problems and to deal with them, often before anyone noticed. Personal control for the occupants was not high, with just infrared 'zappers' for the lights and limited ability to change workstation position.

To test the possible influence of response time, a new variable was added to the Building Use Studies questionnaire in 1995 (Fig. 11.3). The relationship between perceived speed of response to complaints and perceived productivity is shown in Fig. 11.4 and people's perception of 'quickness' (the speed with which occupants think that heating, cooling, lighting, ventilation and noise needs are met) in Table 11.3.

With the usual interpretational caveats relating to small, non-random samples, the results in Figs 11.4 and 11.5 and Table 11.3 are strong grounds for developing this line of analysis further. The association between speed of response and productivity in Fig. 11.4 is positive and significant. Eight out of eleven buildings in Table 11.3 show significant positive associations between perceived quickness of response and perceived productivity. Figure 11.5 indicates that, just like perceived control, the strength of correlation between quickness and productivity increases as the buildings' overall performance decreases. An obvious conclusion from this is that quickness and control are also strongly and significantly associated, and this indeed is the case (for individuals rho = 0.60, $p = 0.0001$; for building means rho = 0.75, $p = 0.0125$).

Response to problems

Have you ever made requests for changes to the heating, lighting or ventilation systems?

Yes *Please describe in box (right)* [1] [2] No

Please give brief details

If yes, how satisfied in general were you with the following?

Speed of response

Please tick

Satisfactory overall	1	2	3	4	5	6	7	Unsatisfactory overall

Effectiveness of response

Please tick

Satisfactory overall	1	2	3	4	5	6	7	Unsatisfactory overall

Figure 11.3 Response time question used in Building Use Studies surveys

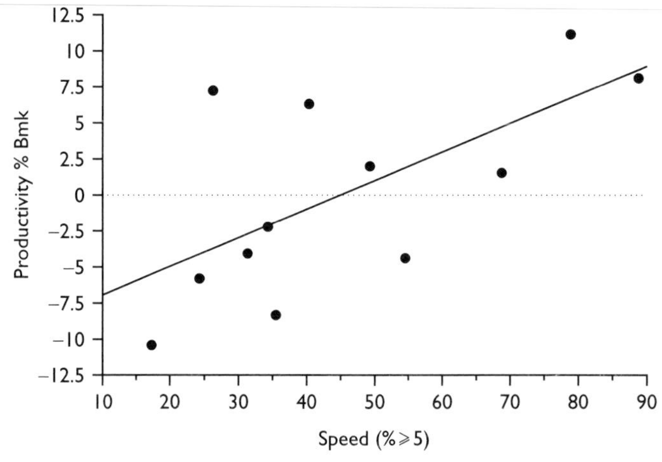

Productivity % Bmk = −8.69 + .2 × Speed (%⩾5); R^2 = .42

Interpretation: The horizontal axis shows the percentage of staff complaining about heating, cooling and ventilation systems who thought that the speed of response by management was satisfactory (a score of over 5 on a seven-point scale). The vertical axis has the perceived productivity for all the staff in the building (including those who did not complain). Perceived productivity and perceived speed of response are significantly associated.

Figure 11.4 Relationship between perceived speed of response in dealing with heating, lighting and ventilation complaints and perceived productivity for twelve study buildings

As measures of response time and personal control are themselves related, are we dealing with two sides of the same coin? To some extent, yes, because responsive control delivers rapid response by definition. But in some buildings a lack of individual control facilities is more than compensated by the excellence of the facilities management arrangements.

Conversely, if designers try to add control in a complicated building which already lacks management resource then their efforts may well be defeated as there will be an inability to manage the added complexity which will induce further chronic failures.

Most buildings tend to have poor levels of perceived control *because* they also have relatively low levels of building management: it has been incorrectly assumed at briefing and design stages that building services technology will automatically deliver what the occupants require without undue extra management intervention or, alternatively, that management will be superhuman.

As the Probe studies show, these assumptions are wrong. The buildings that came out best overall either managed technological complexity with high levels of expertise (e.g. Bordass *et al.*, 1995a) or deliberately rid

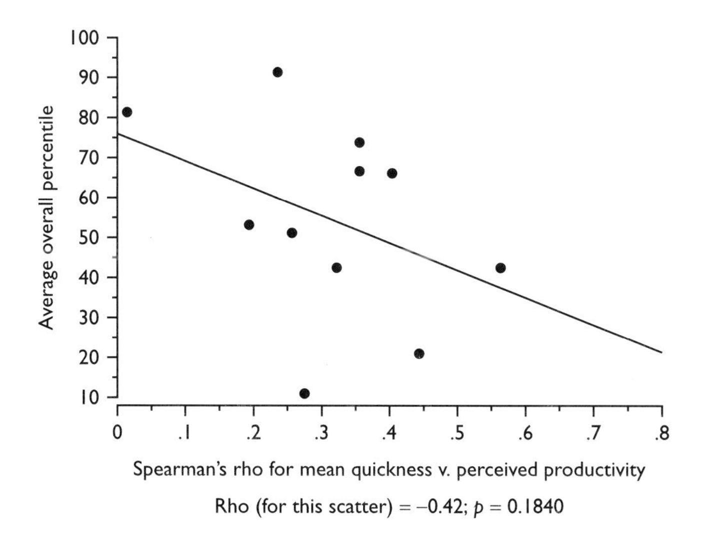

Interpretation: See also Tables 11.1 and 11.3, and Fig. 11.2. As buildings get better (vertical axis), the relationship between perceived quickness and perceived productivity seemingly weakens. This scatter is approaching significance but is **not** significant. A larger sample would clarify this either way.

Figure 11.5 Relationship between perceived quickness and perceived productivity

Table 11.3 Relationships between quickness and perceived productivity for office workers in eleven UK buildings surveyed in 1996–97

Building	*Type*	*Average overall percentile*	*Spearman's rho (corrected for ties) between mean quickness and productivity*	*p value*	*Significant association between perceived productivity and mean quickness?*
A	AC	52	0.25	0.0433	Yes
B	AC	43	0.32	0.0001	Yes
C	NV	81	0.01	0.9084	
D	NV	12	0.27	0.0961	
E	NV	66	0.40	0.0805	
F	AC	67	0.35	0.0025	Yes
G	MM	91	0.23	0.0274	Yes
H	ANV	43	0.56	0.0001	Yes
I	ANV	22	0.44	0.0004	Yes
J	NV	54	0.19	0.0005	Yes
K	NV	74	0.35	0.0176	Yes

Interpretation: See Table 11.1 for definitions. Column 4 has Spearman's rho for mean quickness and perceived productivity. Mean quickness, like mean control, is a composite variable made up from respondents' perceived view of the 'quickness' with which heating, lighting, cooling, ventilation and noise control meet their needs.

themselves of gratuitous complexity (e.g. Standeven *et al.*, 1996b) and a dependence on management.

So designers and managers should consider both personal control and response time implications, rather than think that they are the same. Building Use Studies (unpublished post-occupancy study) finds that when something goes wrong occupants give building managers the benefit of the doubt for a honeymoon period of up to three days, then get upset or give up!

The implication is that real-time responsiveness is something to be considered in the briefing and specification processes, and that different response time standards could be set for different occupier needs. For example, glare and severe overheating need to be dealt with and corrected immediately, whereas a three-day threshold could be used for the replacement of components which directly affect interfaces – simple things such as blinds, chairs, luminaires.

Building depth

The third cluster is building depth. The crucial depth-of-space threshold is some 15 m from wall to wall, around the normal limit of natural ventilation. In the past, we have found that:

- the deeper buildings get, overall satisfaction and productivity tend to go down
- a depth of about 12 m across the building seems about optimal for human performance variables
- shallower plan forms tend to cost about £50/m^2 more, assuming similar cost levels per unit area of envelope and for building services. However, shallower-plan buildings may lend themselves to cheaper, more domestic envelope construction and cheaper services. Unfortunately, cost calculations often find it difficult to consider such trade-offs; economic calculations tend to be more precise at minimising envelope-to-floor area ratios than building services costs, about which they tend to be less well informed (William Bordass Associates, 1992).

Looking at differences in perceived productivity between naturally ventilated (i.e. less than 15 m across) and air-conditioned buildings (usually, but not always, deeper than 15 m) in the current Building Use Studies dataset, the mean perceived productivity is minus 0.19 per cent for NV and minus 4.25 for AC ($p = 0.0097$). This comparison is based on 40 buildings, but does not include either mixed-mode or advanced naturally-ventilated, which are harder to classify by depth.

This does not necessarily mean that naturally-ventilated buildings are better than air-conditioned ones. The pointers are that occupants prefer natural ventilation as the default – in winter, spring and autumn – and air-conditioning, not surprisingly, in the hot, humid parts of summer (Nicol and Kessler, 1998; Oseland *et al.*, 1997). Depth of space is also a correlate for other variables which affect human performance. Many of these have been assessed, although not necessarily in working buildings or conclusively. They include:

- occupants' preferences for window seats (studies usually show that people with window seats tend to be more comfortable (e.g. Nicol and Kessler, 1998; Figure 11.6), but this effect tends to decrease as overall building performance improves (Bordass *et al.*, 1995a)
- ill-health, with the statistical association of chronic, building-related ill-health symptoms (such as dry eyes or stuffy nose), with larger buildings leading to wider speculation about the role of air conditioning as a cause (Wilson and Hedge, 1987).

Building depth is also a correlate for complexity. Buildings have allometric (size) properties which make them disproportionately more complicated as they get bigger. This is a matter not just of building services such as mechanical ventilation and air-tempering, which are always needed with depths greater than 15 m, but also of spatial and behavioural complexity – there are many more activities and much greater likelihoods of conflicts in bigger

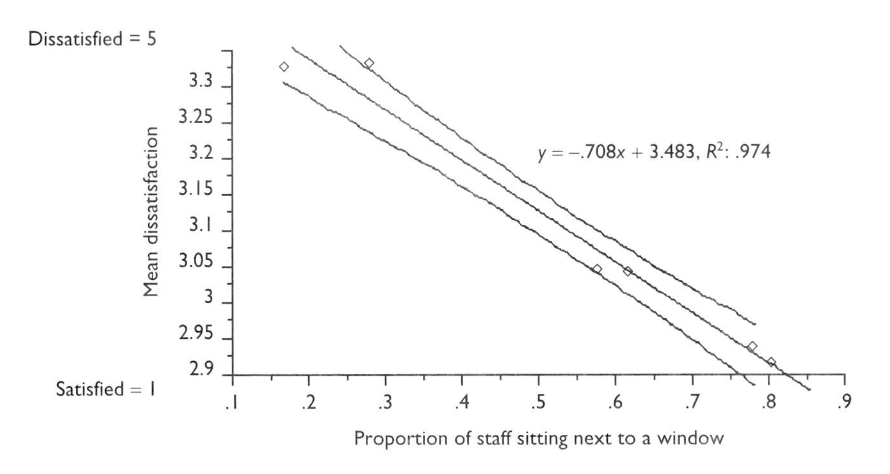

Interpretation: This comes from a survey of six London office buildings carried out by Building Use Studies Ltd for a private client in 1991. The horizontal axis shows the percentage proportion of staff with window seats; the vertical axis the mean scores for overall dissatisfaction on a 5-point scale (a method now discontinued on surveys). For these six buildings, there is a significant and strong relationship: the greater the number of staff with window seats, the less dissatisfaction.

Figure 11.6 Dissatisfaction and window seats

floorplates with higher populations, and a higher dependence on technology and management.

Katsikakis and Laing (1993) is a rare example of a study that measures actual occupational densities, and compares them with design densities. A selection of London offices have been measured for occupied densities, which turn out almost invariably lower than design densities, some substantially so (these findings have more recently been confirmed with a bigger sample by Gerald Eve Research and RICS (1997)). Figure 11.7 is a secondary analysis of data in Katsikakis and Laing (1993) (excluding two buildings with very low occupant densities). It shows how the measured-to-design density proportion varies significantly with both the amount of primary circulation space and the amount of support space. The more primary circulation and support there is, the higher is the measured-to-actual ratio (i.e. the measured density of staff drops with more circulation and support space).

Does this mean that occupiers are compensating for greater complexity by being much more generous with floorspace, or are standards just going up,

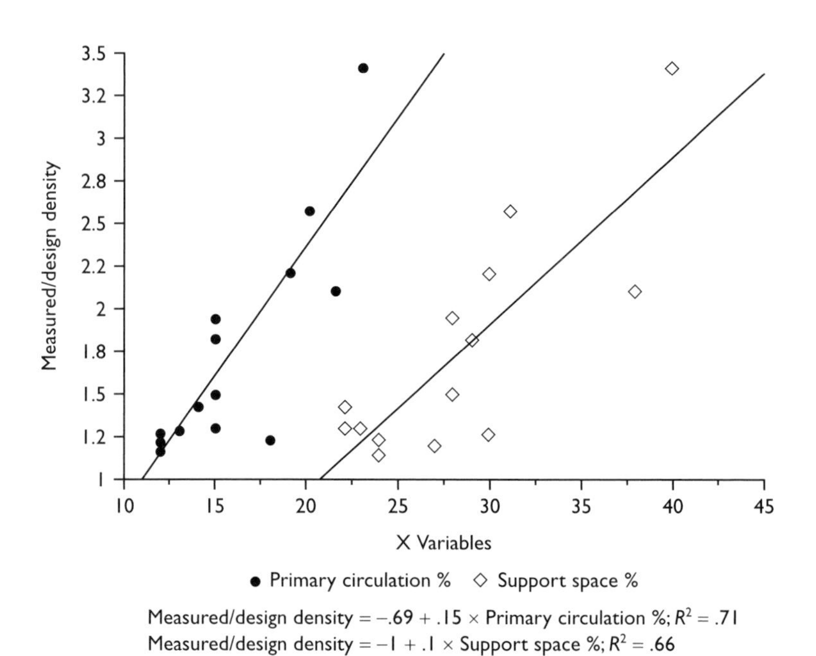

Interpretation: This is a secondary analysis of data first published by Katsikakis and Laing (1993). Two buildings with extremely low densities have been removed from the original dataset of 16 buildings.

Figure 11.7 Ratio of measured to design density by primary circulation and support space for fourteen office buildings

so that office staff are getting the best of both worlds – lower densities and more support space? Unfortunately we do not really know, and do not have productivity data to match with findings on density. As with many other aspects of this tantalising subject, we need a little more information!

Building depth introduces a double-edged effect. As buildings get bigger, they are able to perform more functions and pack more people in, but the penalty is increased operational complexity which creates a greater likelihood of failure – especially chronic performance problems – which increases the cost of management to reduce relative risk. On the other hand, people do not like working at high densities (with the exception, perhaps, of financial dealers, who seem immune!), but higher densities are often perceived to be needed to 'save' on office costs. Katsikakis and Laing (1993) possibly show that when this trade-off is made in reality, building users opt for lower densities because this gives them sufficient degrees of freedom to deal with the consequences of dysfunctional conflicts. What do you prefer: an aircraft with 70 per cent of the seats filled or completely full up?

Workgroups

The fourth variable cluster relates to **workgroups**: remarkably, along with personal control, one of the least understood topics in modern buildings (Trickett, 1991). In offices, perceptions of productivity are higher in smaller and more integrated workgroups. Like control, this will be obvious from personal experience: given an unrestricted choice most will opt for their own room, for instance, or a small workgroup with close colleagues. This said, there are few research data to back it up and, like density and size, this variable needs more work. For example, on the rare occasions when Building Use Studies has looked at workgroup dynamics, productivity has not been measured as well (usually because the client did not want to).

Table 11.4 has preliminary data from a rare case building where both productivity and workgroup topics were studied. In this case, workgroup size could explain differences in overall comfort (smaller is better), but there was no association between size of workteam and overall perceived productivity for the individuals in the study building.

Our confidence in including workgroups as a killer variable comes from work in 1987 (Wilson and Hedge, 1987). Room size is a correlate of perceived control for temperature, lighting and ventilation (Figs 11.8 and 11.9), with perceived control declining with workgroups bigger than about five people. As perceived control is a correlate of perceived productivity, it is fairly safe to assume that workgroup size is also a contributory factor (but so are technical factors).

From a design and management point of view, workgroups are seen as desirable both for space-saving reasons (possibly spurious, see above) and

Table 11.4 Overall comfort scores by size of workteam for a single office building

	DF	*Sum of squares*	*Mean square*	*f value*	*p value*
WKTEAM ≤ 5	1	11.76	11.76	5.63	.0183
Residual	302	631.24	2.09		

Model II estimate of between-component variance: .06
52 cases were omitted due to missing values.

	Count	*Mean*	*Std Dev.*	*Std Err.*
Five or fewer	143	4.58	1.51	.13
Six or more	161	4.19	1.39	.11

52 cases were omitted due to missing values.

Interpretation: These are data for size of workteam for an office building studied in 1996. The analysis of variance shows that workteams of fewer than four people were more comfortable than those with more than four people. The analysis, though, **does not** take account of any possible effects of grade.

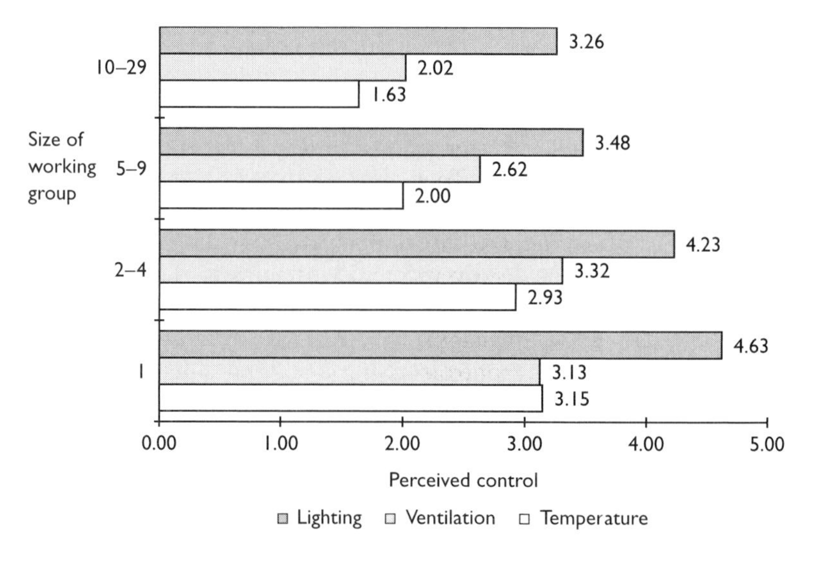

Interpretation: These data are from the Office Environment Survey (Wilson and Hedge, 1987) showing means of values for perceived control for lighting, ventilation and temperature. The OES used a sample of 50 UK office buildings.

Figure 11.8 Size of working group and perceived control no. 1

for better communication between colleagues. There is always a trade-off to be considered between the risk of degrading performance in open plan and cutting people off from each other by putting them in their own rooms. Designers and managers both tend to opt for the open plan approach, but for different reasons. The evidence we have indicates (but does not prove!)

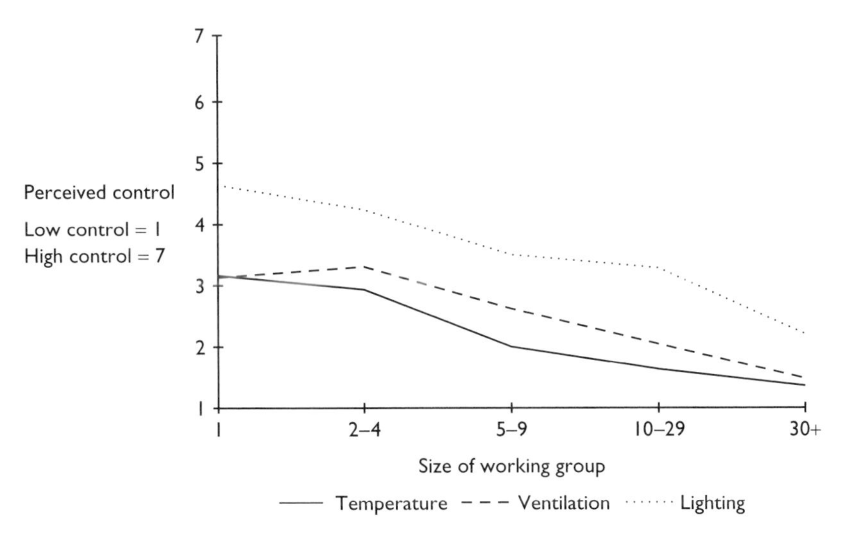

Figure 11.9 Size of working group and perceived control no. 2

that well-integrated workgroups of four to five people will probably be acceptable, but the risks of lower productivity in bigger workgroups can increase substantially thereafter. To support the claimed business benefits it is therefore necessary to put in a much higher level of expertise in building and services design, and facilities management. While we all know this to some extent, the degree of improvement necessary can be much higher than one would think.

A key reason for this is the 'mapping' between the workgroup's activities, the available environmental controls and zones of services. Where the relationship is one-to-one (i.e. everything coincides as it should in a single room), the sole occupant will have full control over lighting, blinds, ventilation, heating, cooling, privacy and noise, and is able to fine-tune to suit needs exactly. Here there will be perceived productivity benefits for the occupant, who will be able to prevent undesirable impacts such as distracting noise or overheating. Supervising managers may be rather less keen on individuals with high levels of privacy and productive interactions may also be reduced, but modern technology is permitting close work integration with less face-to-face contact.

As workgroups get larger, three characteristics affect the occupants, as follows:

1 The mapping between environmental controls, services zones and activities disintegrates (for example, the lighting may be switched for the whole floor rather than for workgroups alone).
2 Occupants have to consider their colleagues' wishes when they want to

make changes. As a result, the likelihood that everyone will be satisfied with the prevailing settings will reduce as the workgroups get bigger. This is an inevitable consequence of (a) size and allometry; (b) differences in individual preference ranges.

3 Long-distance effects become important: for instance, glare from a remote window, possibly even through a glass partition, or draughts of uncertain origin owing to complex movements of air in both naturally ventilated and air-conditioned spaces.

The best strategies for the designer and managers are thus to:

- keep workgroups as small and well-integrated as possible
- make sure that zones of activities map onto the service zones especially for productivity killers such as irrelevant noise, glare and draughts
- keep sources of unwanted distraction to a minimum – we have found that up to 60 per cent of staff sitting in open plan offices can be located directly next to a source of random distraction such as the end-doors which may squeak and bang when closing, the photocopier or the tea/coffee area
- not interfere with sources of wanted information, i.e., information that is needed and relevant to worktasks within earshot and lines of sight, so that people receive reinforcing, relevant data by default
- design and manage the overall worksetting so that the default (i.e. normal) setting is reasonably comfortable, safe and healthy and does not rely on excessive amounts of technological or management input to make it work acceptably.

Conclusions

We have dealt with productivity in the workplace from the perspective of things within the control of building designers and facilities managers. Necessarily this means missing out aspects of productivity largely or entirely outside the influence of building professionals. These include considerations such as workplace stress which we once tested to see whether stress was building-related, and found that for most intents and purposes it was not (Leaman *et al.*, 1990), management attitudes and job satisfaction (Whitley *et al.*, 1996).

Although productivity begs many definitional and methodological questions, we think that the available data tell a clear story which designers and managers can practically incorporate. Buildings, especially offices, work best for human productivity when there are:

- many opportunities for personal control, providing a background for healthy, comfortable and safe operation as well as adaptive comfort

- a rapid response environment, not necessarily only for personal control, but for the many other aspects of a building's operation that might compensate for absence of personal control, such as an excellent complaints monitoring and feedback system
- shallow plan forms, preferably demanding less technically complex and less management-intensive systems (with the added benefit of better energy performance)
- activities which properly fit the services which are supposed to support them, not only in spatial capacity, but for the zoning and control of heating, cooling, lighting, ventilation, noise and privacy.

In some contexts, these will not be possible or desirable. This is perfectly acceptable, but clients, designers and managers then need to appreciate the extra levels of core and support services that will be needed to produce good performance.

Of course, like everything else with buildings, the attributes above are all really aspects of the same thing. Ideally, simple, shallow plan forms, small work rooms, robust and manageable controls and domestic levels of servicing work best. In fact, Raw and Aizlewood (1996) show that building-related chronic illness was significantly lower (and perceived comfort higher) in homes than in offices – just what would be expected from our findings.

However, such 'ideal' design forms (characteristic of offices in the UK up to the 1960s) have long since been superseded. The relentless trend is now towards intensification (and diversification) of building use (Leaman, 1996), with much greater attention paid to:

- risk/value payoffs, not just in rental or property investment terms as in the past, but for the wider canvas of human and environmental resources as well
- business benefits and consequential environmental disbenefits which have to be managed if overall performance improvement is to be achieved
- design strategies linked far more closely to business missions to improve strategic advantage in the market place
- greater interest in 'generic' spaces and forms of servicing which allow rapid switching between different occupier activities.

Bigger and more complex buildings demand subtler strategies for managing this complexity and different design strategies and technologies to support them. Where this is successful, performance gains are possible, but where management does not properly compensate for the extra diligence that technology needs, chronic problems usually result.

The trend towards mixed-mode buildings (with mixtures of natural ventilation and air-conditioning) is a case in point. Treating early findings from Probe and other recent studies very circumspectly, it seems that mixed mode can offer the best of both worlds – better occupant satisfaction and better environmental performance – and occupants can detect the differences. Studies by Rowe and colleagues in Sydney (Rowe *et al.*, 1998 and personal communication) (Table 11.5) suggest that mixed-mode offices not only give performance advantages through better thermal comfort and better perceived ability to perform work (i.e. better productivity), but are also better for perceived air quality and overall satisfaction with workplace. As with British studies, the work also confirms that this delivers better perceived control and leads to much improved energy efficiency. By monitoring the switching behaviour in mixed-mode buildings, Rowe has shown that a control-rich, naturally-ventilated environment is the preferred default – even in subtropical conditions in Australia – and this preference is abandoned only on the minority of occasions when both temperature and humidity exceed tolerance thresholds.

Although we can be upbeat and report these findings optimistically, we have also tried to show where things work and where they don't. Unfortunately, the design and construction industries are much more coy, especially about failures. There is an inclination, even in research and development agencies who should know better, towards reporting just the good news and forgetting about the downsides, which often turn out to be the very things that affect human productivity the most. Many of the issues we

Table 11.5 Performance means by ventilation type for twelve Australian office buildings (Rowe, personal communication)

	Ventilation type	*Overall satisfaction with workplace*	*Impact on ability to perform work*	*Thermal comfort*	*Air quality*
1	MM	4.0	3.7	3.5	3.4
2	AC	2.0	1.9	3.3	2.4
3	AC	3.5	3.4	3.1	3.2
4	AC	3.3	3.4	3.2	2.9
5	AC	2.6	2.8	2.6	2.5
6	AC	2.8	2.9	3.0	2.9
7	NV	3.1	3.3	2.7	3.1
8	AC	2.4	2.5	2.3	1.7
9	AC	3.0	3.1	2.9	2.3
10	AC	3.1	3.2	3.1	2.7
11	MM	3.7	3.7	4.4	4.0
12	NV	2.0	2.9	2.5	2.6

Notes
Performance means are for five-point scales: 1 = low; 3 = average; 5 = high. See also Fig. 11.2 for ventilation types.

have dealt with in this paper have been known about for generations. Poor human productivity in buildings is a function not just of our four 'killer' variables but also poor professional feedback, lack of integration in design processes, lack of care for the primary occupants, weak or non-existent briefmaking, and the convenient but disturbing tendency to forget the bad news.

Our experience with monitoring and troubleshooting studies of UK buildings is that the key to success with building performance lies with managing downsides effectively. Generally this involves:

1. understanding contexts, especially by bringing ruling constraints to the fore at briefing stage, and making sure that everyone shares assumptions early
2. identifying possible downsides and knowing risks for what they really are, so that we are not 'optimising the irrelevant' (Bordass, 1992)
3. keeping technology within thresholds of affordable manageability, so that the inevitable revenge effects can be identified and coped with before they develop into insidious chronic defects
4. taking occupants' complaints seriously and dealing with them quickly and sensitively.

References

Asbridge, R. and Cohen, R. (1996) Probe 4: Queen's Building. *Building Services J.*, Apr., 35–38.

Baker, N. (1996) The irritable occupant: recent developments in thermal comfort theory. *Architect. Res. Q.*, 2, winter.

Baker, N. and Standeven, M. (1995) Adaptive opportunity as a comfort parameter. *Proceedings of the Workplace Comfort Forum*, London.

Bordass, W. (1992) Optimising the irrelevant. *Building Services*, 1992, Sept.

Bordass, W., Bromley, A. and Leaman, A. (1995a) *Comfort, control and energy efficiency in offices*. BRE Information Paper, IP3/95, Garston.

Bordass, W., Bunn, R., Leaman, A. and Ruyssevelt, P. (1995b) Probe 1: Tanfield House. *Building Services J.*, Sept., 38–41.

Bordass, W., Cohen, R. and Standeven, M. (1997) Technical review of engineering and energy issues: Probe office buildings. *Buildings in Use '97: how buildings really work*, London.

Bordass, W., Field, J. and Leaman, A. (1996) Probe 7: Homeowner's Friendly Society. *Building Services J.*, Oct., 39–43.

Bordass, W. and Leaman, A. (1997) From feedback to strategy. *Buildings in Use '97: how buildings really work*, London.

Brill, M. (1986) *The Office as a Tool*. The Buffalo Organisation for Social and Technological Innovation (BOSTI).

Building Use Studies Ltd (1992/93) *User and Automated Controls and Management of Buildings*. Building Use Studies Ltd, London, unpublished report for Building Research Establishment, 1992 (Phase 1), 1993 (Phase 2).

Building Use Studies Ltd (1996) Post-occupancy survey findings, unpublished, 1996. Unlike the Probe studies, which are in the public domain, many other surveys, such as this one, are for private clients only.

Bunn, B. (1993) Fanger: Face to Face. *Building Services*, June.

CIRIA (1994) *Dealing with Vandalism*, CIRIA Special Publication 91, Construction Industry Research and Information Association, London, 31–32.

Cohen, R., Leaman, A., Robinson, D. and Standeven, M. (1996a) Probe 8: Anglia Polytechnic University Learning Resource Centre. *Building Services J.*, Dec., 27–31.

Cohen, R., Ruyssevelt, P., Standeven, M., Bordass, W. and Leaman, A. (1996b) The Probe method of Investigation. *CIBSE National Conference*, Harrogate.

Cohen, R., Standeven, M. and Bordass, W. (1997) Technical review of engineering and energy issues: Probe non-office buildings. *Buildings in Use '97: how buildings really work*, London.

Davies, A. and Davies, M. (1995) *The Adaptive Model of Thermal Comfort: patterns of correlation*. London: CIBSE.

De Dear, R. *et al.* (1993) *A Field Study of Occupant Comfort and Office Thermal Environments in a Hot–Humid Climate*, Final report, ASHRAE RP-702.

Eley, J. (1996) Proving an FM point: One Bridewell Street. *Facilities Management World*, Sept.

Energy Efficiency Office (1991) *Best Practice Programme: Good Practice Case Study 21: One Bridewell Street, Bristol*. Garston: BRESCU.

Gerald Eve Research and RICS (1997) *Overcrowded, Under Utilised or Just Right?*

Humphreys, Rev. M. (1976) Field studies of thermal comfort compared and applied. *Building Services Eng.*, 44, 5–27.

Humphreys, Rev. M. (1992) Thermal comfort in the context of energy conservation, in Roaf, S. and Hancock, M. (eds), *Energy Efficient Building*, 3–13. Oxford: Blackwell Scientific Publications.

Katsikakis, D. and Laing, A. (1993) *Assessment of Occupant Density Levels in Commercial Office Buildings*. London: Stanhope Properties.

Leaman, A. (1996) Space intensification and diversification, in Leaman, A. (ed.), *Buildings in the Age of Paradox*, 44–59. IoAAS, University of York.

Leaman, A. (1997) Probe 10: Occupancy survey analysis, *Building Services J.*, May, 21–25.

Leaman, A. and Bordass, W. (1995) Comfort and complexity: unmanageable bedfellows? *Workplace Comfort Conference*, London, 1995.

Leaman, A. and Bordass, W. (1997) Design for manageability, in Rostron, J. (ed.), *Sick Building Syndrome: concepts, issues and practice*, 135–149, London.

Leaman, A., Bordass, W., Cohen R. and Standeven, M. (1997) The Probe occupant surveys. *Buildings in Use '97: how buildings really work*, London.

Leaman, A., Roys, M. and Raw, G. (1990) *Further Findings from the Office Environment Survey: Stress*. BRE, Garston.

Lloyd-Jones, D. (1994) Windows of Change. *Building Services*, Jan.

Lorsch, H. and Abdou, O. (1994a) The impact of the building indoor environment on occupant productivity: Part I: Recent studies, measures and costs. *ASHRAE Trans.*, 741–749.

Lorsch, H. and Abdou, O. (1994b) The impact of the building indoor environment on occupant productivity: Part 2: Effects of temperature, *ASHRAE Trans.*, 895–901.

Lorsch, H. and Abdou, O. (1994c) The impact of the building indoor environment on occupant productivity: Part 3: Effects of indoor air quality, *ASHRAE Trans.*, 902–912.

McIntyre, D. (1980) *Indoor climate*, 154. London: Applied Science Publishers.

Marmot, M. (1997) Report on perceived control over work, *Br. Med. J.*, July.

Nicol, J. and Kessler, M. (1998) Perception of comfort, in relation to weather and indoor adaptive opportunities, *ASHRAE Trans.*, Jan.

Oseland, N. (1996) Productivity and the indoor environment, *Fourth International Air Quality Conference*, Mid-Career College, London, 19.

Oseland, N. (1997) *The relationship between energy efficiency and staff productivity*. Building Research Establishment internal CR90/97, Garston.

Oseland, N., Brown, D. and Aizlewood, C. (1997) Occupant satisfaction with environmental conditions in naturally ventilated and air conditioned offices, *CIBSE Mixed Mode Conference*, May.

Pepler, R. and Warner, R. (1968) Temperature and learning: an experimental study. *ASHRAE Trans.,* **74**, 211–219.

Raw, G. and Aizlewood, C. (1996) *The Home Environment Survey: a pilot study of acute health effects in homes*. Building Research Establishment, Garston, 109/1/3.

Raw, G., Roys, M. and Leaman, A. (1993) Sick building syndrome, productivity and control. *Property J.*, Aug.

Robson, C. (1993) *Real World Research*. Oxford: Blackwell.

Rowe, D., Forwood, B., Dinh, C. and Julian, W. (1998) Occupant Interaction with a Mixed Media Thermal Climate Control System Improves Comfort and Saves Energy. *AIRAH Meeting*, Sydney.

Standeven, M. and Cohen, R. (1996) Probe 5: Cable and Wireless College. *Building Services J.*, June, 35–39.

Standeven, M., Cohen, R. and Bordass, W. (1995) Probe 2: 1 Aldermanbury Square, *Building Services J.*, Dec. 29–33.

Standeven, M., Cohen, R. and Bordass, W. (1996a) Probe 3: C&G Chief Office. *Building Services J.*, Feb. 31–34.

Standeven, M., Cohen, R. and Leaman. A. (1996b) Probe 6: Woodhouse Medical Centre. *Building Services J.*, Aug., 35–38.

Tenner, E. (1996) *Why Things Bite Back: New technology and the revenge effect.* London: Fourth Estate.

Trickett, T. (1991) Workplace design: its contribution towards total quality. *Facilities*, **10**, no. 9.

Vischer, J. (1989) *Environmental Quality in Offices.* New York: Van Nostrand Rheinhold.

Whitley, T., Dickson, D. and Makin, P. (1996) The contribution of occupational and organisational psychology to the understanding of sick building syndrome, *CIBSE/ASHRAE Joint National Conference*, Harrogate, 1996, 133–138.

William Bordass Associates (1992) *Specification and Design for Building Depth*, internal note.

Wilson, S. and Hedge, A. (1987) *The Office Environment Survey*. London: Building Use Studies.

Chapter 12

Individual control at each workplace: the means and the potential benefits

David P. Wyon

Introduction

We all have days when we feel that the office environment is reducing our productivity. Whether the problem we experience is due to the lighting, noise, temperature, air quality or some other physical factor; if we can't change anything, the only remaining option is to change everything by going to a different location to work. Even in the rare circumstance when this is an option, it has many obvious disadvantages. Providing alternative locations is expensive; books, papers and equipment at one location are not available at another; moving between alternative locations takes time; and variable locations makes 'teaming' more difficult. Working at home requires either a rare degree of trust, or a shift toward piecework accountability and ultimately self-employment, which suits only a tiny minority of office workers. The best solution is a workplace where you can change something. In this context, anything is better than nothing, and more is better. Indoor environmental control is traditionally provided on a group basis. The problem is that individual differences are such that a 'good' indoor environment is accepted as one where 80 per cent are satisfied. The remaining 20 per cent are expected to endure conditions which may adversely affect their work, their comfort, and their health. The solution is to provide some means of individually adjusting each occupant's microclimate. It will be shown that this is perfectly possible, and very beneficial.

Following a general discussion of the degrees of freedom that are currently available to office workers, the current consensus on acceptable levels of temperature and noise in offices will be set out, taking account of recent experimental evidence for interactions between these factors. New estimates of the range of individual control of temperature required to ensure comfort for a given proportion of a group will be given, based on recent experimental evidence. Estimates will then be made of the degree to which fan noise may be increased to extend the range of individual thermal and air quality control, without negating the purpose by increasing the overall percentage dissatisfied. Published research on performance under cold and heat stress will be shown to be capable of predicting quite accurately the beneficial effects of individual control on group average performance of office work in a real-world situation in which one particular type of work could be identified as critical for productivity.

However, even if environmental effects on the performance of specific types of office task could be accurately predicted, it would still be difficult to predict the 'bottom-line' productivity of office workers. At lower levels in the job description hierarchy, the percentage of an employee's time spent on a given task may be known quite exactly. In very simple job descriptions, such as copy-typing, it approaches 100 per cent. At levels only slightly above this it becomes increasingly difficult to identify which tasks are being performed, which are critical for productivity, and how much time is spent on each. Individual control must then be assumed to benefit productivity if it is capable of providing optimal conditions for individual comfort and health.

The need for individual control

User-initiated changes are made necessary by a large number of '**drivers for change**'. Examples of these will be given, beginning with several that occur in the visual environment. Visual information can be a stimulus or a distraction, yet visual information can be essential for effective teaming. View-out through windows allows occupants to focus on distant objects as well as near ones, helping them to avoid eye-strain. View-in and the consequent lack of privacy can be stressful. The non-visual effects of any lighting arrangement are positive or negative depending on the task, e.g. bright lighting raises levels of mental arousal, which is fine for some tasks, but not for others. Daylight changes all the time, and glare, direct or reflected, generally occurs only at certain times of day and with certain materials, such as paper with a glossy finish.

Similarly, there are drivers for change in the acoustic environment. Auditory information can be a stimulus or a distraction, yet it may be essential for effective teaming. The non-informational effects of noise, i.e. those that are a function of noise level rather than of the distracting effect of unwanted acoustic information, include an effect of noise on levels of mental arousal. Noise increases arousal. Arousal affects occupants' distribution of attention. Cue-utilisation, or breadth of attention, is reduced at high levels of arousal. Thus noise, like bright lighting, may or may not be beneficial for the task in hand.

Indoor air quality may also be a driver for change. Thresholds of irritation vary with health and between people. Unwanted odour can be as distracting as noise. Episodic dust contamination and periodic cleaning will alter the required clean air supply rate.

Indoor air temperatures can change occupant requirements. Thermal discomfort is very distracting. Mental performance is greatly reduced by moderate levels of heat stress, and manual dexterity is measurably reduced even by moderate levels of cold stress. Even in standard clothing, neutral temperatures vary between individuals over a 10 K range, and differences between individuals' clothing and activity levels have large effects on neutral temperature. The intensity of many subclinical symptoms such as sick building syndrome (SBS) is affected by air temperature, and this may also lead to a requirement for change.

Delegation of control

The means by which users are allowed to change something to cope with the above '**drivers for change**' are not nearly as important as the principle that users should be enabled and empowered to initiate change themselves. The more '**degrees of freedom**' that can be designed into a workplace, the better. The principle of '**bringing the user back into the loop**' is far more important for health, comfort and productivity than optimising uniform conditions to accord with group average requirements, yet indoor environmental research has always concentrated on the group and ignored individual choice. If users can optimise their own work environment in multiple ways, designers do not even need to know what group average requirements are; it is sufficient to know the range that is required on each dimension.

Delegation of control to users obviously means that they must take on some new responsibilities. They must understand the way the building works and the consequences of their actions, so they must be given **insight**. They must learn to use the control delegated to them, and as learning cannot take place without feedback, they must be given online **information**. Only when they have both insight and information can they be given **influence**. The 3I principle of user empowerment is herewith formulated and states that all three – insight, information, and influence – must be provided. Providing any one of them, or even any two, will fail. This principle applies to a surprisingly wide range of human activities, from indoor environmental optimisation to accident prevention in complex systems.

Existing means of delegating control to the individual

Task lighting is common in Europe, but less common in North American and Asian offices. Access to view-out is a rare privilege of rank in many North American office buildings, while it has been elevated almost to a human right in Scandinavian office design. Adjustable solar shading should always accompany proximity to windows, but often does not. Adjustable visual screening can make it possible to trade off view-out against view-in. Provision of these four degrees of freedom addresses several of the drivers for change listed above.

Doors which can be closed are really the only effective way of reducing the noise problems listed above. However, in order to facilitate teaming, i.e. to signal accessibility and allow view-in, office doors are often left open unless they are transparent. Glass doors and internal walls, as used to admit daylight to core common areas in the **combi** offices found in Sweden, resolve this conflict and restore the additional degree of freedom that permits doors to be closed to reduce noise distraction. Combi offices incorporate a new design concept that places small individual offices around an office building's perimeter, each with a window and a door. They have solid walls between them, but floor-to-ceiling glass interior walls with glass doors in order to share daylight and view-out with a central common area. Functions such as storage, printing, copying, break facilities, and team-space are situated in the central common area of the building. The design was originally conceived as a retrofit conversion of the much-hated **open landscape** offices that had been introduced from the United States, but it has been so successful that it is now used even in new office construction. Cubicles, which in the USA are the most common retrofit solution to the environmental problems of open landscape offices, leave much to be desired acoustically: it has been found that adjustable background levels of 'white noise', when provided as part of a '3I initiative', are used by as many as 40 per cent of US office workers in cubicles, presumably to reduce the distracting effect of nearby conversations. The additional background noise is switched off automatically when the user is absent, although acoustic telephone signals are allowed to continue to annoy and distract the occupants of nearby cubicles. The obvious solution to this acoustical problem, a flashing light signal instead of the ubiquitous warbling signal, does not seem to have occurred to anybody.

Free-standing air cleaners are often an appropriate means of providing optional breathing zone filtration (BZF) in cubicles and rooms. Both this function and ducted local air-supply vents may be found as features of desk-mounted microclimate control units, sometimes known ambiguously as **environmentally responsive workstations**, but here termed **individual microclimate control devices** (IMCDs). Local air-supply vents effectively move the user upstream and provide improved air quality even when there is a high degree of recirculation. Entrainment of room air in the plume from the vent usually means that even providing 100 per cent fresh air may not achieve very much better ventilation efficiency in the breathing zone. Where possible, operable windows provide an important additional degree of freedom in this respect.

The ducted local air supply vents mentioned above can also provide a high degree of local air-temperature control. However, an equivalent effect on individual heat balance can be achieved by non-ducted desk-mounted IMCDs even in an isothermal, well-mixed space by means of adjustable local air velocities and thermal radiation from heated panels close to the body. Users provided with these two degrees of freedom can adjust the equivalent temperature they experience by as much as 3 K above and below actual room air temperature. This is shown below to be sufficient to bracket 99 per cent of individual neutral temperatures, greatly reducing complaints of thermal discomfort. The reduction in the cost of dealing with complaints alone can make this approach cost-effective, as each response to a 'hot or cold call-out' is currently estimated to cost about $100. However, it is important to ensure that increasing the cooling power by raising local air velocities does not inconvenience close neighbours, irritate the eyes or blow papers off horizontal surfaces, and that heated panels are placed where they are most effective (where their angle-factor to the body is greatest). Desk fans and private-initiative heaters do not usually fulfil these criteria. Having the means to adjust local equivalent temperature to current personal requirements is probably the least common and the most appreciated degree of freedom of all. Thermal complaints always seem to come top of everybody's list.

The expected benefits of delegating control

When users are provided with some of the above degrees of freedom and are empowered to use them effectively in accordance with the 3I principle, it may confidently be assumed that health, comfort and productivity will increase. Users learn very quickly to use whatever degrees of freedom are provided, although it must be said that the experience is often unfamiliar, as very few of the degrees of freedom listed above are in fact provided in conventional offices. The most important exception is openable windows – in older buildings, in Europe – which may explain their popularity and the widespread resistance to their elimination. As users find they have remedies for many of their complaints, they complain less often. The reduction in the cost of handling complaints can be surprisingly large.

Delegation of control in the office of the future

The future workplace will contain many, and perhaps all, of the degrees of freedom described above and identified as being available today. In addition, IMCDs will have occupancy sensors and will be linked to the building control system to provide high-grade additional control information: whether workplaces are occupied, and if so, whether the users are within the control band delegated to them. This is a higher grade of information than is available today from wall-mounted thermostats in zones which may or may not be occupied. Energy conservation strategies can be implemented in unoccupied zones, and zone set-points can be adjusted adaptively if any users are taking maximum heating or cooling power from their units. The control delegated to the users creates a wider 'dead-band', which simplifies the control problem and makes it possible to conserve energy by allowing the building to 'float' more naturally with changes in thermal load and in outdoor climatic conditions.

Building control systems will have access to many more sources of information in the future workplace. They will be able to read thermal sensors on every desk, i.e. sensors located in the occupied zone instead of on the corridor wall, as is currently the norm. The use of existing telephone and computer links for this purpose will make it unnecessary to install dedicated wiring, which will reduce installation costs considerably. Carbon dioxide sensors in the return ducts from each zone or space will provide immediate and distributed information on occupancy, signalling intrusion in spaces expected to be unoccupied at that time, and accurately quantifying occupancy to ensure an adequate fresh air supply at all times. Even smokeless, smouldering fires can be detected by carbon dioxide sensors. They can be discriminated from human occupancy by the time course of their source strength, or by carbon monoxide sensors, making it possible to sound the fire alarm reliably at a much earlier stage than is detected by smoke and flame sensors.

In future workplaces, users will be able to access the building control computer at any time by telephone or modem, to obtain online information, adjust set points, register complaints and request maintenance. The system will be able to give them an intelligent response, reporting current operating conditions, giving advice, explaining that there are conflicts between users in the same area, recording complaints and confirming the dispatch of maintenance personnel. Automatic complaint logs of this dialogue will record all system information relevant to a complaint in ways that are prohibitively difficult and expensive to achieve today. Automatic analysis of incoming complaints will make it possible to optimise building operation continuously in the medium term, keeping users firmly in the loop. Loss of perceived control and lack of confidence in building management are major factors in the downward spiral of problem buildings. Future workplaces will quite simply have the communication channels to the user that are needed to reverse this spiral.

Estimating how much thermal microclimate control is required

International Standard ISO7730 includes the PMV equation, which makes it possible to predict the air temperature for neutral heat balance of a group of subjects with known activity level (i.e. metabolic rate) and clothing insulation (i.e. Clo-value), taking account of mean radiant temperature, air humidity and air velocity. In practice, the uncertainties involved in estimating the mean activity level and mean effective clothing insulation of a real-life group, taking account of the usually completely unknown mean insulation value of the chairs on which they sit, mean that these values usually have to be 'reverse-engineered' from the mean temperature at which such a group is known to be thermally neutral. The equation can be used to predict the adjustments necessary to maintain thermal neutrality for a group if one of the six factors should change by a given amount, but it should be noted that the approach takes no account of the spread of clothing insulation values that always occurs within a group, of the extent to which different individuals compensate for habitual differences in metabolic rate, skin fold thickness and preferred skin temperature by choosing to dress differently at the office, or of individual differences in the width of the thermal range which is regarded as acceptable. Subjects in the predicted mean vote (PMV) experiments wore standard clothing and adjusted the air temperature until they personally were in thermal neutrality. Subjects never wore their own clothing, and never performed actual office work in their own idiosyncratic way. Percentage of people dissatisfied (PPD) values should therefore not be used to determine the control range of IMCDs.

The inter-individual standard deviation (SD) of neutral temperature about the group mean which was obtained in the PMV experiments was of the order of 0.8–1.1 K, while it was determined to be 2.6 K by Grivel and Candas (1991) in an attempt to replicate the PMV experiments in France: their subjects wore standard clothing but could vary their metabolic rate at will. In an experiment by Wyon and Sandberg (1996) on discomfort due to thermal gradients, over 200 office workers wore their own habitual office clothing while they performed office work in a test room. They were randomly assigned to nine thermal conditions over a range of thermal conditions equivalent to 2.8 K as measured empirically, using a thermal manikin to estimate resulting total heat loss in each condition. Age, gender and thermal gradient were equivalent at each level of operative temperature. If their SD had been 2.6 K, 30 per cent would have been too hot even in the coldest condition, and 30 per cent would have been too cold even in the warmest condition: in fact, only 7 per cent and 18 per cent respectively were dissatisfied with their total heat loss. The difference shows that clothing is used adaptively to reduce individual differences, and is compatible with SD = 1.17 K. This is currently the best available estimate of the range of individual preferences under realistic office conditions. Table 12.1 uses this

Table 12.1 Range of individual control required for a given percentage to be satisfied

% Comfortable	*Range, K*	*Range, F*
85	3.4	6.1
90	3.9	7.0
95	4.6	8.3
99	6.0	10.8

value to predict the range of individual control of whole-body equivalent operative temperature which would be necessary to achieve different proportions of comfortable individuals.

It should be noted that this may vary between populations. The necessary control range will be increased by the imposition of a dress code which makes it more difficult to adjust clothing insulation between individuals. Table 12.1 does not predict the necessary supply air temperature range of an IMCD, as the cooling effect is strongly dependent on exactly how the plume strikes the body, i.e. how much of the body is affected, which parts were affected, and the resulting air velocity and turbulence close to the body. In practice the ability of a given design of IMCD to affect whole-body heat balance must be determined empirically, by experiment. This has already been done for a desk-mounted IMCD, using a thermal manikin (Wyon and Larsson, 1990). The range of individual adjustment of whole-body equivalent operative temperature was 3.8 K with supply air temperature equal to air temperature, and 4.9 K with supply air temperature 8 K below ambient, sufficient according to the data underlying Table 12.1 to ensure comfort for 89 per cent and 96 per cent of a group, respectively. Moisture evaporation may reasonably be expected to increase the cooling effect of the air stream sufficiently to extend the individual control range to over 6.0 K, sufficient to ensure comfort for 99 per cent of a group. Human experiments in the field, taking account of both evaporative cooling and user behaviour, could provide a quantitative estimate of the range of control of an IMCD, in terms of equivalent room temperature change, if carried out in the following way: subjects who had become familiar with the IMCD would be asked if they were willing to turn it off for a period of about 90 minutes. Those who agreed would then be assigned at random to one of two conditions: (1) IMCD turned off for 90 minutes; or (2) IMCD remaining in operation. The hand skin temperature of each subject would be measured at the beginning and end of the period. In warm conditions, it is to be expected that the mean hand skin temperature of Group 1 would increase in relation to that of Group 2, whereas in cool conditions it would decrease. By repeating the experiment over a range of room temperature conditions, the room temperature bias equivalent to the cooling and heating benefit of the IMCD, in terms of the resulting hand skin temperature actually achieved, would be derived empirically.

The ASHRAE *Handbook of Fundamentals* (1997) suggests that an acceptable percentage comfortable would be 80 per cent, but does not attempt to predict the degree of individual control that would be necessary to ensure that any given percentage could be comfortable, merely reproducing the PMV/PPD approach of ISO7730 for predicting individual differences in thermal comfort sensation. The limitations of this approach have been set out above. However, a European Union Prestandard (1994) defines three levels of thermal environmental quality in Section A.1.3, ‘Categories of thermal environment'. These are defined to ensure 85 per cent, 90 per cent or 95 per cent comfortable in terms of whole-body equivalent operative temperature. The EU Prestandard does not address the provision of individual control in any quantitative way, but states (p. 13) that 'it is an advantage if some kind of individual control of the thermal environment can be established for each person in a space. Individual control of the local air temperature, mean radiant temperature or air velocity may contribute to balance the rather large differences between individual requirements and therefore provide fewer dissatisfied'.

The Scandinavian HVAC Association had already recommended three levels of thermal quality (SCANVAC, 1991), corresponding to 80 per cent, 90 per cent and >90 per cent comfortable in terms of whole-body operative temperature. In contrast to the EU Prestandard, SCANVAC suggests that a range of individual adjustment of ±2 K would be necessary to ensure >90 per cent comfortable. SCANVAC did not have access to the then confidential development data of Wyon and Larsson (1990) showing that this was indeed an achievable range even without sweating or cooling of the supply air by more than 1.5 K below ambient. Table 12.1, which is based on results that have become available since 1991, predicts that an individual adjustment range of ±2 K about the group optimum would indeed ensure >90 per cent comfortable, but with a small margin. The basis for a reasonably good consensus on this important point may thus be said to exist already.

In summary, it is estimated that 99 per cent of a group would be thermally comfortable if the equivalent room temperature provided by their microclimate could be individually adjusted over a range of 6.0 K, 95 per cent with 4.6 K, and 90 per cent with 3.9 K. Dress codes increase these ranges. Office cubicle walls will probably be required for floor-, ceiling-, or wall-mounted IMCDs, to deflect air streams and thereby concentrate their effect locally, while furniture-based IMCDs have been shown to be capable of satisfying 99 per cent of a group even without them. A standard approach to the assessment of individual adjustment is proposed, allowing widely different solutions, including openable windows, to be quantitatively and empirically compared.

Acceptable levels of fan noise for individual thermal control

An experiment by Clausen *et al.* (1993) at the Technical University of Denmark quantified subjective preferences for different combinations of noise, temperature and IAQ. Subjects made their preference judgement on the grounds of first impressions only, and the noise was recorded road traffic noise. The results in Table 12.2 were obtained for the percentage dissatisfied at different noise levels.

Table 12.2 Subjective percentage dissatisfied with road noise in the DTU experiment

dBA	*% Dissatisfied*
35	—
40	5
45	10
50	20
55	35
60	55

Other results of the DTU experiment indicate that unless each change of 1 K in operative temperature in the desired direction can be achieved by means of an increase in noise level of less than 3.9 dBA, the overall degree of subjective discomfort will actually increase. This appears to be the first direct comparison of the subjective annoyance from noise and heat stress that has ever been made. However, the data used to compile Tables 12.1 and 12.2 may be used to equate the increase in percentage dissatisfied when the noise level increases with the decrease in percentage dissatisfied when the range of individual control increases (in the hypothetical case where increased fan noise is necessary to increase the range of individual control of operative temperature), as shown in Table 12.3.

An assumed linear relationship between these equivalent levels of dBA and control range (dBA = 68.6 – 6.2 × range K) accounts for 99.9 per cent of the variance, and the coefficient of regression indicates that increasing the range of individual control by 1 K would be 'worth' 6.2 dBA, if fan noise may be equated with road traffic noise. The DTU figure of 3.9 dBA per K is rather lower and as it was based on a direct comparison, must be regarded as more reliable until field studies prove otherwise. A behavioural study of the trade-off between thermal discomfort and noise could be carried out as follows: the set point of a room thermostat would be linked to the volume control of recorded fan noise in such a way that noise will increase linearly from zero at 27°C to a maximum at 21°C; for different fixed values of the maximum noise level ranging from 70 dBA down to 35 dBA, subjects would be allowed to select their preferred set-point on the basis of several days' work in the room; this study would verify or disprove the trade-off predicted in Table 12.3 and would provide a better basis for the design of future IMCDs. A pilot experiment recently performed under tropical conditions in the Philippines (Santos and Gunnarsen, 1997) indicates that subjects are indeed capable of systematically trading-off temperature against noise, although the trade-off function obtained in this experiment would not apply to unacclimatised subjects elsewhere.

Table 12.3 Levels of noise and thermal control range for a given percentage dissatisfied

% Dissatisfied	*dBA*	*Range, K*
5	40	4.58
10	45	3.84
20	50	3.00
35	55	2.18

Estimating the productivity impact of providing individual control

Published research on group performance decrement in response to thermal changes is assumed to be valid for individuals, if individual neutral temperature is substituted for group average neutral temperature. The findings of a review of thermal effects on performance (Wyon, 1993) may be simplified for the present purpose as follows:

1 **Thinking:** the performance of mental tasks requiring concentration was typically reduced by 30 per cent at 27°C (6 K above group average neutral temperature 21°C), in comparison with 20°C. Individual performance is therefore assumed to be 100 per cent at temperatures up to individual neutrality, to decrease linearly to 70 per cent over the next 6 K, and to remain at 70 per cent at higher temperatures.
2 **Typing:** individual performance of routine, well-practised office work such as typing is assumed to be 100 per cent at temperatures up to individual neutrality, to decrease linearly to 70 per cent over the next 4 K, and to remain at 70 per cent at higher temperatures.
3 **Skill:** individual performance of skilled manual work is assumed to be 100 per cent at temperatures down to 6 K above individual neutrality, and to decrease linearly with temperature to 80 per cent at temperatures 12 K or more below individual neutrality.
4 **Speed:** the speed of individual finger movements is assumed to be 100 per cent at temperatures down to 6 K above individual neutrality, and to decrease linearly with temperature to 50 per cent at temperatures 12 K or more below individual neutrality.

As in Table 12.1, individual neutral temperatures are assumed to be distributed normally about the group mean with SD = 1.17 K. Providing individual control makes it possible for each individual to approach thermal neutrality to the extent to which control has been delegated. It is assumed that an individual would refrain from using this possibility if performance would thereby be reduced, and would not go beyond neutrality even if performance would thereby be increased. The impact of individual control on group average performance is small for subjects whose individual neutral temperature is close to the group average, and large for subjects who would prefer temperatures very different from the group average. Table 12.4 shows group average performance at room temperatures ranging from 3 K below group average neutral temperature to 6 K above, with no individual control (IC = 0 K), using the above assumptions. Column 6 is the unweighted mean of all four task types, i.e. assuming an equal amount of time is regularly spent on each type of task. The effect on group average performance of providing ±3 K of individual control (IC = ±3 K) is shown in Table 12.5 for the same range of room temperatures.

The values in parentheses in Table 12.4 are the values observed in the

Table 12.4 Group average performance for four task types, with IC = 0 K (no individual control), at room temperatures ranging from 3 K below to 6 K above group average neutral temperature. The values in parentheses are the actual experimental results on which the interpolations are based

K	*Thinking*	*Typing*	*Skill*	*Speed*	*Mean*
−3	100.8	102.1	90.3	75.7	92.2
−2	100.6	101.7	91.4	78.6	93.1
−1	(100.0)	(100.0)	92.6	81.4	93.5
0	98.1	95.1	93.7	84.3	92.8
+1	94.6	86.5	94.9	87.1	90.8
+2	89.9	76.8	96.0	90.0	88.2
+3	84.7	(70.0)	97.1	92.9	86.2
+4	79.4	67.2	98.3	95.6	85.1
+5	74.3	66.6	99.3	98.1	84.6
+6	(70.0)	66.5	(100.0)	(100.0)	84.1

Table 12.5 Performance improvement with IC = ±3 K as a percentage of Table 12.4 reference values, at room temperatures ranging from 3 K below to 6 K above group average neutral temperature

K	*Thinking*	*Typing*	*Skill*	*Speed*	*Mean*
−3	0.0	0.0	3.4	8.6	3.0
−2	0.2	0.4	3.5	8.6	3.1
−1	0.8	2.1	3.4	8.6	4.0
0	2.7	7.0	3.4	8.6	5.4
+1	6.0	15.2	3.4	8.5	8.3
+2	10.1	23.2	3.3	8.1	11.2
+3	13.4	25.1	2.9	7.1	12.1
+4	15.2	19.3	2.1	5.4	10.5
+5	15.6	10.2	1.2	3.3	7.6
+6	14.7	3.5	0.6	1.4	5.1

experiments. The cold conditions in the skill and speed experiments are off scale, 12 and 18 K below the upper reference temperature shown. The calculated values are expressed as a percentage of performance in the reference condition in each experiment. The values represent the performance levels that may be expected at different room temperatures in conventional buildings with no individual control.

The values in Table 12.5 may be directly added to the corresponding values in Table 12.4 to obtain the performance levels that may be expected when individual control corresponding to room temperature changes of ±3 K is provided. As in Table 12.4, 'mean' performance in column 6 is the unweighted mean of the values in columns 2–5. It should be noted that this is an appropriate estimate of the overall impact on productivity only if an equal amount of time is spent on all four types of task, which will not often be the case.

A field experiment on individual control: the West Bend Mutual study

The rate at which insurance claims were processed at the West Bend Mutual insurance company was shown to increase by 2.8 per cent when individual control was operative, in comparison with when it was installed but inoperative (Kroner *et al.*, 1992). This productivity metric is mainly determined by decision-making, i.e. thinking, and is close to the above prediction of 2.7 per cent for IC = ±3 K. In the report of this field experiment the West Bend management state that 'general productivity benefited by between 4 and 6 per cent', which is between the values predicted for thinking and for typing, and very close to the productivity improvement predicted when an equal amount of time is spent on all four types of task (5.4 per cent improvement). The figure of 8.1 per cent dissatisfied with the thermal environment (voting 1 or 2 on a scale from 1 to 5) was down from 50 per cent dissatisfied in the old building and would be expected if individual control of ±2.2 K were available, if room temperature were set to the group average neutral temperature, and if the standard deviation of individual neutral temperature were as has been assumed above (SD = 1.17 K). Field experience thus supports the present theoretical predictions.

Summary

It has been shown that individual control equivalent to ±2 K can be achieved and would satisfy >90 per cent, while individual control equivalent to ±3 K would be necessary to satisfy 99 per cent. A new method has been described for calculating the expected effect on group average productivity of providing individual control. The provision of individual control equivalent to being able to change room temperature in the range ±3 K about the actual value, if used by each individual to approach conditions providing individual thermal neutrality whenever this can be done without decreasing performance on the task in hand, would increase group average performance by up to 7.0 per cent, depending on the nature of the task, even when room temperature was equal to group average neutral temperature. Averaging with equal weighting across the four very different tasks considered, all relevant to office work or light industrial work, the improvement in productivity achieved at the group average neutral temperature is 5.4 per cent if occupants use this strategy. The performance improvement due to individual control increases at room temperatures above group average neutral temperature. The outcome in terms of bottom-line productivity depends on the proportion of each person's time for which each task is critical – which becomes increasingly difficult to estimate at higher levels in the job hierarchy – but it should be remembered that this advantage occurs in addition to the comfort and motivational advantages of enabling 99 per cent of a group to achieve thermal comfort when room temperature is set to the group average neutral temperature. The calculated impact on productivity is a conservative estimate, as comfort in itself has not been assumed to improve performance, although it may well do so. If an IMCD uses a fan to provide additional cooling effect under individual control, the noise increase should be less than 3.9 dBA per K equivalent, or no overall subjective benefit will be experienced. As it is not particularly difficult to reduce the noise level of the small fans required for individual microclimate control, productivity improvements have been calculated assuming that there will be no negative effects of fan noise on performance.

References

ASHRAE (1993) *Handbook of Fundamentals*. Atlanta, GA: ASHRAE.

Clausen, G., Carrick, L., Fanger, P.O., Sun Woo Kim, Poulsen, T. and Rindel, J.H. (1993) A comparative study of discomfort caused by indoor air pollution, thermal load and noise. *Indoor Air*, **3**, 255–262.

European Union Prestandard (1994) *Ventilation for Buildings – Design Criteria for the Indoor Environment*. CEN Ref. prENV 1752:1994 E Final Draft.

Grivel, F. and Candas, V. (1991) Ambient temperatures preferred by young European males and females at rest. *Ergonomics*, **34**(3), 365–378.

Kroner, W., Stark-Martin, J.A. and Willemain, T. (1992) *Rensselaer's West Bend Mutual Study: using advanced office technology to increase productivity*. Center for Architectural Research, Rensselaer Polytechnic Institute, Troy, NY.

Santos, A.M.B. and Gunnarsen, L. (1997) Trade-off between temperature and noise, air velocity and window area during chamber tests. *Proceedings of Healthy Buildings/IAQ '97*, **2**, 41–46.

SCANVAC (1991) *Classified Indoor Climate Systems*. Stockholm: Swedish Indoor Climate Institute.

Wyon, D.P. (1993) Healthy buildings and their impact on productivity. *Proceedings*

of the 6th International Conference on Indoor Air Quality and Climate – Indoor Air '93, **6**, 3–13.
Wyon, D.P. (1998) Individual control at each workplace for health, comfort and productivity. *INvironment*, **4**(1), 1–6.
Wyon, D.P. and Larsson, S. (1990) *Individual Control of the Microclimate for Sedentary Work* (in Swedish). Laboratory report, Autocontrol AB, Gothenburg, Sweden. (Thermal manikin test results from an early prototype of the desk-mounted Climadesk IMCD are available as 'EHT Profile Diagrams').
Wyon, D.P. and Sandberg, M. (1996) Discomfort due to vertical thermal gradients. *Indoor Air*, **6**(1), 48–54.

Prior publication

This chapter is based on a paper to Indoor Air '96 at Nagoya, Japan, 21–26 July 1996, and on the content of a lecture at the Workplace Comfort Forum 'Creating the productive workplace' in London, 30 October 1997 (Wyon, 1998).

Chapter 13

Creating high-quality workplaces using lighting

Jennifer A. Veitch

Introduction

Many people have read about the Hawthorne experiments on illumination (Roethlisberger and Dickson, 1939; Snow, 1927). The investigators set out to understand the effects of lighting on the performance of workers assembling electrical products. They pre-selected a set of employees to participate in the tests and moved their work area to a specially prepared space, where they worked under a variety of lighting conditions. The results surprised everyone: regardless of the direction of the lighting change (even when lighting levels dropped), the work output of the employees increased. Even when the investigators gave the appearance of having changed the lighting, but had in fact simply taken out and replaced the same lamps, performance increased.

The results led to an important series of studies concerning the relationships between employers and employees (Roethlisberger and Dickson, 1939). On closer analysis the investigators realised that the special experimental set-up, separate from other employees, and the knowledge that they were participating in work that might benefit their working conditions, were powerful motivators to the participants in the lighting experiments. The investigators, and many others, concluded that the physical environment at work was relatively unimportant to workers' performance. Management–employee relations seemed to be the important consideration.

This conclusion rests on an oversimplification: the assumption that there is a direct cause-and-effect relationship between physical conditions and human behaviour. Because they did not find that lighting levels improved performance, the researchers assumed that light levels were irrelevant to performance. Instead, psychologists now know, the effects of physical conditions on human behaviour are mediated by complex cognitive processes. In the Hawthorne experiments, one likely explanation for the results is that people interpreted their special workroom and the changing experimental conditions as indications of managers' concern for their welfare. The employees' interpretation probably led to the improved work output. It was not the case that the physical environment was irrelevant to their performance, rather that its effect was mediated by the participants' expectations and beliefs.

One consequence of the Hawthorne experiments was a drought in psychologists' interest in lighting and other physical conditions (Gifford, 1997); today, we are only beginning to understand how the lit environment influences mental states and processes that, in turn, determine work performance, satisfaction, and other important outcomes. The motivation for much of this research, for lighting and for other indoor conditions, is economic: buildings cost less than employees, so any environmental condition that decreases individual performance (either in quantity or quality), increases absenteeism, or contributes to turnover, is more expensive to organisations than the capital and operating costs of better indoor environments (Woods, 1989). Productivity, the organisational outcome most sought after, is a complex concept defined in different ways in various disciplines, but is not synonymous with individual performance (Pritchard, 1992). One commonality in definitions of productivity is the concept of efficiency. Productivity is an index of output relative to inputs. Using the efficiency definition, a poor indoor environment decreases organisational productivity both by reducing revenue and by increasing costs.

The research literature concerning micro-level environmental conditions and productivity does not lend itself to direct calculations of the consequences for organisations (Rubin, 1987). Simple, direct promises of the form 'This lighting design will improve productivity by 15 per cent' are not possible, however desirable they might be to marketing agents. However, by understanding the psychological and organisational processes that influence employees in their working environments, we can develop advanced design recommendations that support employee workspace needs. This is the approach taken here.

Veitch and Newsham (1998a) have presented a model for lighting–behaviour research that includes six categories of human needs addressed by lighting. These are visibility, task performance, social behaviour and communication, mood and comfort, aesthetic judgements, and health and safety. The literature documents several psychological processes thought to mediate the relationship between lighting conditions and these behaviours, of which three are discussed here: visibility, arousal and stress, and positive affect. They are presented in order of increasing psychological complexity, following the path from the luminous stimulus through visual processing, physiological responses, and interpretative, cognitive processes. In the concluding section, we turn to the integration of this information with other considerations such as energy efficiency, architecture, and costs, to produce a model for achieving good-quality lighting in workplaces. Good-quality lighting, supportive of human needs, contributes to conditions that sustain organisational productivity.

From light to vision

Light is visible electromagnetic radiation between 380 and 780 nm in wavelength (Commission Internationale de l'Éclairage, 1987). We see light directly from its sources – the sun, fire, electric lamps – and we see objects when light is reflected from them. Light enters the eye through the pupil, is focused by the lens, and is detected by photoreceptive cells on the retina (for more detail about the visual system, see Dowling (1987), or Spillman and Werner (1990)). The focal point for a correctly focused eye is a point on the retina called the fovea, where specialised cells (cones) detect colour in red, green, or blue bands. These cells respond selectively to the wavelength composition of the light, and higher centres in the brain interpret the signals to perceive colour. Elsewhere on the retina, other photoreceptors (rods) detect light and dark. Neural impulses from the retinal receptors pass through the optic chiasm to the lateral geniculate nuclei (LGN), and terminate in the primary visual cortex.

Lighting researchers have focused much of their attention on visibility, with the result that we have an excellent understanding of what is needed to make objects visible. Four variables have the greatest effects: the age of the viewer, task size, task/background contrast, and task luminance.[1] Objects are more easily viewed when they are larger, have higher contrast, and at higher luminance. Rea and Ouellette (1991) produced a model of relative visual performance (RVP) that allows precise predictions of visual performance given these input conditions. The model includes the effects of the decreasing visual acuity that occurs with age.

The model shows asymptotic relationships. For any given task luminance and task size, there is a range of task contrasts for which RVP is high and nearly constant; however, below a certain value (which depends on the luminance and size), RVP drops drastically. Rea and Ouellette (1991) have called this a 'plateau and escarpment' model, for this reason. Likewise, for a given task contrast and size there is a range of task luminances that produces consistently high RVP. The asymptotic relationship between visual performance and luminance means that, as Boyce (1996) said, 'To put it bluntly . . . for many visual tasks, lighting is unimportant to visual performance' (p. 44). There is a broad range of acceptable light levels that provide adequate quantity of illumination to see in most workplaces.

Certain industrial tasks, however, will require special attention to the task characteristics. Some details are so small or have such low contrast that increasing the task luminance will not sufficiently increase task visibility (for example, sewing with black thread on black cloth). In these cases, magnifiers and directional lighting (to increase contrast by using relief) will be necessary parts of lighting design. For more details, see common recommended practice documents such as the CIBSE *Code for Interior Lighting* (Chartered

Institution of Building Services Engineers, 1994) or the IESNA *Lighting Handbook* (Rea, 1993).

In any workplace, lighting conditions that reduce task contrast will adversely affect visibility. Reflected images of luminaires in VDT screens and veiling reflections (regular luminaire reflections superimposed on diffuse lighting) are examples of glare that reduces task contrast. Changes in the lighting can prevent such problems, as can changing certain task characteristics. Parabolic-louvred luminaires, designed to reduce glare from reflected images in VDTs, lead to better performance on computer-based tasks and lower ratings of glare (Veitch and Newsham, 1998b). Computer-based task performance is also better when the display uses a light background and dark characters (Sanders and Bernecker, 1990; Veitch and Newsham, 1998b).

The two retinal receptor types have different spectral sensitivities and functions. Rod vision dominates at low light levels (below 0.034 cd/m^2); this sensitivity function is called the **scotopic sensitivity curve**, and peaks at 505 nm (blue-green). Cone vision dominates at light levels typical of interiors, and is described by the **phototopic sensitivity curve**, which peaks at 550 nm (yellow-green). It is because scotopic vision is principally based on rod receptors that our colour vision is so poor at night. Some researchers believe that both scotopic and photopic vision influence visual performance even at high light levels. They argue that scotopic processes control pupil size such that a higher concentration of short-wavelength light (around 505 nm) reduces pupil size sufficiently to increase the resolution of small details (Berman *et al.*, 1993, 1994), much like a smaller aperture on a camera. However, this notion is controversial and other researchers have not repeated these results (Halonen, 1993; Rowlands *et al.*, 1971).

Fluorescent lighting is a complex system that includes the lamps, the luminaires that hold them, an optical system (usually either a lens or a louvre), and a ballast to create the needed circuit conditions (voltage, current, and waveform) for starting and operating the lamp. Conventional ballasts used a magnetic core–coil system that resulted in oscillations at twice the rate of the AC electrical supply (thus, luminous modulation of 100 Hz in the UK and Europe, 120 Hz in North America). Most people cannot perceive this modulation as flicker, but there is evidence of neural activity in response to modulation at rates as high as 147 Hz (Berman *et al.*, 1991). This response occurs not only at the retinal level, but in the LGN and the visual cortex as well (Eysel and Burandt, 1984; Schneider, 1968). Mounting evidence suggests that low-frequency flicker interferes with visual processing, disrupting the eye movements in reading, causing visual fatigue, and hampering visual performance and computer-based task performance (Veitch and McColl, 1995; Veitch and Newsham, 1998b; Wilkins, 1986). Energy-efficient, high-frequency electronic ballasts, which use integrated electronic circuitry to operate lamps at rates between 20,000 and 60,000 Hz, alleviate this problem because the modulation rate exceeds the capacity of human physiology. Functionally, they produce a constant luminous output.

The sensory system, from retina to visual cortex, is but the first step in a sequence of complex processes leading to the perception of objects, colour, depth, movement, and other features. Without these interpretative mechanisms, the images delivered by the sensory system would be so much noise, like a grainy, blurred photograph. Lighting has few direct effects on perceptual processes, but lighting influences on the visual image can interact to produce visual illusions. For example, depth perception on stairs requires a contrast contour of either colour or shadow. Some visual illusions caused by lighting can be dangerous if they obscure important details such as the depth of a stair or the movement of machinery (stroboscopic effect). Detailed consideration of these processes is beyond the scope of this chapter; however, excellent general works are available (e.g. Gregory, 1978).

Activation, arousal, and stress

Arousal is a general state of mental and physical activation (Landy, 1985). Arousal theory holds that there is an inverted-U function between arousal levels and behaviours, with an optimal arousal level for each behaviour. For task performance, the curve is believed to shift up for simple tasks (that is, the optimal arousal level is higher if the task is easier). **Stress** is the name for a set of physiological and hormonal changes that arise in response to threatening or unpleasant events, called **stressors**. Stressors can include environmental conditions such as direct glare, or loud noise; life events, such as bereavement; or emotional states, such as conflict. Chronic exposure to stressors can lead to unpleasant health effects, such as high blood pressure. Stress is linked to arousal because the response to such events can include a heightened state of neurological activation. Thus, in lighting, the goal is to create luminous conditions that would lead to optimal arousal, while avoiding conditions that might act as stressors.

Light exposure is well known to suppress melatonin secretion and to control circadian rhythms (Hill, 1992).[2] Melatonin induces sleep; its suppression leads to wakefulness. Growing insight into these mechanisms has led to the development of schedules and technologies to increase the light exposure of night-shift workers, successfully improving their adaptation (Boyce *et al.*, 1997; Czeisler *et al.*, 1990). Both timing and intensity of illumination (illuminance[3]) influence the outcome. Some photobiologists believe that most people suffer from a deficit of light exposure, even day-shift workers (Brainard and Bernecker, 1996), but there is no consensus on what the necessary daily light dose might be, nor on the need for electric lighting to provide it. Moreover, to increase lighting levels substantially, as would be required for biological effectiveness, would risk serious energy consumption and glare-control problems.

Arousal explanations for lighting effects are difficult to test because of a logical flaw in arousal theory. To test the theory, one needs to determine the relationship between the independent variable (lighting levels, for example) and arousal levels and, in addition, the relationship between arousal and the performance of various tasks. One needs to know the optimal arousal level for optimal task performance and one needs to know which lighting conditions will cause that arousal level. Usually, however, the optimal arousal level is inferred after the fact based on the physical conditions that led to the best performance. For example, if I conducted a study of lighting level effects that found that performance was best under 800 lx and worse under both 400 and 1200 lx, then I might conclude that 800 lx creates the optimal arousal level and that the study supports arousal theory. However, the study does not test the theory at all, because one could equally well argue, if performance under 400 lx was best, that both 800 and 1200 lx produced too high an arousal level and that the experimental conditions captured only part of the inverted-U curve.

We do not know the optimal arousal level for most tasks, nor is there a clear relationship between illuminance and general arousal levels (Veitch and Newsham, 1996). None the less, the belief that increasing illuminance will increase arousal and improve task performance is long-standing, and underlies many of the changes in lighting practice over this century (Pansky, 1985). Many experiments in the laboratory and the field have attempted to address this belief, with very mixed results. Gifford *et al.* (1997) conducted a meta-analysis (a quantitative review that combines the results of many studies into one general conclusion) of the literature concerning illuminance and office task performance. Many studies could not be included in the review because their published reports included too few details. Of the studies that could be included, contrasts between low (average 70 lx) and medium (average 486 lx) illuminance levels did not produce significant effects on task performance; however, contrasts between low and high illuminance (average 1962 lx) produced a (statistically significant) average correlation of 0.25 between illuminance and task performance, which is a small- to medium-sized relationship.

Arousal is one possible explanation for this relationship, but other explanations are equally plausible. For example, the higher illuminance levels might have improved visibility; 70 lx is quite dark for paper-based work. Novelty is another possible explanation. On closer analysis, Gifford *et al.* (1997) found that studies that provided more than a 15-minute adaptation period showed a smaller effect. This suggests that a high illuminance can improve office task performance as compared to a very low one, but the effect might not last. Most people, of course, spend considerably longer than 15 minutes in their offices. Once they adapt to conditions, illuminance probably does not influence performance, provided the level is adequate for seeing task details. More light is not necessarily better light.

Another notion about the arousal effects of illuminance is that higher light levels lead to louder conversation and more communication. Sanders *et al.* (1974) observed louder communication in naturally-occurring groups in a university corridor near areas where the lamps were on than in delamped areas (over the range 10–270 lx). They concluded that higher arousal associated with the brighter areas caused the effect, but because the areas were continuous along one corridor it is impossible to determine to which lighting conditions the speakers had adapted when the measurements were taken. In contrast, Veitch and Kaye (1988) reported that groups of female university students conversing about fictional job candidates were louder under low illuminance (400 lx) than high illuminance (1274 lx). The authors speculated that the unusual nature of the dim lighting condition caused the louder speech; conversely, students in a brightly lit classroom are usually not expected to speak loudly. In fact, one could argue that the unusual lighting was more arousing than the bright lighting; but this leaves one with a circular argument in which any outcome can be explained using arousal theory. Good tests of arousal theory require unambiguous predictions about the expected effects on physiological systems (Blascovich and Kelsey, 1990; Venables, 1984), but most lighting researchers lack the expertise to make such predictions and their required measurements.

Whether or not lighting conditions influence arousal, it is clear that they can act as stressors. A bright light shone directly on the face during interrogation is a classic example of a deliberately created stressful, threatening situation. Glare, either directly from light sources in the field of view or by reflection in glossy surfaces, is the less extreme instance of light as a stressor. Discomfort glare is a well-known phenomenon that has a physiological basis (Berman *et al.*, 1994). Very high luminances in the field of view, or very highly non-uniform luminance distributions, can cause discomfort; laypeople believe that glare can cause headaches (Veitch and Gifford, 1996a). Veiling luminances can also be unacceptable even if the degree of contrast reduction does not reduce visual performance (Bjørset and Frederiksen 1979).

Low-frequency flicker might also constitute a stressful lighting condition. In a field experiment in which participants were unaware of the changes in lighting conditions, reports of headaches and eyestrain dropped dramatically when fluorescent lights were run on high-frequency electronic ballasts instead of low-frequency magnetic ballasts (Wilkins *et al.*, 1989). Lindner and Kropf (1993) also found that increasing the operating rate of fluorescent lighting decreased the complaint rate for eyestrain, headache, and other symptoms.

Beliefs, expectations, and emotions

Physiological responses and the behaviours they trigger are modified by other, higher cognitive processes and existing states. What we already know and believe influences the information we attend to and can bias our thinking and perception. We pay more attention to information that is vivid and personal, and, we remember more accurately the information that is consistent with our beliefs (Norman, 1976). In judging probabilities of success or failure, we assign greater weight to information that is more easily recalled (Tversky and Kahneman, 1974). If we have selectively attended to cues or stimuli around us, or if our beliefs are based on inaccurate information or misunderstanding, then we can be led by these biases to make incorrect decisions.

Knowledge about laypeople's beliefs and expectations concerning lighting has implications for workplace lighting design. For example, people distrust fluorescent lighting on principle, believing that it can cause adverse health effects ranging from headache to melanoma (Stone, 1992; Veitch and Gifford, 1996a; Veitch *et al.*, 1993). Except for headache and eyestrain associated with flicker, discussed above, these fears are groundless (Stone, 1992). However, these fears influence lighting choices: people who feared the ill-effects of fluorescent lighting were less likely to purchase compact fluorescent lamps for their homes (Beckstead and Boyce, 1992). Successful attempts to implement novel lighting technologies and designs will require an understanding of existing beliefs about lighting in order to direct information and education towards removing such biases.

Another reason for understanding existing beliefs and preferences about lighting is to create luminous conditions that match what the occupants want and expect. Obtaining these lighting conditions is believed to lead to a pleasant emotional state that psychologists call **positive affect**. Positive affect theory (Baron, 1994) states that environmental conditions that create positive affect lead to better performance, greater effort, less conflict, and greater willingness to help others. Experiments in which positive affect was induced using fragrance have supported this theory (Baron, 1990; Baron and Thomley, 1994). Application of positive affect theory to lighting requires, first, that we know which luminous conditions people prefer. Conclusive proof of the theory would accrue with evidence that performance and other outcomes were consistently better under the preferred conditions than under non-preferred conditions. The scientific lighting literature currently shows limited support (Baron *et al.*, 1992; Knez, 1995), but in these studies positive affect was not consistently related to the experimental lighting conditions.

One reason for this could be the wide individual differences in preferred lighting conditions. Although many studies have found that people prefer illuminance levels that are higher than current recommended practice (e.g. Boyce, 1979; Halonen and Lehtovaara, 1995; Begemann *et al.*, 1994), the variability from one person to another is also striking. Halonen and Lehtovaara (1995) concluded that it was so great that it would be impossible to design an automated daylight-linked control system that would suit the majority of individuals. Several investigators have attempted to identify preferred luminance levels for walls and luminance ratios in workplaces, but a comparison of them shows marked differences in the preferred levels (Loe *et al.*, 1994; Tregenza *et al.*, 1974; Ooyen *et al.*, 1987; Miller *et al.*, 1995). The tendency for lighting researchers to base their conclusions on small samples, short exposures, and poorly controlled research designs can explain these inconsistent results. Cultural variations both between nations and across time might also explain the discrepancies in the literature; the foregoing studies include samples from the UK, Sweden, The Netherlands, and the USA.

Individual, manual controls or lighting design choices are two strategies that might address the problem of individual differences in lighting preferences. When one does not know which conditions will create positive affect, these would allow people to self-select. Many lighting designers and researchers alike believe that people with personal controls will be more satisfied and will work more productively (e.g. Barnes, 1981; Simpson, 1990). Not only will desired luminous conditions result, the reasoning goes, but the state of perceived control is itself desirable.

The psychological literature generally supports that hypothesis, but with some exceptions (Burger, 1989). Perceived control does not lead to desirable outcomes when the individual fears that making the wrong choice will cause a loss of face, when it appears that an expert is more likely to make a better choice, or when one thinks that choosing will lead to failure. In one experiment, making a lighting choice in front of the experimenter and another participant led to poorer and slower performance on a creativity task than occurred for people who had no lighting choice (Veitch and Gifford, 1996b). Furthermore, Wineman (1982) suggested that in a demanding, stressful workplace, providing additional choices about physical conditions could add to the job load in undesirable ways. Individualised controls that people do not understand how to use, that do not work as intended, or that demand responses from overworked employees are unlikely to achieve satisfactory results. Likewise, providing lighting choices without providing background information about what luminous conditions they will create or how they work is unlikely to create perceived control, especially if the choices are inconsistent with employees' existing beliefs and expectations.

Although the precise levels for luminance, luminance ratios, and illuminance remain in debate, there also exist certain consistent, general preferences. People prefer brighter vertical surfaces to dark ones (e.g. Ooyen *et al.*, 1987; Sanders and Collins, 1996). They also prefer daylight when it is available, and many believe that it is superior to electric light (Heerwagen and Heerwagen, 1986; Veitch *et al.*, 1993; Veitch and Gifford, 1996a). Preferences are also greater for luminous conditions described as 'interesting', which are usually non-uniform but use luminance to reinforce the design features (Hawkes *et al.*, 1979; Loe *et al.*, 1982, 1994).

Few studies have set out explicitly to test the positive affect theory, and none has clearly demonstrated that luminous conditions create positive affect. Baron *et al.* (1992) found that although there were no effects of illuminance or lamp type on direct measures of positive affect, the pattern of results was consistent with the effects of other conditions known to create positive affect. For instance, participants rated the appearance of a room lit with warm-white lamps at 150 lx as more pleasant than the other conditions, and also performed better on a word categorisation task in that room. Preferred conditions also were associated with a preference for resolving disagreements with cooperative strategies rather than conflict. These effects are similar to behaviour in people whose positive affect was induced by the receipt of a small gift. Knez (1995) attempted to replicate the work of Baron *et al.*, with mixed success. He observed inconsistent effects of lighting conditions on measures of positive and negative affect, but the conditions that caused better affect (more positive or less negative affect) were associated with better performance outcomes.

Comparisons between lighting systems provide no clearer results. A field comparison of office areas retrofitted with either parabolic-louvred luminaires or suspended lensed direct/indirect luminaires found that the employees preferred the lensed direct/indirect luminaires and rated their work performance as better than the employees whose area received the parabolic louvred luminaires (Hedge *et al.*, 1995). Katzev (1992) compared performance and satisfaction in four identical enclosed offices lit with different lighting systems. A recessed direct/indirect system was one of the more preferred systems (preference rankings depended on the way preference was assessed), and in that room reading comprehension was highest. However, typing performance was lowest in that room. It is possible that positive affect influences certain work behaviours more than others, or alternatively that lighting preferences differ for various tasks, but these questions remain for future research to answer.

The future of lighting research lies in deepening our understanding of the role of cognitive processes in lighting–behaviour relationships. The interplay of beliefs, expectations, affect and perceived control is complex, making research in this area appear unnecessarily abstract for practical application. Only when we have any understanding of how luminous conditions influence work performance, communication, perceptions, and social behaviours will we begin to be able to make precise predictions about the likely effects of a particular design on occupants in the same way that visual performance can be precisely predicted. This understanding can only come from thorough investigations into these fundamental processes.

Conclusions

'Hide the source, light the walls; don't cause glare problems' is how one well-known designer summarised his views on lighting quality for offices at a seminar on office lighting (M. Kohn, personal communication). This chapter has expanded on that good advice using examples drawn from the scientific lighting literature. A set of general guidelines for workplace lighting that meets occupant needs is provided in Table 13.1 (refer to Chartered Institution of Building Services Engineers (1994) or Rea (1993) for more detailed guidance).

The complexity of the lighting–productivity equation increases when one considers that lighting installations must meet both immediate and long-term needs of several groups. There are the needs of occupants, which this chapter has emphasised; the economic needs of clients; the needs to integrate lighting in a pleasing way with architecture and to respect building codes and standards; and, not least, the long-term need to protect the environment. Lighting quality requires energy efficiency (Figure 13.1). Although energy-efficient lighting installations carry a premium in initial costs, interactions with other building systems (particularly cooling systems) can reduce the payback period for energy-efficiency investments by lowering annual energy costs (Newsham and Veitch, 1997).

Although the literature is not conclusive on such topics as preferred luminance ratios or desired wall luminance, it does support the argument that investments in lighting pay off in the long run. Lighting contributes to

Table 13.1 Lighting guidelines for productive workplaces

Mediating process	*Guidelines*
Visibility	• appropriate horizontal and vertical illuminances for tasks and viewers • control unwanted light (glare), both direct and reflected • use high-frequency ballasts for fluorescent lights
Arousal, activation and stress	• investigate light exposure for aiding night-shift workers • avoid creating stressors: direct glare, excessive luminance contrast • use high-frequency ballasts for fluorescent lights
Beliefs, expectations, affect	• learn end-users' expectations and beliefs about lighting • educate users before implementing new technologies or designs • create interest by integrating luminance variability with architecture • keep vertical surfaces bright • use daylighting and windows where possible • consider individual controls, but educate users and maintain systems

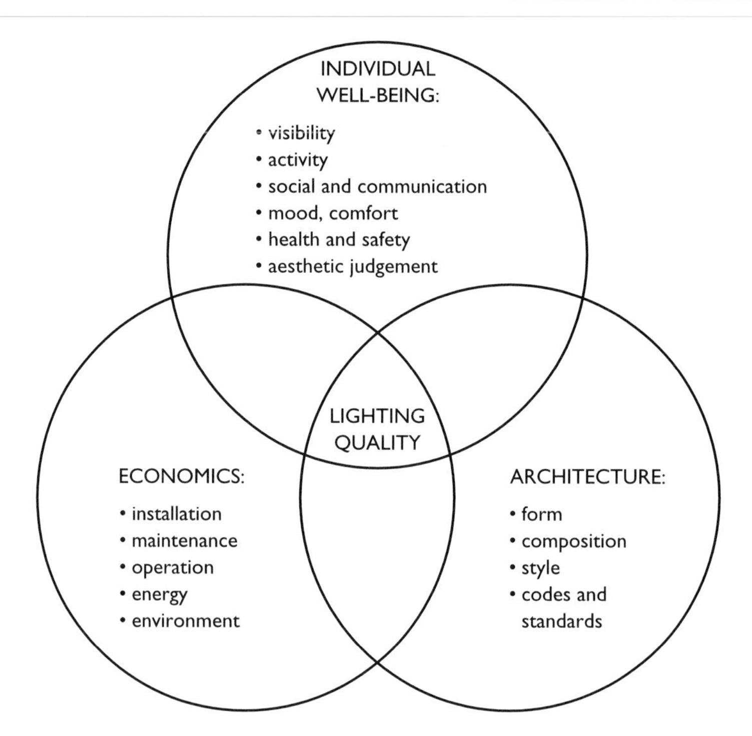

Figure 13.1 Lighting quality: the integration of individual well-being, architecture and economics

environmental satisfaction and to individual performance at levels sufficient to support capital and operating expenses for lighting (e.g. Veitch and Newsham, 1998b; Wilkins *et al.*, 1989). Lighting is a popular target for such investment because it is a small portion of the building investment, which itself is a minor expense in comparison to employee compensation (Woods, 1989). Fisk and Rosenfeld (1997), in calculating the potential economic consequences of investments in indoor environment improvements, estimated that the performance effects now reported could translate into productivity increases between 0.5 per cent and 5 per cent, which is within the range of acceptable return on investment.

Clearly, the goal of creating high-quality lighting to improve organisational productivity is more challenging than the Hawthorne researchers realised when they changed workplace lighting by increasing the wattage of incandescent lamps. Good-quality lighting demands simultaneous resolution of requirements that sometimes conflict, and coordination with other building systems. Difficult or not, the effort is worth while, for in achieving this goal, everyone benefits.

References

Barnes, R.D. (1981) Perceived freedom and control in the built environment. In J.H. Harvey (ed.), *Cognition, Social Behavior, and the Environment*, 409–422. Hillsdale, NJ: Erlbaum.

Baron, R.A. (1990) Environmentally induced positive affect: its impact on self-efficacy, task performance, negotiation, and conflict. *J. Appl. Soc. Psychol.*, **20**, 368–384.

Baron, R.A. (1994) The physical environment of work settings: effects on task performance, interpersonal relations, and job satisfaction. In B.M. Staw and L.L. Cummings (eds), *Research in Organizational Behavior*, vol. 16, 1–46. Greenwich, CN: JAI Press.

Baron, R.A., Rea, M.S. and Daniels, S.G. (1992) Effects of indoor lighting (illuminance and spectral distribution) on the performance of cognitive tasks and interpersonal behaviors: the potential mediating role of positive affect. *Motivation and Emotion*, **16**, 1–33.

Baron, R.A. and Thomley, J. (1994) A whiff of reality: positive affect as a potential mediator of the effects of pleasant fragrances on task performance and helping. *Environ. and Behav.*, **26**, 766–784.

Beckstead, J.W. and Boyce, P.R. (1992) Structural equation modeling in lighting research: an application to residential acceptance of new fluorescent lighting. *Lighting Res. and Technol.*, **24**, 189–201.

Begemann, S.H.A., Aarts, M.P.J. and Tenner, A.D. (1994) Daylight, artificial light, and people. *Annual Conference of the Illuminating Engineering Society of Australia and New Zealand*, Melbourne, November.

Berman, S.M., Bullimore, M.A., Jacobs, R.J., Bailey, I.L. and Gandhi, N. (1994) An objective measure of discomfort glare. *J. Illuminating Eng. Soc.*, **23**(2), 40–49.

Berman, S.M., Fein, G., Jewett, D.L. and Ashford, F. (1993) Luminance-controlled pupil size affects Landolt-C task performance. *J. Illuminating Eng. Soc.*, **22**(2), 150–165.

Berman, S.M., Fein, G., Jewett, D.L. and Ashford, F. (1994) Landolt-C recognition in elderly subjects is affected by scotopic intensity of surround illuminants. *J. Illuminating Eng. Soc.*, **23**(2), 123–130.

Berman, S.M., Greenhouse, D.S., Bailey, I.L., Clear, R. and Raasch, T.W. (1991) Human electroretinogram responses to video displays, fluorescent lighting and other high frequency sources. *Optometry and Vision Sci.*, **68**, 645–662.

Bjørset, H.-H. and Frederiksen, E. (1979) A proposal for recommendations for the limitation of the contrast reduction in office lighting. *Proc. 19th Session of the Commission Internationale de l'Éclairage*, Kyoto, Japan (CIE Publication no. 50, 310–314). Paris: Bureau Centrale de la CIE.

Blascovich, J. and Kelsey, R.M. (1990) Using electrodermal and cardiovascular measures of arousal in social psychological research. In C. Hendrick and M.S. Clark (eds), *Research Methods in Personality and Social Psychology*, 45–73. Newbury Park, CA: Sage.

Boyce, P.R. (1979) Users' attitudes to some types of local lighting. *Lighting Res. and Technol.*, **11**, 158–164.

Boyce, P.R. (1996) Illuminance selection based on visual performance – and other fairy stories. *J. Illuminating Eng. Soc.*, **25**(2), 41–49.

Boyce, P.R., Beckstead, J.W., Eklund, N.H., Strobel, R.W. and Rea, M.S. (1997) Lighting the graveyard shift: the influence of a daylight-simulating skylight on the task performance and mood of night-shift workers. *Lighting Res. Technol.*, **29**, 105–134.

Brainard, G.C. and Bernecker, C.A. (1996) The effects of light on human physiology and behavior. In *Proceedings of the 23rd Session of the Commission Internationale de l'Éclairage, New Delhi, India, November 1–8, 1995*, vol. 2, 88–100. Vienna: Bureau Centrale de la CIE.

Burger, J.M. (1989) Negative responses to increases in perceived personal control. *J. Personality and Soc. Psychol.*, **56**, 246–256.

Chartered Institution of Building Services Engineers (1994) *Code for Interior Lighting*. London: CIBSE.

Commission Internationale de l'Éclairage (1987) *International Lighting Vocabulary*, CIE Publication no. 17.4. Vienna: Bureau Centrale de la CIE.

Czeisler, C.A., Johnson, M.P., Duffy, J.F., Brown, E.N., Ronda, J.M. and Kronauer, R.E. (1990) Exposure to bright light and darkness to treat physiologic maladaptation to night work. *New Engl. J. Med.*, **322**, 1253–1259.

Dowling, J.E. (1987) *The Retina. An Approachable Part of the Brain*. Harvard, CT: Belknap Press.

Eysel, U.T. and Burandt, U. (1984) Fluorescent light evokes flicker responses in visual neurons. *Vision Res.*, **24**, 943–948.

Fisk, W.J. and Rosenfeld, A.H. (1997) Estimates of improved productivity and health from better indoor environments. *Indoor Air*, **7**, 158–172.

Gifford, R. (1997) *Environmental Psychology: Principles and Practice*, 2nd edn. Boston, MA: Allyn & Bacon.

Gifford, R., Hine, D.W. and Veitch, J.A. (1997) Meta-analysis for environment–behavior research, illuminated with a study of lighting level effects on office task performance. In G.T. Moore and R.W. Marans (eds), *Advances in Environment, Behavior, and Design*, vol. 4, 223–253. New York: Plenum.

Gregory, R.L. (1978) *Eye and Brain: the Psychology of Seeing*, 3rd edn. New York: McGraw-Hill.

Halonen, L. (1993) *Effects of Lighting and Task Parameters on Visual Acuity and Performance*, NTIS No. PB94–179231. Espoo, Finland: Helsinki University of Technology.

Halonen, L. and Lehtovaara, J. (1995) Need of individual control to improve daylight utilization and user's satisfaction in integrated lighting systems. *Proc. 23rd Session of the Commission Internationale de l'Éclairage*, New Delhi, India, 1–8 November 1995 (vol. 1, 200–203). Vienna: Bureau Centrale de la CIE.

Hawkes, R.J., Loe, D.L. and Rowlands, E. (1979) A note towards the understanding of lighting quality. *J. Illuminating Eng. Soc.*, 8, 111–120.

Hedge, A., Sims, W.R. and Becker, F.D. (1995) Effects of lensed-indirect and parabolic lighting on the satisfaction, visual health, and productivity of office workers. *Ergonomics*, **38**, 260–280.

Heerwagen, J.H. and Heerwagen, D.R. (1986). Lighting and psychological comfort. *Lighting Des. Appl.*, **16**(4), pp. 47–51.

Katzev, R. (1992) The impact of energy-efficient office lighting strategies on employee satisfaction and productivity. *Environ. and Behav.*, **24**, 759–778.

Knez, I. (1995) Effects of indoor lighting on mood and cognition. *J. Environ. Psychol.*, **15**, 39–51.
Landy, F.J. (1985) *Psychology of Work Behavior*. Homewood, IL: Dorsey.
Lindner, H. and Kropf, S. (1993) Asthenopic complaints associated with fluorescent lamp illumination (FLI): the role of individual disposition. *Lighting Res. Technol.*, **25**, 59–69.
Loe, D.L., Mansfield, K.P. and Rowlands, E. (1994) Appearance of lit environment and its relevance in lighting design: experimental study. *Lighting Res. Technol.*, **26**, 119–133.
Loe, D.L., Rowlands, E. and Watson, N.F. (1982) Preferred lighting conditions for the display of oil and watercolour paintings. *Lighting Res. Technol.*, **14**, 173–192.
Miller, N., McKay, H. and Boyce, P.R. (1995) An approach to a measurement of lighting quality. *Proc. of the Annual Conference of the Illuminating Engineering Society of North America, New York, 29–31 July*. New York: IESNA.
Newsham, G.R. and Veitch, J.A. (1997) *Energy-efficient lighting options: predicted savings and occupant impressions of lighting quality. CLIMA 2000 Conference*, Brussels, 30 August–1 September.
Norman, D.A. (1976) *Memory and Attention*. New York: Wiley.
Ooyen, M.H.F. van, Weijgert, J.C.A. van de and Begemann, S.H.A. (1987) Preferred luminances in offices. *J. Illuminating Eng. Soc.*, **16**(2) 152–156.
Pansky, S.H. (1985) Lighting standards: tracing the development of the lighting standard from 1939 to the present. *Lighting Des. Appl.*, **15**(2), 46–48.
Pritchard, R.D. (1992) Organizational productivity. In M.D. Dunnette and L.M. Hough (eds), *Handbook of Industrial and Organizational Psychology*, 2nd edn, vol. 3, 443–471. Palo Alto, CA: Consulting Psychologists' Press.
Rea, M.S. (ed.) (1993) *Lighting Handbook: Reference and Application*, 8th edn. New York: Illuminating Engineering Society of North America.
Rea, M.S. and Ouellette, M.J. (1991) Relative visual performance: a basis for application. *Lighting Res. Technol.*, **23**, 135–144.
Roethlisberger, F.J. and Dickson, W.J. (1939) *Management and the Worker*. Cambridge, MA: Harvard University Press.
Rosenthal, N.E. (1993) Diagnosis and treatment of seasonal affective disorder (clinical conference). *J. Am. Med. Assoc.*, **270**, 2717–2720.
Rowlands, E., Loe, D.L., Waters, I.M. and Hopkinson, R.G. (1971) Visual performance in illuminance of different spectral quality. Paper presented at the *17th Session of the Commission Internationale de l'Éclairage*, Barcelona.
Rubin, A. (1987) *Office Design Measurements for Productivity – a Research Overview* (National Bureau of Standards Report NBSIR 87-3688). Washington, DC: Department of Commerce.
Sanders, M., Gustanski, J. and Lawton, M. (1974). Effect of ambient illumination on noise level of groups. *J. Appl. Psychol.*, **59**, 527–528.
Sanders, P.A. and Bernecker, C.A. (1990) Uniform veiling luminance and display polarity affect VDU user performance. *J. Illuminating Eng. Soc.*, **19**(2), 113–123.
Sanders, P.A. and Collins, B.L. (1996) Post-occupancy evaluation of the Forrestal Building. *J. Illuminating Eng. Soc.*, **25**(2), 89–103.
Schneider, C.W. (1968) Electrophysiological analysis of the mechanisms underlying critical flicker frequency. *Vision Res.*, **8**, 1235–1244.

Simpson, M.D. (1990) A flexible approach to lighting design. *Proc. CIBSE National Lighting Conference*, Cambridge, 8–11 April, 182–189. London: Chartered Institution of Building Services Engineers.

Snow, C.E. (1927) Research on industrial illumination. *The Tech Eng. News*, 8(6), 257, 272–274, 282.

Spillman, L. and Werner, J.S. (1990) *Visual Perception: the Neurophysiological Foundations*. San Diego, CA: Academic Press.

Stone, P.T. (1992) Fluorescent lighting and health. *Lighting Res. Technol.*, **24**, 55–61.

Tam, E.M., Lam, R.W. and Levitt, A.J. (1995) Treatment of seasonal affective disorder: a review. *Can. J. Psychiatr.*, **40**, 457–466.

Tregenza, P.R., Romaya, S.M., Dawe, S.P., Heap, L.J. and Tuck, B. (1974) Consistency and variation in preferences for office lighting. *Lighting Res. Technol.*, **6**, 205–211.

Tversky, A. and Kahneman, D. (1974) Judgment under uncertainty: heuristics and biases. *Science*, **185**, 1124–1131.

Veitch, J.A. and Gifford, R. (1996a) Assessing beliefs about lighting effects on health, performance, mood and social behavior. *Environ. and Behav.*, **28**, 446–470.

Veitch, J.A. and Gifford, R. (1996b) Choice, perceived control, and performance decrements in the physical environment. *J. Environ. Psychol.*, **16**, 269–276.

Veitch, J.A., Hine, D.W. and Gifford, R. (1993) End users' knowledge, preferences, and beliefs for lighting. *J. Interior Des.*, **19**(2), 15–26.

Veitch, J.A. and Kaye, S.M. (1988) Illumination effects on conversational sound levels and job candidate evaluation. *J. Environ. Psychol.*, **8**, 223–233.

Veitch, J.A. and McColl, S.L. (1995) On the modulation of fluorescent light: flicker rate and spectral distribution effects on visual performance and visual comfort. *Lighting Res. Technol.*, **27**, 243–256.

Veitch, J.A. and Newsham, G.R. (1996) Determinants of lighting quality II: Research and recommendations. *104th Annual Convention of the American Psychological Association*, Toronto, ERIC Document Reproduction Service no. ED408543.

Veitch, J.A. and Newsham, G.R. (1998a). Determinants of lighting quality I: state of the science. *J. Illuminating Eng. Soc.*, **27**(1), 92–106.

Veitch, J.A. and Newsham, G.R. (1998b) Lighting quality and energy-efficiency effects on task performance, mood, health, satisfaction and comfort. *J. Illuminating Eng. Soc.*, **27**(1), 107–129.

Venables, P.H. (1984) Arousal: An examination of its status as a concept. In M.G.H. Coles, J.R. Jennings and J.A. Stern (eds), *Psychophysiological Perspectives: Festschrift for Beatrice and John Lacey*, 134–142. New York: Van Nostrand Reinhold.

Wilkins, A.J. (1986) Intermittent illumination from visual display units and fluorescent lighting affects movements of the eyes across text. *Human Factors*, **28**, 75–81.

Wilkins, A.J., Nimmo-Smith, I., Slater, A. and Bedocs, L. (1989) Fluorescent lighting, headaches and eye-strain. *Lighting Res. Technol.*, **21**, 11–18.

Wineman, J.D. (1982). The office environment as a source of stress. In G.W. Evans (ed.), *Environmental Stress*, 256–285. New York: Cambridge University Press.

Woods, J.E. (1989) Cost avoidance and productivity in owning and operating buildings. *Occupational Med.*, **4**, 753–770.

Notes

1 Luminance is the quantity of luminous flux propagated in a given direction from a point on a surface. Colloquially, this is what is generally meant when we speak of the brightness of an object, although this use confuses the photometric quantity and the sensation of brightness, which depends on the state of adaptation of the eye as well as the luminance of the object (Rea, 1993).
2 Phototherapy using bright light exposure is a common treatment for seasonal affective disorder, but this use has no bearing on general lighting practice. Interested readers are directed to Rosenthal (1993) and Tam *et al.* (1995) for reviews.
3 Illuminance is the technical term for the area density of luminous flux incident on a surface; colloquially, we speak of 'light levels'. Lighting recommendations are specified in terms of illuminance largely because this value is in the control of the lighting specifier, although the visual system sees luminance, which is the product of illuminance and reflectance. Reflectances of walls, furnishings, floors and ceilings determine luminance but are usually not under the control of the person choosing the lighting.

Part 4

Concentration and thinking

Chapter 14

Attention and performance in the workplace

Roy Davis

Most of our activities in the workplace involve exercising skills of various kinds. Control of these activities is organised at different levels in the nervous system, from relatively low levels (such as keyboard skills) to relatively high levels (such as overall planning of a project).

Attention, the conscious awareness of the task in hand, may be focused at different levels. It is important for comfort and efficiency that it is focused at an appropriate level. For example:

- if one is preparing a computer spreadsheet one doesn't want to be distracted by a sticky key or a malfunctioning mouse
- if one is leading a discussion on company policy one doesn't want to be disturbed by an erratic air-conditioning system.

In this chapter I shall look briefly at the characteristics of skilled tasks from the point of view of their attentional demands, and give a few examples to show where better design could lead to improvement in both comfort and efficiency.

At the outset it is worth noting that it is not just in the workplace that we exercise our skills, but in all our everyday transactions with things and with people. As well as going to work we play games involving all kinds of mental and physical effort for **enjoyment**. The activity of playing the game is intrinsically satisfying. No external reward is necessary.

Sometimes the activities we carry out at work are intrinsically satisfying; often they are not. The work may be carried out solely for the reward or pay external to the task. Pay is clearly important for satisfaction at work, but there is no reason why we shouldn't attempt to make the skills we use at work as satisfying as those we use in play.

Hence my theme is not just a matter of making work comfortable, which was the intent of many of the old ergonomists, but that of making it more **enjoyable**.

The nature of skills

Skills involve coordinated goal-directed activities. The criterion of success is whether the goal is achieved, **not** whether a routine sequence of actions is accurately repeated (Bartlett, 1947).

The essence of a goal-directed system is that relevant information is taken in, a course of action is formulated, the action is carried out and, if the system is working as it should, it receives information back about the results of its action – the extent to which the goal has been achieved.

When we engage in skilled activities we bring into action all kinds of goal-directed systems at different levels. For example, the neuromuscular system that enables us to grasp a pencil without either letting it slip or crushing it depends on information circulating at the level of the neural control loops in muscles and joints, whereas the neurophysiological system that enables us to point the pencil accurately at a target depends on different kinds of information at a different level (the control loops involved in eye–hand coordination).

Higher-order cognitive decision-making systems require the input and analysis of other kinds of information. Their proper functioning may depend on the integrity of the lower-order systems, but the higher-order systems do not (usually) access the kind of information which the lower systems are processing.

Attentional control may be moved between levels, but for the most fluent performance it is best to direct attention at the highest level required by the task.

On the other hand, during training, attention may be directed at quite low levels of task organisation. For example:

- how to control the limbs when making a stroke at tennis/cricket
- how to control the fingers when developing keyboard skills.

The result of training is to make actions at lower levels almost **automatic**. One no longer has to think about them. Attention is left free to deal with organisation at a higher level.

In learning a skill we build up at each level an internal (neural) representation or model of the way we interact with the external situation. Provided there is no radical change in the situation, the model allows **prediction** or even **anticipation** of the next move, without collecting and analysing further information. So the system becomes quicker to react and generally more efficient. Attention is left free to deal with organisation at a higher level so that one can think in terms of **strategy**, rather than **moves** (Fitts and Posner, 1967).

When, however, something goes wrong, if the external situation changes, or one is disturbed by events, or otherwise stressed, attention is diverted back to the level where the inconsistency or conflict occurs. The internal representation may have to be modified, the model reprogrammed, all of which may disrupt the task in hand, giving rise to feelings of discomfort and dissatisfaction.

There is an old distinction (which in some languages is preserved by using different words) between:

- **knowing that** – propositional knowledge, which can be made **explicit**, and verbalised
- **knowing how** – procedural knowledge, which is often **implicit** and cannot be put into words.

Procedural knowledge is very characteristic of skilled activities. Most people who can ride a bicycle find it difficult, if not impossible, to describe how they do it. This is one of the reasons why the best performers are not necessarily the best instructors.

Not only may we be unable to describe the constituents of skilled activities which we carry out very competently, we may be **unaware** of them while we are carrying them out. For an experienced driver the actions involved in changing gear become almost automatic; they are carried out below the level of awareness. Only if something goes wrong is attention directed towards them. Furthermore, when the task in hand is being carried out smoothly, it may be disrupted if one tries to attend to what is going on at the level of constituent activities.

> The centipede was happy, quite,
> Until the toad in fun
> Said, 'Pray, which leg comes after which?'
> This raised his doubts to such a pitch
> He fell exhausted in the ditch,
> Not knowing how to run.

These characteristics of skilled tasks have material consequences for the design of the workplace:

- **for training the novice** – identifying and using appropriate information; making it easy to deal with, without having to direct attention towards it
- **for expert performance** – recognising that highly organised processing of information is going on below the level of awareness and that decisions/actions may be controlled from this level (indeed, the smooth execution of the task in hand depends on such organisation).

Now let me turn from talking about the characteristics of skilled activities to give a few examples of principles which can be applied in the workplace.

The use of information in training

It is important to recognise that the information or 'cues' used by an expert at the job are not necessarily those used spontaneously by a beginner. During training one must direct the attention of the novice towards the appropriate cues.

For example, in training keyboard skills the novice may rely on vision for locating the keys, whereas the expert relies on touch and proprioceptive information from the muscles and joints, leaving vision free for reading text and monitoring the progress of the task.

This has very general application. For example, in all kinds of social interaction 'cues' are generated, such as a slight change in facial expression, or in tone of voice, Unless attention is drawn to them a person may be unaware of producing such cues. A skilful participant will be sensitive to the cues he/she is producing and in noticing the cues generated by the other participant(s). Social skills of this kind are also susceptible to training.

Different ways of presenting information

The way in which we deal with information depends a great deal on the way it is presented to us. For a simple example, contrast the way in which the time of day may be presented by an analog clockface as compared with a digital display (see Fig. 14.1).

Both show the same time of day, but the analog display is usually read as 'twenty to eleven', whereas the digital display is usually read as 'ten-forty'. Whereas a digital display gives a clear read-out of the instantaneous value, an analog display may have an advantage for judging intervals, estimating rates of change, or reading off the time one has to one's next appointment. One **sees** that there are twenty minutes to elapse before 11 o'clock, rather than having to subtract 10.40 from a mental representation of 11.00 hours. If circumstances lead to errors being made, the type of error generated by misreading a digital display will be different from that generated by misreading an analog display. For example, one may confuse visually similar figures in a digital display, whereas one may mistake the hour hand for the minute hand in an analog display.

There is a long and tragic history of aviation accidents attributed to pilots misreading their height above ground level as a result of confusing the pointers in the 3-pointer altimeter, a design which was used for years in civil and military aircraft (Fitts, 1951; Green, 1983).

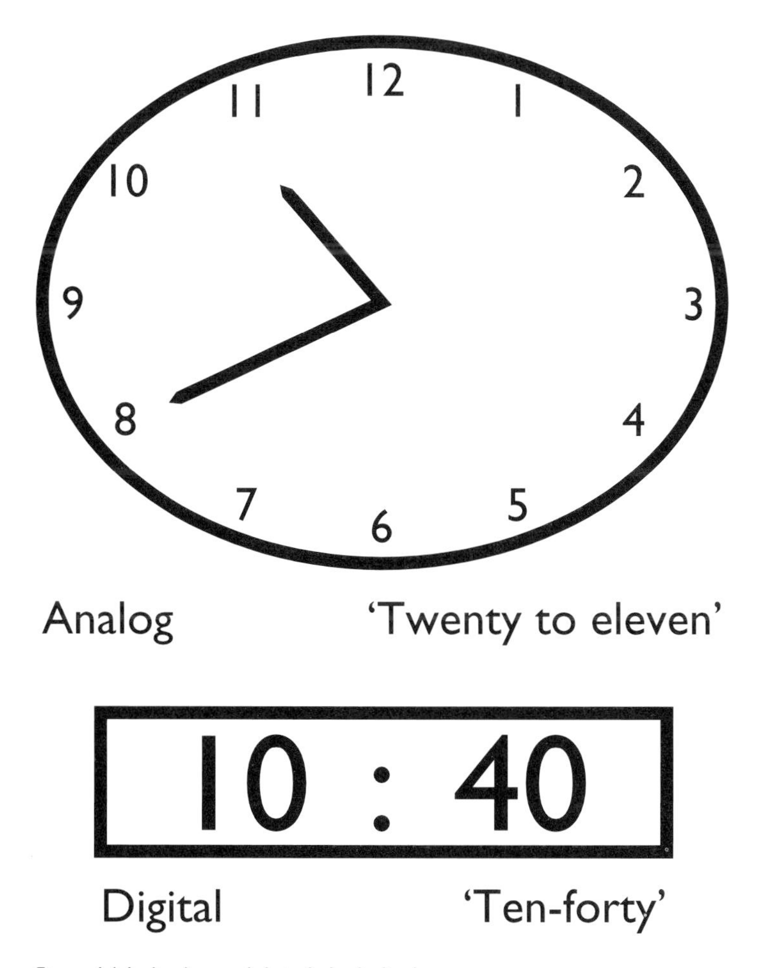

Figure 14.1 Analog and digital clock displays

Using visual symbols/icons to convey information or instructions

With the expansion of international trade and travel there is an increasing use of symbolic displays to provide information and to give instructions. A visual symbol may be used to supplement, or replace entirely, a written notice, which by its nature is language-specific. If the symbol is well designed and generally understood its use can save much time and trouble. If it is inappropriate it may be confusing, even hazardous.

Figure 14.2 shows traffic signs taken from the *Highway Code* (Department of Transport, 1996, 1999) which, presumably, are generally understood. Note the sign prohibiting the passage of bicycles (Fig. 14.2(iv)),

Figure 14.2 Traffic signs from the *Highway Code* (1996)

which is derived from the sign prohibiting the passage of all vehicles (Fig. 14.2(iii)).

Figure 14.3(i) and (ii) shows signs that are also in common use. Referring to the *Highway Code* one finds that (ii) is given as the sign prohibiting the passage of pedestrians (similar in style to the sign prohibiting the passage of bicycles), but one quite often sees the sign shown in (i) used for the same purposes! In fact, in earlier versions of the *Highway Code* (e.g. 1968 edition), it is this sign, with the diagonal bar, that is used to indicate 'no pedestrians'.

What is one to make of this change in use in the *Highway Code*? It is widely accepted that the diagonal bar indicates prohibition of the activity depicted, for example, in 'no smoking' signs, and in the current *Highway Code* signs for prohibiting left, right and U-turns (Fig. 14.2(ii)). The sign

Figure 14.3 Alternative signs prohibiting the passage of pedestrians

without the diagonal bar is interpreted differently by different people, and in different contexts. It creates confusion because the way it is used conflicts with the expectations held by many people. Has it contributed to any accidents? It is difficult to tell, because the kind of information it conveys tends to be processed automatically. People are not aware of its potential ambiguity unless their attention is drawn directly to it.

Interest in the relationship between signs and what they symbolise goes back to classical antiquity. Plato's dialogue *Cratylus* introduces a discussion on the principles of naming, and raises many issues still unresolved today. To what extent is there a 'natural' relationship between a symbol and what it represents? What is the basis for it? Does it depend on properties of the world common to everyone's experience, or is it specific to certain cultural, social and language groups (Brown, 1958)?

There is evidence that some relationships are found universally, whereas others are culturally dependent. The way in which arbitrary shapes are matched with 'nonsense' sounds is consistent over a wide range of culture and language groups. For example, there is good evidence for universal agreement that the sound produced by saying 'TAKETE' is a more appropriate match to the shape in Fig. 14.4(ii), than the sound of either 'MALUMA' or 'ULOOMU', which are thought to be more appropriate to the shape in Fig. 14.4(i) (Köhler, 1929; Davis, 1961). On the other hand, associations such as those between colour and mood may be culturally determined (Davis, 1995). Of course as cultural influences extend internationally, there is a tendency towards universal interpretations of signs and symbols which were originally culture-specific.

This is not the place to attempt to sort out these issues, but I will give one example to illustrate their application, in the design of visual displays for

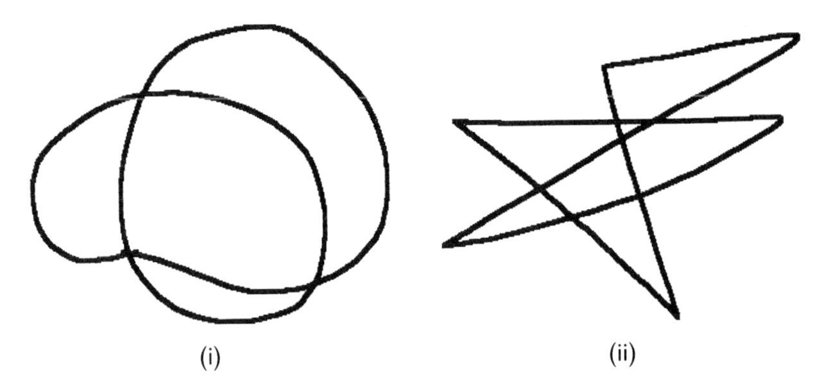

Figure 14.4 Shapes similar to those used by Köhler (1929) and Davis (1961)

aircraft pilots. 'Head-up' displays project information about the state of the aircraft (air-speed, altitude, compass bearing, etc.), so it is seen through the windscreen in front of the pilot, superimposed on the landscape in the background. For pilots of military aircraft, 'helmet-mounted displays' are being developed on similar principles (Stinnett, 1989). The display may also be used to provide information for the pilot about features in the surroundings which may not be seen, perhaps because of bad weather conditions. This may require a symbolic representation of some features, such as other aircraft in the vicinity, and it is important that these symbols are interpreted unambiguously and as quickly as possible.

Figure 14.5 shows two alternative displays. One is intended to represent a flight of aircraft identified as 'friendly', the other a flight not identified, or potentially hostile. If one asks almost anyone in the general population to choose between the two displays as to which best indicates 'friendly' and which indicates 'hostile', there will be almost universal agreement. One may argue about the reason why, but one neglects these relationships at one's peril.

Stimulus–response compatibility

Our consideration of skilled activities has shown that we build up expectations about the way in which the artefacts we construct relate to things in the external world, and I have given examples of the kind of expectations people have when interpreting visual signs and symbols.

Another kind of expectancy may be described in terms of **stimulus–response compatibility**. When one is required to select a response on the basis of information presented in some kind of display, the way in which the information is presented may suggest that one kind of response is more appropriate than another.

Figure 14.6(i) is an example of high stimulus–response compatibility,

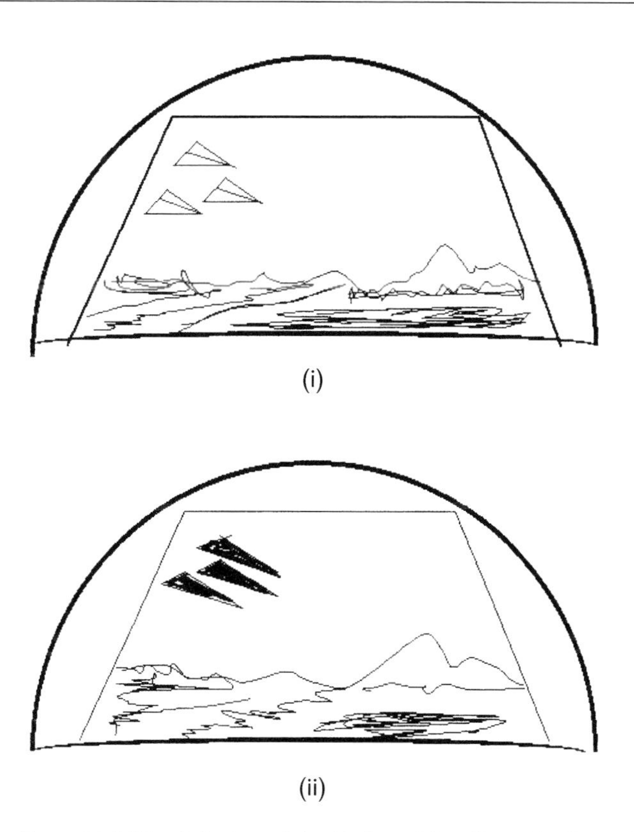

Figure 14.5 Visual displays of the kind used for pilots of military aircraft

simple spatial correspondence The button to be pressed is immediately opposite the signal light (although it could have been designed otherwise). In Fig. 14.6(ii) the relationship between display and control is more problematic; it rather depends on the assumptions about what it is one is controlling. In Fig. 14.6(iii) the most compatible relationship between the numerical display and the response panel seems obvious, but perhaps only for those who are accustomed to using numerals in the context of a general habit of reading text from left to right? In Fig. 14.6(iv) one would generally move a control to the right to move a display value upwards (and both would be described as an 'increase') but this is not always the case.

The controls of gas and electricity cookers show an interesting variation. To increase the heat from an electric cooking ring the control knob must be rotated clockwise, whereas to increase the heat from a gas ring the control knob must be rotated anticlockwise. Although many people have carried out these actions on both types of cooker many times in their lives, their knowledge about which way the knobs should be turned may remain totally implicit, and when questioned they may be unable to say (Oborne, 1987).

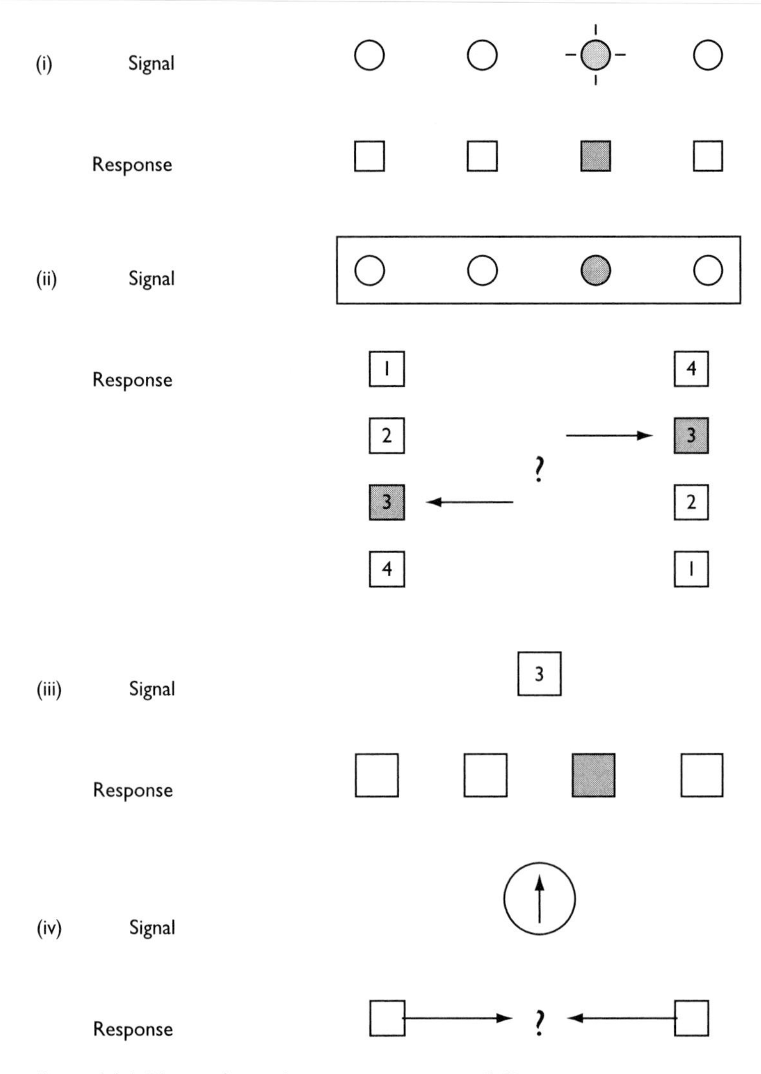

Figure 14.6 Types of stimulus–response compatibility

Natural objects often show '**affordances**'. By their very look they invite certain actions (Gibson, 1979). Artefacts can be designed to show such affordances. A door handle can visually suggest 'push me' or 'pull me'. When the design fits the function all goes well, but all too often it does not.

Figure 14.7 is a diagram of a passenger exit door on the Barcelona Metro, Line 1. This is a very pleasant line to travel on, but the exit doors present a problem. They open automatically once the door latch is released, but to do this the passenger has to push the latch in the direction shown as 'a'. This is

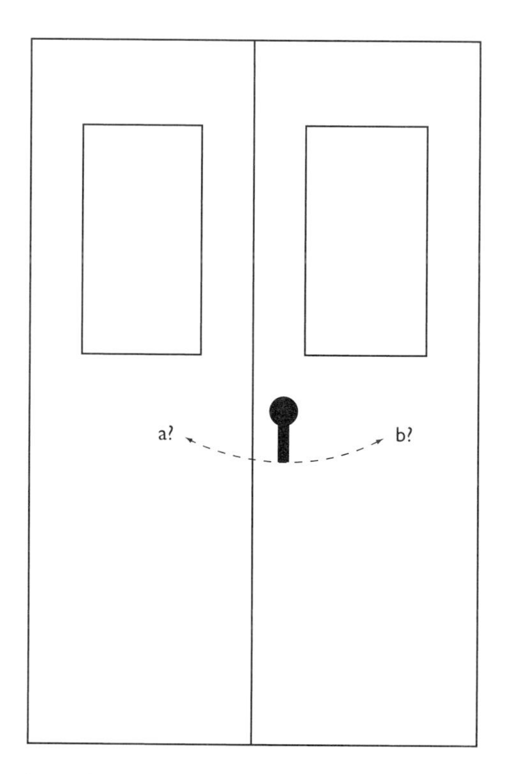

Figure 14.7 Door latch, Barcelona Metro, Line 1

contrary to expectation, and against the movement of the door as it starts to open. It is quite common to see visitors to the city struggling with the door latch and sometimes failing to get out at their chosen station!

A similar problem arose with the latch used to lock the toilet door on some British Rail InterCity trains a few years ago. Figure 14.8(i) shows, from inside the door, the latch in the 'locked' position. It became obvious that because of the way the latch was fitted, many passengers regarded this as the 'open' position and confusion arose when trying to get out. The design was subsequently modified by engraving an arrow on the latch handle to show which way to move it (Fig. 14.8(ii)). But perhaps it could have been solved by fitting the latch the other way round in the first place (Fig. 14.8(iii))?

I use these examples to illustrate a simple point. If one gets the design right, if it fits in with people's expectations, the action at this level is carried out 'automatically' without having to divert attention. If expectations are contravened, attention has to be redirected, and one's whole pattern of thought and behaviour may be disturbed. The point may be simple, but how often is it ignored in practice?!

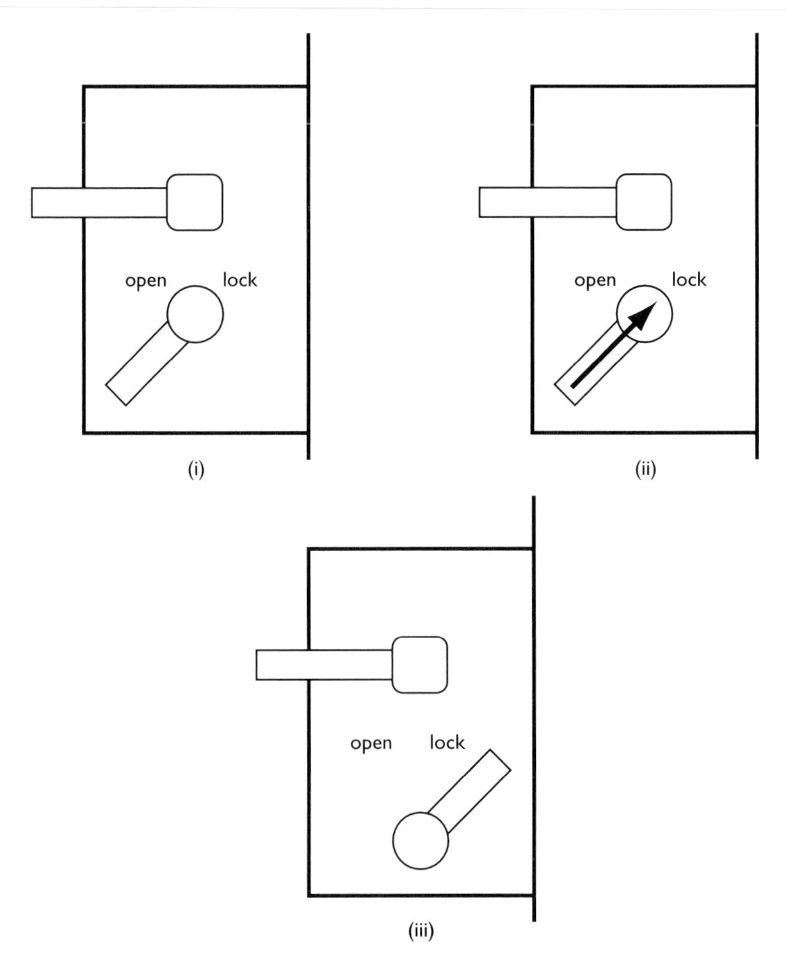

Figure 14.8 Door latch, British Rail InterCity toilets

Before leaving this theme may I give one more example of recent folly, in the design of numeric keypads.

Figure 14.9(i) shows the arrangement used for telephones. Figure 14.9(ii) shows the arrangement used for calculators and computer keypads. Why should they be different? Many people reveal on casual questioning that they are unaware of the difference. Does it matter? It may! Tapping in well-rehearsed codes tends to become an almost automatic procedure. If one transfers the habit of entering a code from one keyboard to another there is a potential source of error. If one has to divert attention to observe precisely which keyboard layout is in front of one, it slows one down and may be distracting. It is a further hazard in trying to use the computer for telephone dialling.

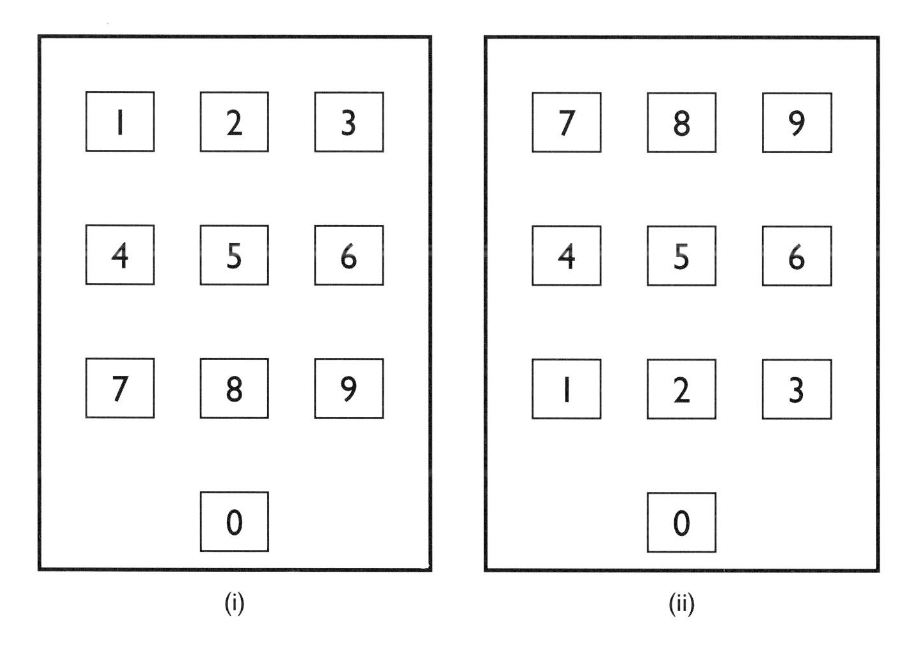

Figure 14.9 Numeric keypad layouts for telephone, calculator/computer, cash-point

When it comes to keypads at bank cash-points, the situation becomes even more confusing. In Britain the telephone layout seems to have gained dominance for cash-points, but in continental Europe one can find either arrangement in different banks in the same street. I have even seen one arrangement used by a particular bank for a publicity display in their window and the other arrangement used for its own cash-point at the entrance to the bank.

Relating psychological variables to physical variables

We have seen how, when using **one** device to control **one** variable in the physical world, unfortunate design can cause difficulties for the user. When several variables have to be controlled it can become even more confusing. I take an example from Donald Norman (1990) of a situation which at one time or another confronts all travellers – that of using the hotel shower.

What you want to control ('the psychological variables') are:

- the temperature of the water (how hot?)
- the flow (how much?)

The way the system usually works is that 'hot' water is produced at a fixed temperature and the flow of 'hot water' is controlled by one tap. The flow of 'cold water' is controlled by another tap, and one has to regulate the temperature of the shower by mixing the flows from the 'hot' and the 'cold' taps.

So the 'physical variables' to be controlled are:

(a) the flow of water in the 'hot' tap
(b) the flow of water in the 'cold' tap.

The temperature and the flow of water interact, so that in order to change one of them one inevitably has to change the other, with consequent problems of adjustment, aggravated by the delay between the user's action and the system's response. (This delay is something which always has to be taken into account in the design of control and communication systems, whether physical or human, or a mixture of both (Wickens, 1992).)

There are, of course, various designs of 'mixer' taps, some with elaborate multi-function controls, but many of them seem to aggravate the problem, leaving the unaccustomed user even more baffled.

One recommendation we can derive from this example is that the mapping of physical variables onto psychological variables should be as straightforward as possible. People find it very difficult to cope with interaction of this kind between the variables they are trying to control. In Donald Norman's terms the system should be as **transparent** as possible to the user (Norman and Draper, 1986).

The controls should feel good. They should inspire confidence. The level of control should be appropriate. The system should be responsive, and if the **quality of control** is good, people will enjoy using it.

How to take account of the user?

Most designers nowadays are well aware of the importance of taking account of the user, both during the design process and after the user's experience of the finished product. However, a word of caution may be needed!

One of the commonest ways of taking account of the views of potential users is by administering questionnaires, perhaps supplemented by interviews, open-ended or structured, with varying degrees of constraint. In most cases the information from the user is in the form of verbally expressed responses (whether spoken or written). This information is very valuable, but it may not be enough.

I will pass over the fact that in an interview, or when answering questionnaires, people may not say what they mean, for all kinds of reasons, which are well documented. More significant for my theme today is the fact that much of the information which people have stored inside them, which they use in work and play and in all everyday activities, is not accessible to verbal description. It is implicit procedural knowledge, not explicit propositional knowledge.

Hence it becomes very important to watch what people do, and analyse their behaviour, as well as to listen to what they say.

I think it was John Christopher Jones (1992) who said that design is the process of making external what is normally internal, or, if I may paraphrase, making explicit the implicit. People may not be able to say, perhaps not even realise what they would like, but they know a good thing when they see one, or rather after they have experienced one.

So what a real designer has to do is dig out from the users their implicit knowledge, and put it to use, not just as it is, but extrapolate from it, to anticipate what the users are going to need, going to like, in the future. This requires a combination of skills and insight which not many people possess, but it is a goal we can set ourselves, and train our students to accomplish.

References

Bartlett, F.C. (1947) The measurement of human skill, *Br. Med. J.*, **1**, 835–838, 877–880.

Brown, R. (1958) *Words and Things*. Glencoe, IL: The Free Press.

Davis, R. (1961) The fitness of names to drawings, a cross-cultural study in Tanganyika. *Br. J. Psychol.*, **52**, 3, 259–268.

Davis, R. (1995) The basis for intersensory relations. Cognitive or emotional? *Proc. 8th Congress, European Society for Cognitive Psychology*, Rome.

Department of Transport (1996) *The Highway Code*. London: HMSO.

Fitts, P.M. (1951) Engineering psychology, in *Handbook of Experimental Psychology* (ed. S.S. Stevens). New York: Wiley.

Fitts, P.M. and Jones, R.E. (1961) Psychological aspects of instrument display. Analysis of 270 'pilot-error' experiences in reading and interpreting aircraft instruments, in *Selected Papers on the Design and Use of Control Systems* (ed. H.W. Sinaiko). New York: Dover.

Fitts, P.M. and Posner, M.L. (1967) *Human Performance*. Belmont, CA: Brooks/Cole.

Gibson, J.J. (1979) *The Ecological Approach to Visual Perception*. Boston, MA: Houghton Mifflin.

Green, R.G. (1983) Aviation psychology. *Br. Med. J.*, **286**, 1880–1882.

Jones, J.C. (1992) *Design Methods*, 2nd edn. New York: Van Nostrand Reinhold.

Köhler, W. (1929) *Gestalt Psychology*. New York: Liveright.

Norman, D.A. (1990) *The Design of Everyday Things*. New York: Doubleday.

Norman, D.A. and Draper, S.W. (1986) *User Centred System Design*, Hillsdale, N.J.: Lawrence Erlbaum.

Oborne, D.J. (1987) *Ergonomics at Work*, 2nd edn. Chichester: Wiley.

Stinnett, T.A. (1989) Human factors in the super cockpit, in *Aviation Psychology* (ed. R.S. Jensen). Aldershot: Gower Technical.

Wickens, C.D. (1992) *Engineering Psychology and Human Performance*, 2nd edn. London: Harper Collins.

Chapter 15

Concentration and attention: new directions in theory and assessment

David A. Schwartz and Stephen Kaplan

Introduction

The human being is a highly adaptable animal. It can survive and even accomplish things under a wide range of circumstances. While this adaptability is an admirable asset it can at times be a serious liability. The reason for this is simple. People who adapt to an unfriendly environment – for example a poor work situation – may appear not to be handicapped by the inadequacy of the setting. In general this is an illusion: there are costs, but they are frequently hidden. The costs of an unfriendly work environment, for example, might appear at home rather than on the job. Or they may be reflected in work productivity, but not recognised as related to problems in the setting. Employee characteristics such as irritability and distractibility might be attributed to the person rather than to the setting. To the extent that costs are hidden and unrecognised, careful analysis will not be carried out and corrective action will not be taken.

Given these problems, the main purposes of this chapter are: (1) to make visible these often hidden consequences – both conceptually and as they express themselves in the workplace; (2) to describe possible interventions to eliminate these unfortunate consequences; and (3) to examine some ways of assessing the effectiveness of these interventions. Our intention is to introduce the conceptual underpinning of this approach in as intuitive a way as possible. For a more rigorous treatment of the conceptual issues and a description of the considerable empirical work that supports this approach, see Cimprich (1993) and Kaplan (1983, 1995).

What constitutes mental effort?

A useful way to think about the often hidden costs of an unfriendly environment is in terms of mental effort. Mental effort, in turn, is importantly related to the difficulty of what we are trying to do. At every moment of our waking lives we find ourselves attending to or thinking **something**, but not all forms of thought are equally taxing. Compare, for example, the experience of daydreaming with the experience of studying train schedules. They differ certainly with respect to content, but also with respect to how much mental effort we need to expend in the process. Another example: compare the experience of sitting in a cafe watching the parade of passers-by with the experience of trying to remember where you left your keys when you emptied your pockets last night. Again, the two experiences differ with respect to content, but also in terms of the amount of mental effort each demands. If we were to consider a large number of such comparisons, we would discover that two factors seem to distinguish those situations that demand mental effort from those during which our thoughts seem to unfold effortlessly.

Goals and obstacles

First, we tend only to exercise mental effort in the service of some goal. The nineteenth-century psychologist William James wrote that 'We never make an effort to attend to an object except for the sake of some remote interests which the effort will serve' (1892, p. 88). When we have no particular goal or purpose in mind, we let the outer world's impressions and the shifting contents of memory and imagination carry us where they will. Once we undertake to achieve some desirable future state, however – that is, once we set ourselves some goal – certain phenomena in the world or bits of information in our minds take on greater salience than do others. And because of their greater relevance for the accomplishment of our aims, these salient objects and ideas attract our attention and we dwell on them more than we might otherwise, plucking them, as it were, from the stream of ongoing thought.

A goal, therefore, seems a necessary condition for mental effort. It is not, however, a sufficient condition. Goal-directed thought, like goal-directed action, is effortful only when obstacles exist to its completion. Examples of physical obstacles, such as impassable mountains and unfordable rivers, come readily to mind, but obstacles may also be psychological in nature. Suppose, for example, a man hiking down a trail comes to a point at which the path branches into two forks. No physical barrier blocks his path, but none the less he finds that he can't proceed. Why? Because he doesn't know which fork to take. Uncertainty, therefore, is one kind of psychological obstacle.

While the lack of information can paralyse action, a surfeit of information can prove equally disruptive. This latter kind of psychological obstacle occurs whenever one confronts multiple sources of information, whether in the world or in one's memory, all of which impinge on awareness more or less simultaneously. In our lives we face this kind of psychological obstacle more or less all the time, for at any given moment there exist numerous sights, sounds, smells and so on to which we could attend, as well as an ample storehouse of knowledge we could draw upon. Only a small subset of the information available to us, however, is likely to be relevant to any particular goal we might undertake. It's useful, for example, to draw upon one's knowledge of aerodynamics when attempting to sail a boat, but not when choosing a wine to drink with dinner. It is useful to focus on a street sign when finding one's way in an unfamiliar neighbourhood, but not when attempting to cross the street safely. Thus, effective functioning requires the ability to select, both from our knowledge and from the world, the information most relevant to our goal, and to stave off the interference from the rest. Let us, therefore, define concentration as this **process of selecting from among multiple sources of information those most relevant to some goal, and of managing the interference from non-selected sources.**

Note that while we are often aware of ourselves selecting one type of information and rejecting another, nothing in the proposed definition requires consciousness of the selection process. When our powers of concentration are operating effectively, in fact, we may be totally unaware that we are constantly rejecting information irrelevant to our momentary goals. When our powers of concentration begin to flag, however, we may find it increasingly difficult to ignore irrelevant information, and so increasingly may notice things that we previously had unconsciously suppressed. A concrete example may serve to illustrate the point. A colleague sometimes leaves his office door open when he works at his desk. The desk faces the door, and the door opens on to a long hall that accommodates considerable pedestrian traffic over the course of the day. He has found that in the morning, while working on some task, he hardly notices the steady passage of people in the hall outside. By the end of the afternoon, however, his head reflexively jerks up each time a person passes by, and he finds it necessary to close the door in order to sustain attention to the work at hand. An undergraduate student recently described to us an experience that may be another example of a decline over time in the ability to screen out interference. He typically studies during the evening in his dormitory room, and finds that the music emanating from the adjacent room seems to grow steadily louder as the evening progresses.

Mental fatigue

The examples above suggest, and experimental studies confirm, that certain limits exist on our capacity to concentrate, such that after periods of sustained mental effort we find it more difficult to focus our attention or to manage interference in pursuit of some goal. The reader no doubt has experienced the phenomenon of 'mental fatigue' in his or her own life, and so understands intuitively both the feelings and the consequences associated with this state of mind. The fatigued state is, in fact, so familiar that most people probably accept it simply as a fact of life without questioning why the human mind should be susceptible to this sort of cognitive impairment. And yet, why should the mind tire? If, as we have argued, achieving goals often requires the ability to select from among multiple sources of information and to manage interference, why should evolution not have equipped us with the capacity to concentrate without wearying for hours, for days, for however long it might take to achieve a given goal? Nobody knows for certain, but comparisons with other species may offer some clues.

Anyone who has observed trained sheep dogs herding a flock, for example, can attest to the intensity of concentration and stamina these animals bring to their work. Hundreds of generations of selective breeding have produced a dog that will perform physically and intellectually demanding work for hour after hour without tiring. One handler has written that male border collies working stock will neglect to eat and will even ignore females in heat. Much as an employer or manager might admire this kind of workaholism on the part of an employee, the process of natural selection clearly would not favour the ability to concentrate tirelessly on tasks that serve no intrinsic survival or reproductive purpose. Perhaps in the future corporations will attempt to engineer fatigue-resistant employees just as the British highlanders have engineered the traits of the border collie. In the meantime, we shall have to design our workplaces around the limitations of a creature that is easily distracted not only by food and sex, but by many other types of interference as well.

Mental effort and mental fatigue in the workplace

On the job, the sources of interference that tax our powers of concentration can be either intrinsic to a given task or extraneous. Intrinsically demanding tasks are those that, for example, require us to consider several ideas at once, to generate and evaluate alternatives, to analyse complex data sets, to plan and organise, to bring order to a confusing mass of information, to coordinate the work of several people, and so on. What all these tasks have in common is that they present a worker with a problem that often has no obvious or routine solution, a problem which by virtue of its complexity defies the application of any simple algorithm. When faced with such a situation, one must explore the problem space, conceive and compare alternatives, take account of various constraints, select a particular interpretation or course of action, and proceed to the next stage of the process. In a word, intrinsically demanding tasks require one to **think**, and thinking, as we all know, is hard work.

Is the banishment of effort a useful strategy?

Suppose, now, a consultant proposes that your company could increase productivity by decomposing all intellectually demanding tasks into a sequence of simple if–then rules that direct an employee's thought along prescribed pathways, thus freeing the worker from having to exercise any judgement. Should you embrace this proposal? Probably not. First, one usually cannot anticipate all possible situations that may arise on the job for which some action may be needed. Thus, there is no getting away from the need for considered judgement. And even were it possible to engineer tasks in a way that allowed for their mindless performance, we ought not to do so, because when we remove opportunities for exploration, for creativity and experimentation, and for wrestling with uncertainty, we undermine the very conditions that nurture excellence on the job. Moreover, we end up removing those features that make intellectually demanding work so satisfying and which thus attract and retain talented people to our organisations. In their perceptive analysis of outstanding companies, Peters and Waterman (1982) emphasised the importance of what they called 'productivity through people'. In successful companies, Peters and Waterman observed, managers challenged employees to take on increasing responsibility in the workplace and to explore possible innovations that might improve the company's performance. Thus, far from freeing the employee from the challenges of thinking and decision, there are strong arguments for moving in the opposite direction. Let us, therefore, take the intrinsic demands on our powers of concentration as a given.

Sources of job interference

Extraneous sources of interference, however, are another matter entirely. Extraneous sources of interference include such things as random noise, interruptions, lack of privacy, poor lighting, that we can well do without. If we assume that both the extraneous and the intrinsic sources of interference tax the same limited mental capacity, then the more mental effort one expends managing extraneous interference, the less mental effort one can devote to the intrinsic demands of the task at hand. Thus, the more inhospitable the work environment, the more quickly will people's effectiveness and productivity decline.

Consequences of fatigue

This decline can manifest itself in many different ways. For example, laboratory experiments have found that when people become mentally fatigued they behave more impulsively and exhibit greater distractibility. On the job, such lapses of attention and of impulse control can undermine employees' effectiveness, with severe, even lethal consequences. Supporting this concern is a study of the relationship of mental lapses to auto accidents (Larson and Merritt, 1991). The authors examined the driving records of individuals who reported experiencing frequent absent-mindedness and found that these people had caused a significantly higher level of accidents. Translating this example to a work setting, one needs only to consider some of the occupations for which lapses in attention could be particularly costly. Research by Yoshitake (1978) provides thought-provoking examples. He found that people in certain occupations were particularly likely to experience 'difficulty of concentration'. The fatigue-prone employees included 'helicopter pilots, air-traffic controllers, key punchers, bank clerks, broadcasting personnel, incinerator-plant operators, pharmacists and research workers' (p. 232).

Fatigued people also tend to think and act in stereotyped, habitual, or routine ways. They fail to notice subtle cues that make one course of action more appropriate than another, attending instead to what is most obvious and performing in ways that are most familiar. In their analysis of the central role of innovation in effective organisations, Peters and Austin (1985) emphasise the importance of going round the formal structure and the usual ways of doing things. Clearly this would be difficult for individuals suffering from fatigue.

Finally, fatigue induces impatience and irritability, which can increase feelings of tension and hostility among coworkers and hamper progress on collaborative projects. A fatigued workforce thus is likely to experience more on-the-job accidents, commit more careless errors, conceive and/or implement fewer innovations, and experience more interpersonal tensions than will a less fatigued workforce. All of this, of course, translates into lost earnings, lost opportunities, and probably lost personnel. This would seem an ample incentive for employers to identify ways to nurture employees' cognitive functioning on the job.

Strategies of intervention

There are two categories of intervention for keeping mental fatigue levels under control. The first involves the elimination or mitigation of sources of extraneous interference. There are a wide range of possible interventions in this category. They can be as simple as installing sound-dampening materials to reduce noise levels or as complicated as improving lines of communication to reduce worker uncertainty. The second category involves interventions that actually reduce fatigue. Since these events restore the individual to a pre-fatigued level of effectiveness, they are referred to as **restorative**. In the context of the workplace, a particularly helpful restorative intervention has been to provide a window with a view that includes natural elements (Kaplan, 1993). A modest literature is available that provides both conceptual and empirical guidance that could prove helpful in understanding, identifying, and implementing appropriate interventions (Kaplan, 1995).

Assessing the effect of an intervention

Whatever changes one decides to introduce, some method for assessing whether the intervention has achieved its desired effect is essential. Ultimately, the goal is to determine whether the workplace changes have improved your organisation's productivity. 'Productivity', however, means different things to different types of organisation. A manufacturing concern will assess it differently to an educational or social service institution. In addition, concentration ability is only one of several factors that influences how well the members of an organisation perform their jobs. Motivation and knowledge, for example, also matter a great deal. Thus, we will confine ourselves in this chapter to a discussion of tasks psychologists have devised to assess basic cognitive processes, leaving the measurement of productivity to those more familiar with the particular kind of work a given organisation performs.

Although the tasks we describe below may seem dull and unrelated to the type of work employees do on the job, we believe they are useful tools for assessing changes in a person's ability to concentrate; that is, to select goal-relevant information and to manage interference posed by irrelevant information in memory or in the task environment. As we have seen, this abstract ability is a prerequisite for successful performance across a wide range of activities and settings.

The tasks

By way of summary, in our discussion thus far we have proposed that concentration is the process of suppressing cognitive interference ('mental noise', if you will) in the service of some goal. We have further suggested that sources of interference that tax our limited powers of concentration can be either intrinsic or extraneous to one's task, but that both types of interference draw upon the same limited cognitive capacity. We infer, therefore, that reducing the amount of extraneous interference workers confront on the job, or providing opportunities for mental restoration, should improve their ability to suppress intrinsic interference, enabling them to work better for longer periods of time. Each of the following tasks serves as a measure of concentration because successful performance requires coping with some form of intrinsic interference.

Hidden figures test (Witkin et al., *1971)*

Each item of this timed paper-and-pencil task consists of a complex geometric figure within which a simpler geometric pattern is embedded (Fig. 15.1). Some versions of the task specify which of several simple target patterns is embedded in each complex figure, while others do not specify.

This is a test of your ability to tell which one of five simple figures can be found in a more complex pattern. At the top of each page in this test are five simple figures lettered A, B, C, D, and E. Beneath each row of figures is a page of patterns. Each pattern has a row of letters beneath it. Indicate your answer by putting an X through the letter of the figure which you find in the pattern.

NOTE: There is only one of these figures in each pattern, and this figure will always be right side up and exactly the same size as one of the five lettered figures.

Now try these 2 examples.

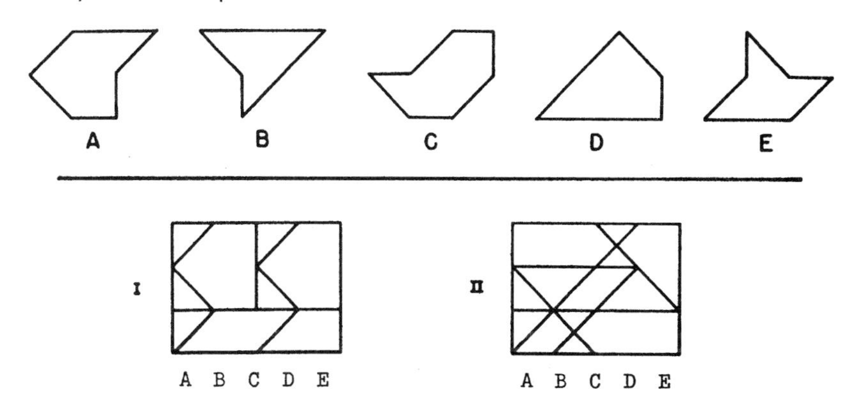

The figures below show how the figures are included in the problems. Figure A is in the first problem and figure D in the second.

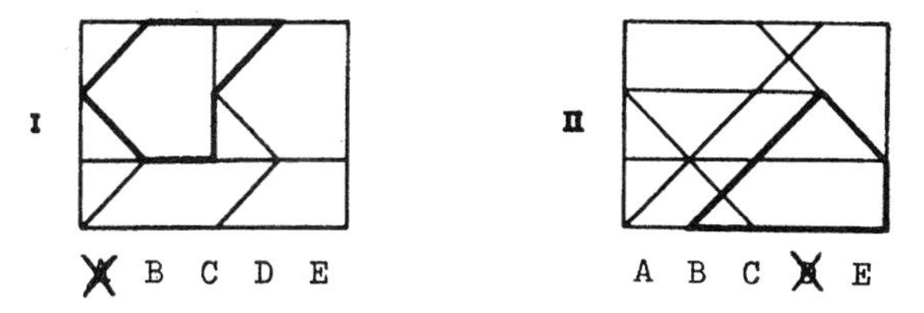

Your score on this test will be the number marked correctly minus a fraction of the number marked incorrectly. Therefore, it will **not** be to your advantage to guess unless you are able to eliminate one or more of the answer choices as wrong.

You will have **10 minutes** for each of the two parts of this test. Each part has 2 pages. When you have finished Part 1, **stop**. Please do not go on to Part 2 until you are asked to do so.

DO NOT TURN THIS PAGE UNTIL ASKED TO DO SO.

Figure 15.1 Hidden figure task (© 1962 Educational Testing Service)

The participant's task is to locate and trace the outline of the simple pattern embedded in each complex figure. The participant's score is the number of items completed correctly within the allotted time.

Psychologists have traditionally used this task to study supposedly stable individual differences in a cognitive factor, known variously as 'field-dependence', 'flexibility of closure', and, most transparently, 'perceptual disembedding skill'. While some researchers have interpreted performance on the task as a measure of intelligence or cognitive style, and have attempted to correlate performance on the task with other psychological variables, it will suffice for our purposes to articulate those aspects of the task that render it a suitable measure of concentration ability. On each item of the task, a person must search for a target object within a visually noisy environment. More technically, in order to locate the target pattern within the complex figure, a person must impose a figure-ground segregation upon a scene that does not lend itself easily to such parsing. Psychologically, the task thus calls upon many of the same processes as does trying to discern the contours of a well-camouflaged animal. Concentration comes into play because the person studies each complex figure with a goal in mind (i.e. the location of the simple target figure) and because achieving the goal requires overcoming the perceptual interference created by the context in which the target is embedded.

Letter cancellation (Diller et al., *1974)*

This task was designed to assess an individual's capacity to sustain vigilance, but it can also be conceptualised as a measure of one's ability to pursue a goal in the face of interference intrinsic to the task. The task consists of page containing rows of letters, and individuals are instructed to locate and mark (e.g. by circling or striking through) every instance of a particular character (Fig. 15.2). The individual's score is the number of targets correctly marked within the allotted time. In our own research we separately record the number of errors of omission and of commission.

The task can be made difficult in several ways. First, one can select a target letter that has few distinctive features, that is, a character that closely resembles one or more other characters. In Fig. 15.2, for example, the target letter *i* closely resembles the letter *j*. In fact the physical form of *i* is wholly subsumed within the form of *j*, thus making discrimination difficult. The task can be made still more difficult by complicating the instructions, such as by making the target a conjunction of two conditions. For example, we might instruct participants to strike through only those instances of the letter *i* that precede a vowel. One can vary the task in many other ways as well. The letter cancellation task, like the hidden figures task, requires an individual to search for a target within a visually noisy context, though in this case the target and its context consist of recognisable symbols rather than arbitrary geometric shapes.

i d f q j q i g i b i j q i f a m j c j y i r v e i t q d i c t i g w c g l o m o j s j e j k t

d k e e v j p q w o v w e g t p o i n l v z t p b y g l i t g r f t c s n o j c q g b m r

p n f m t d j i h d w v s c l i e r h s i c e k a p n i a t z e g f n j j i e z s v j i p c s l

h i z s j i l v c g l j k j c s m a n j i t n j q e i m v t o z h g i a l t c p i d q j i v t h o

g o c j j j q p y v i v s j j n y o p t o j i j g b k t c v l b t o h j o g b t s i v e h i z i

i m t g j c n i l i g v l z p j l e u n g w r b i n t c j p t g n b h q d y e s l i t m r v b

e i t q d i c t i e w c g l o m o j s j e j k t v z t p b y g l i t g r f t c s i o j c q g b m

a p n t t z e g f n j t e z s v j d p c o g w j q e i m v t o z h g i a l t c p i u q j i v t h

r b i n t c j p w c g i o m o j s j e j k t v z t p b y g l i t g i s v r f t c s n o j c q n o

g l j k j c s m a n j i t n i q e i m v t o z h g d k e e v j p q w o v w e g t p o i v l b

v z t p b y g l i t g r f t c s n o j c q g b m y i d f q j q i g i b i j q i f a m j c j y i r j

d j c h d w v s c l b r h s i w r k a p n t e i e g j c n i o i g v l z p j l e u n g w r a d

c g l o m o j s j e j k i v z t p b y g e i t g w c g l o m o j s j e j k t v z t p b y g l i d

n j i t n j q e i m v t o z h g d k e e v j p q w o h i z s j i l v c g l j k j c s m a n j i s

b r h s i w r k a p n t t z e g j c n i e i c v l a p n t t i e g f n j t e z s v j d i k a p n t

v s h i u n y o p i o n p s g b k t c v l b t o h j o n j i t n j q e i m v t o z h i d k e e

e j k t v z i p b y g p i q g w c g l o m o j s i l i t g r f t c s n o j c q g b m i w v s c

b r h s i d k e e v j p q w o v i e g t p o i v l b h j g l o m o j s j e j k t v z t p b l i l

d w v s c l b r h s i w r k a p n t t z e g f h i a s j i l v c g l j k j i e s m a n j n r o i

w c g l o m o j s j e i k t v z t p b y g l i a g r m v t o z h g i a l t c p i d q j i e t p b

Figure 15.2 Letter cancellation task

Category-matching task (Purdum and Schwartz, 1995)

This task, adapted from Melnyk and Das (1992), was developed to assess an individual's ability to select one kind of information over another. Each page of the task contains pictures from five different categories: boats, fish, balls, mammals, and fruit (Fig. 15.3). These pictures are arranged on each page in thirty-six pairs, with four pairs on each of nine lines (Fig. 15.4). In about 20

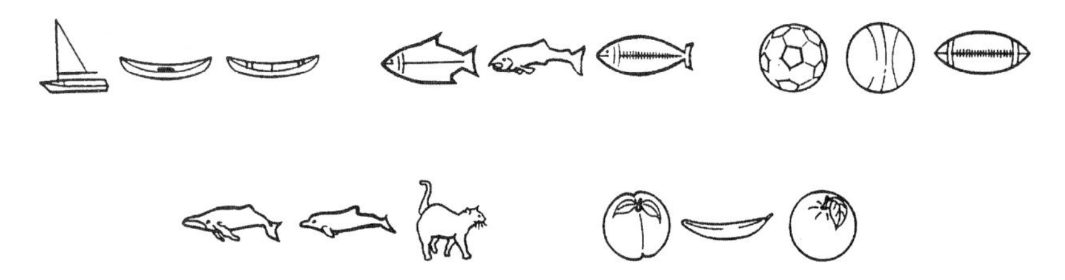

Figure 15.3 Category match task items grouped by category (© 1995 Gary E. Purdum and David A. Schwartz)

per cent of the pairs, both pictures are from the same category; these are the **target** pairs. The remaining pairs, in which the two pictures are from different categories, are the **distractor** pairs. Participants are instructed to draw a circle around each target pair, and only around the target pairs. They are further instructed to work as quickly and as accurately as they can, from left to right across the page beginning with the top row. When administering the task to groups of people, we typically allot the participants 60 seconds to complete as much of the task as they can. When administering to an individual, one has the option of allowing the person to complete all three pages of the task and measuring how long it takes the person to complete the task. Whichever timing method one chooses, the number of correctly circled pairs serves as one score. In addition, each participant receives a score for each of two kinds of errors: errors of **omission**, in which a participant fails to circle a target pair, and errors of **commission**, in which a participant circles a distractor pair.

Note that one member of each category appears physically more similar to members of another category than to members of its own category. For example, the kayak looks more like the banana than it does the sailboat, and the dolphin looks more like the fish than it does the cat. We know from experimental research that people judge physical similarity more quickly and easily than they do categorical similarity. A person completing the task thus will be psychologically predisposed to commit both errors of omission (by judging the dolphin and cat as dissimilar) and errors of commission (by judging the banana and kayak as similar). Close concentration is required to resist these error tendencies.

Melnyk and Das (1992) found that performance on their version of this task discriminated between adolescent students classified as either good or poor attenders on the basis of teacher ratings. They concluded that performance on the exercise functions as a measure of individuals' capacity to inhibit responses to distractors, a capacity roughly synonymous with the ability to concentrate. In our own research (Schwartz, 1994) we found that performance on the task modestly predicted teenagers' self-reports of focus, persistence, self-control, and effectiveness, all aspects of functioning in which we

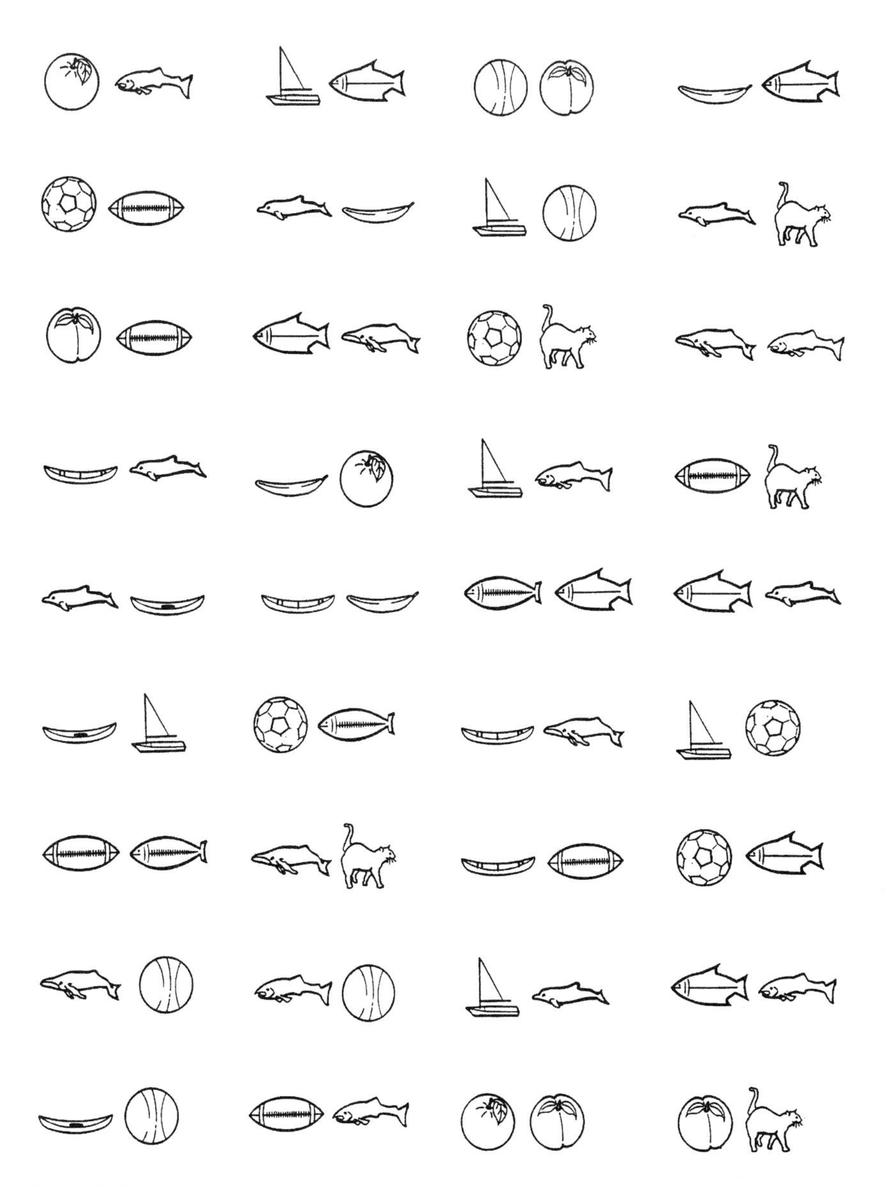

Figure 15.4 Sample page of category match task

believe concentration plays an important role. Furthermore, in a study of concentration ability among University of Michigan undergraduates (Schwartz, 1995), we found that performance on this task correlated negatively with self-reported forgetfulness and mental fatigue.

The importance of multiple measures

There are many other simple pencil-and-paper tasks that one might profitably employ as measures of concentration. In fact, we believe that many of the tasks neuropsychologists use to assess frontal lobe functioning will serve as measures of concentration. Let us conclude by emphasising that when one undertakes the assessment of some workplace intervention on workers' ability to concentrate, it is imperative to administer a battery of several different measures. There is no pure measure of concentration, and performance on any given task is a function of many different factors. Administering only one measure, therefore, leads to a hopeless confounding of concentration ability with the influence of other variables such as perceptual speed, scanning strategies, memory and spatial and verbal fluency. By administering a variety of tasks, and by employing the appropriate statistical analysis, we can study the independent contribution of concentration ability to performance on the various tasks, and how this ability changes in response to our attempts to enhance it.

Conclusions

It is our thesis that mental fatigue plays a profound, albeit often hidden role in undermining productivity in the workplace. Since salaries are often the greatest single budget item, in both the public and private sectors, Romm and Browning (1994) have argued that even a modest increase in productivity can have substantial economic benefit. In that case, mental fatigue seems too expensive to ignore. We have divided possible interventions into two categories, those oriented to limiting fatigue-causing factors and those that reduce fatigue directly. Noise control and the reduction of distractions fall into the first category; providing a view of nature out the window, the second. In neither case can or should the improvements be taken on faith. There are measures that can provide useful indices of productivity, and hence of the degree to which interventions have led to improvement. Ongoing research of this kind promises not only to enhance employee productivity, but very likely, to enhance the quality of workplace life as well.

References

Cimprich, B. (1993) Development of an intervention to restore attention in cancer patients. *Cancer Nursing*, **16**, 83–92.

Diller, L., Ben-Yishay, Y., Gerstman, L.J., Goodkin, R., Gordon, W. and Weinberg, J. (1974) *Studies in Cognition and Rehabilitation in Hemiplegia* (Rehabilitation Monograph No. 50). New York: New York University Medical Center Institute of Rehabilitation Medicine.

James, W. (1892) *Psychology: The briefer course*. New York: Holt.

Kaplan, R. (1993) The role of nature in the context of the workplace. *Landscape and Urban Planning*, **26**, 193–301.
Kaplan, S. (1983) A model of person–environment compatibility. *Environ. Behav.*, **15**, 311–332.
Kaplan, S. (1995) The restorative benefits of nature: toward an integrative framework. *J. Environ. Psychol.*, **15**, 169–182.
Larson, G.E. and Merritt, C.R. (1991) Can accidents be prevented? An empirical test of the cognitive failures questionnaire. *Appl. Psychol. Int. Rev.*, **40**, 37–45.
Melnyk, L. and Das, J. (1992) Measurement of attention deficit: correspondence between rating scales and tests of sustained and selective attention. *Am. J. Mental Retardation*, **96**(6), 599–606.
Peters, T.J. and Austin, N. (1985) *A Passion for Excellence*. New York: Random House.
Peters, T.J. and Waterman, R.H. (1982) *In Search of Excellence*. New York: Harper & Row.
Purdum, G.E. and Schwartz, D.A. (1995) Category match task (unpublished).
Romm, J.J. and Browning, W.D. (1994) *Greening the Building and the Bottom Line: Increasing Productivity through Energy-Efficient Design*. Snowmass, CO: Rocky Mountain Institute.
Schwartz, D.A. (1994) The measurement of inhibitory attention and psychological effectiveness among adolescents. Unpublished Master's thesis, University of Michigan.
Schwartz , D.A. (1995) A naturalistic study of mental fatigue among college undergraduates (unpublished).
Witkin, H.A., Oltman, P.K., Raskin, E. and Karp, S.A. (1971) *A Manual for the Embedded Figures Tests*. Palo Alto, CA: Consulting Psychologists Press.
Yoshitake, H. (1978) Three characteristic patterns of fatigue symptoms. *Ergonomics*, **21**, 231–233.

Part 5

Case studies

Chapter 16

Managerial and employee involvement in design processes

Jean E. Neumann

The problem

During the course of organisational consultancy with clients who require assistance in addressing changes or developments to social systems, I frequently have met or been presented with strong feelings about the design, construction and use of buildings. The people expressing these feelings tend to be the users of the building: managers and employees of a sub-unit for whom the decision to expand, build or renovate their workplace was taken elsewhere in the hierarchy of their larger organisation. In all the cases I have come across, consultation of the users was practically non-existent. Attempts to exert even mild influence were met by resistance from those personnel responsible for the design and building processes. As a result, productivity suffered dramatically post-construction.

As an applied social scientist who has been working with issues of motivation and cooperation since 1971, I find it hard to understand that well-established methods of involvement and consultation seem to be under-utilised in the construction industry. Manufacturing, finance and retail industries have proved the productivity benefits of such methods for at least three decades. The concept of user involvement has been progressively important in information technology industries.

By increasing the involvement of the users of buildings in design and construction processes, construction managers could help avoid the worst problems of productivity caused by the construction industry in carrying out its work. The notion of the construction 'client' needs to be broadened from the joint executives who are commissioning the building to include the managers and employees of the sub-unit(s) who will be using the actual building. Efforts to decrease adversarial relationships within the construction supply network can be expanded to introduce more cooperation with those who are going to be using a building.

Many different actors are engaged already in designing, constructing and using buildings. They hold perceptions characteristic of their identities and positions which can work against appropriate involvement and consultation of users. An understanding of the role of perceptions in creating productive workplaces may help to motivate some willingness to cooperate more actively with users.

My colleagues and I have identified a model of the construction process overlaid with points of potentially useful involvement. Combined with knowledge about how to craft involvement strategies, that model can serve as a guide for thinking through with clients the questions, 'which users should be involved, when, and about what issues?'. Much is known about managerial and employee involvement in decision-making that can be applied to the construction situation.

Even so, construction managers may not be willing to develop their competence in this regard if it means extra meetings within the client system. A determination to contribute to a genuinely productive workplace is needed. Three cases from my own experience illustrate the negative impacts of inadequate involvement on workplace productivity.

The case of 'no windows'

I had been awarded a significant consulting contract with the UK site of a multinational chemicals corporation. The 20-year-old site was in the process of being expanded from one to two facilities and products. Simultaneously, headquarters wanted the UK company to introduce 'group-based working' on the shopfloor and to increase employee quality of working life. In early summer, I arrived to begin consulting on these topics, delighted to be visiting a particularly beautiful part of the country. The manager of the new plant offered to walk me round the construction site. We dressed in protective gear and climbed steps, looking at expansive empty floors and trying to imagine where the operators would be located during their working hours. There were no walls built yet, and I was thrilled by the terrific views surrounding the plant. I asked enthusiastically, 'Where will the windows be?'. The Plant Manager looked puzzled and said, 'There are no windows.' To my look of amazement, he responded, 'Do you think that matters?'.

My determined 'Yes!' seemed to confuse the Plant Manager even more. He took me to speak with the Senior Construction Engineer, on secondment from headquarters. To make a long story short, the Engineer explained that windows could not be put in the chemical plants. Even though it would be six months before the walls were built, such a major decision would hold up construction deadlines. My client system, the UK site, was a part of the Manufacturing Division, whereas the Engineer worked for the Buildings Division. Any major change would have to work its way up five levels of hierarchy to the Engineer's boss and back down again. Attempts to influence the 'no windows' decision on the part of the Plant Manager and his Managing Director failed.

A few years later, managers in the new plant complained that operators tended to cluster at the doors during working hours. These same operators were caught engaging in dangerous acts to entertain themselves during off-shifts. One particularly wild stunt was reported in the national press. The lack of windows seemed to lure workers away from their positions and contributed to an isolated workplace.

The designers, working in the headquarters' Building Division, altered an existing standard blueprint for the new UK plant. Their offices were located in the midst of an ugly, five-mile maze of chemical plants and pipes. Windows were not a priority for them. The organisation of design–build meant that the users of the factory, managed by a separate corporate division, had practically no involvement in design. The Senior Construction Engineer, eventually sympathetic to the problem, confided that his bonus depended on meeting his deadlines and, besides, his family wanted to be gone from the UK by Christmas.

The case of the colour-coordinated floors

Productivity had been reliable for years within the Patents Department at the headquarters of a European mining company. More recently, however, there had been an increase in inter-office bickering and intra-departmental conflict. I was brought in to analyse the problem and work with the partners on a solution. It quickly became apparent that the workload had increased dramatically during the previous year. No new solicitors had been hired, however, because the entire company was suffering a spending freeze. Partners had been told that this freeze was in order to help the company to pay for its new headquarters building.

To a person, all eighty employees reported hating the new building in advance of even seeing it. For decades, the department had resided away from the main headquarters, located in cramped but cosy wooden-panelled rooms winding throughout a Victorian office block. Soon, they were to be moved into a new 25-floor skyscraper. Three qualities of the new building's design upset the solicitors. Firstly, each floor would be mostly open-plan with only a handful of the most senior partners being given offices. In addition to the lack of privacy afforded the majority of employees, special treatment for some violated the department's culture which under-emphasised hierarchy and status. Secondly, the furniture would be standardised chairs, desks and filing cabins. Desks were to be bolted to the floor so the design could not be changed, and no family photos or other personalised touches would be allowed. The solicitors felt this over-control contradicted their rights as individuals, as well as forcing them out of a comfortable, idiosyncratic legal atmosphere into a 'corporate man' culture. Lastly, the colour scheme of each floor was different: carpet, drapes, wall hangings and furniture were made to match the colour for that floor. The Patents Department floor would be red.

A meeting was convened between the four Directing Senior Partners, the Facilities Director and representatives from the construction management firm in charge of the new building. The Partners expressed their concern at the damage the move would have on productivity, and made some suggestions for 'their floor' to minimise the shock anticipated to their departmental culture. While the Facilities Director sympathised, he was concerned that changes would add to an already exorbitant construction bill. One of the architects was less sensitive. After listening in silence for over an hour, he held forth for ten minutes about the aesthetic beauty of the design, the concept underlying it and the general lack of artistic taste in employees. Needless to say, the meeting did not result in any changes to plan.

I rang my client contact three months after the department had moved into the new building. He reported that there had been a 30 per cent turnover in the department leading up to and since the relocation. There was a general feeling that relations between senior and junior partners were colder than before; a more formal atmosphere prevailed as communications were more prescribed in the big open-plan office. 'Quick chats in the corridor' or 'popping into an office for a word' no longer took place. The department had been allowed to hire new solicitors, but those attracted to work in the new building seemed a 'different breed' from those who had worked there before. My client contact had managed, with some effort, to gain permission for photos and plants to be left on desks overnight.

The case of a great building that does not work

A European government department commissioned a new 'operation central' for one of its agencies. As public relations was important for this agency, executives took a decision to make sure that the building 'looked great' and contradicted a 'stuffy reputation'. They hired an internationally known architect to design the building, while managing the construction using multiple contractors and subcontractors well-known to them. The cost of an ambitious design with unusual materials put pressure on the agency to keep building costs as low as possible. Another factor in terms of cost was the date on which the new building could be occupied: the agency would incur additional expenditure for any delays, which the Minister would not allow.

Pressures on costs and pressures on time created havoc among the nearly thirty contractors and subcontractors. They blamed each other for delays and cost-overruns, often suing one another to avoid being blamed by the agency. Many delays were due to such factors as: mismatches between an ordered part (e.g. a window) and the space created for the part; damage to designed materials when building did not go to plan; and difficulty in scheduling specialist contractors in the correct sequence and in relation to other contractors. Despite all these problems, the building was finished on time and attracted pubic comments that it was 'a great building'.

The building opened, however, before the landscaping was completed and most of the operational equipment installed. For several months, employees complained of working on a building site. This inconvenience was compounded by many petty 'hygiene' problems. For example, the majority of doors stuck, requiring a hard push to open or close. The restaurant equipment turned out to be too large for the space designed and built for it. Employees had to improvise tables and cabinets to hold equipment, thus disrupting traffic flows in some areas. Air-conditioning and heating functioned unevenly in many sections of the building. Agency managers reported that morale remained low (and absenteeism high) for months as employees adjusted to the disappointments of the 'great building that does not work'.

Attending to social and organisational perceptions

These three case stories illustrate the main points of my argument. The multiple actors involved in designing, building and using workplaces have different roles and responsibilities. From their positions within occupational groups and organisational contractors, they perceive different priorities and needs for what constitutes a productive and healthy workplace. Often the losers in this difference of perceptions are the users of the buildings. From my professional value system, I think that users of buildings have a right to influence the design of their working environment.

Perceptions held by multiple actors within the design–build process shape the degree to which those buildings make a contribution towards productivity and well-being. The perceptions of an expert in some aspect of construction can often result in a better quality of working environment than lay users could desire or suggest themselves. Typical examples include air quality, lighting, traffic flows, noise pollution and ergonomics. Other aspects of construction, however, are open to wide interpretation and influenced by factors that have little to do with potential workplace productivity as determined by experts or users. In addition to the obvious example of aesthetic values, many elements of initial design conception and specification requirements emerge from negotiation between designers, contractors and those members of the client system who take executive decisions.

It is not unusual for each group of actors to experience their perceived needs and opinions as more worthy than those of the other actors. The Senior Construction Engineer in the chemicals plant considered the negative consequences of having to change his construction schedule as more important than the negative consequences of the chemical operators having to work without windows. Those in the final implementation stages of building the new mining headquarters felt little empathy for the solicitors' attachment to a 'legal culture' working environment. And, the executives of 'operation central' were determined to avoid additional expenditure for their budgets regardless of the discomfort of employees or the chaos and financial losses to contractors.

Social and organisational forces often constitute barriers to recognising and incorporating different perceptions into design and building processes. Financial limits would be mentioned by most clients and construction managers as a shaping force in the designs that they commission. However, the actual organisational structure and decision-making processes within the client system emerge in these three cases as significant as well. The 'discovery' that windows might matter to the Manufacturing Division's mandate for a better quality of working life for its employees stood in direct opposition to the Buildings Division's construction schedule. Neither the Engineer nor the Plant Manager was sufficiently convinced that such a social concern would be welcomed by their divisional vice-presidents to justify raising it. Similarly, the 'discovery' that employee satisfaction might suffer in the face of the colour coordination and open-plan designs for the new headquarters was hidden from the executives involved. The executives of the government agency focused on the appearance of the 'great building' and left more junior staff to struggle with the implications of unreasonable and ineffective decisions.

Actors involved in design, building and use hold different roles and responsibilities. Contained within social identities and organisational boundaries, these roles and responsibilities encourage ways of looking at the numerous decisions necessary in the construction industry. These ways of looking, or perceiving, result in decisions considered positive by one group of actors and negative by others.

I am using the word 'perception' to mean these socially and organisationally supported ways by which people view, feel and generally sense the tasks of designing and building. Perception, in this meaning, does not concern primarily individually oriented 'human factors' or 'ergonomics', notions which emphasise the physiological bases of human sensation and perception. I focus, instead, on ways of thinking and apprehending the world that have been socialised into the occupational and other identities held by the people who make up the groups and organisations responsible for construction (Holti and Standing, 1997).

My purpose is to illustrate how differing priorities and understandings of what is important – embedded in the organisation of design, building and use – often result in conflicting notions of what is a productive and healthy workplace. Further, such differences in perception often carry with them ideas of superiority and other judgements of worth that hinder genuine collaboration and problem-solving.

Sometimes those differences have to do with the design itself. The designers of the chemicals plant did not consider windows necessary or important, even in a facility located in an area of outstanding natural beauty: chemical operators often work without windows. By contrast, they were very interested in the challenge presented by new computerised technologies, and in linguistic difficulties of working with contractors from throughout Europe. The Architect of the mining corporation's headquarters felt strongly about the superiority of his aesthetic sense over that of the solicitors. The Facilities Manager, already overwhelmed by the task of dealing with a 25-floor building, considered consistency easier to manage than the idiosyncratic concerns of a particular department. The ambitious design of the 'great building' superseded usability. The agency's middle managers dealt with their thirty contractors as if they were building a standard design that had been built many times before, not the unique design with many specialist materials and subcontractors that their executive bosses had commissioned for them. When there is relative harmony on the design, difference in perceptions focus on priorities for decision-making and task implementation related to design and build processes.

All three cases indicate one result of inadequate consultation with a broad sample of the user population. Mismatches between users' perceptions and needs and the building design only became apparent when the chemical plant and the mining headquarters were almost built. At that point, schedules and budgets influenced the lack of action most strongly. Individual needs, e.g. the Senior Construction Engineer's bonus and the Architect's professional ideology, may play some role. However, perceptions of organisational decision-making and occupational groups' task flows play a more powerful role.

These aspects of organisational behaviour impact directly on the degree to which a completed building can be turned into a productive workplace by those people who must work within it. Ironically, in these instances, the productivity and well-being of designers and builders tend to take priority over the productivity and well-being of users. Users are then left to adjust themselves or those aspects of the building that can be made flexible.

As the three cases illustrate, some of the adjustments employees make are not in the best interests of their employing organisations. Employees of the chemical plant enacted feelings of isolation by leaving workstations to cluster at doors or in a central control room that held computer 'windows' of all the plant. Solicitors left their jobs in the Patents Department, while those remaining evolved a new way of working with new employees that felt less productive than before. And employees of the 'great building' complained bitterly of petty hygiene problems and enacted low morale through high absenteeism.

Broader, more effective participation of user groups could help avoid some of the worst excesses of design and build processes. Construction management already involves a complicated multiple party network; key players could be forgiven for not wanting to add more. However, mechanisms for effective participation have been tested for decades in other industries – notably manufacturing – and can readily be introduced into construction. Such processes can go far in recognising and incorporating different perceptions.

Managerial and employee involvement in decision-making

'Involvement' and 'consultation' are versions of a broader area of organisational behaviour related to participation in decision-making. The terms and phrases have changed over the years: employee participation, participative management, employee involvement, employee empowerment, teamwork and flexibility, to name a few. But the basic research underlying these seeming fads is strong and stable.

People are more likely to be motivated to implement decisions well that they have had a role in shaping, however minor that role. Many lessons have been learned by those in other industries who have worked extensively with involvement and consultation processes. I think three are of particular importance to the issue of managerial and employee involvement in design and construction processes. These can be summarised as a general principle: involvement in design and construction processes should be genuine, appropriate and interconnected.

Genuine involvement means that managers and employees are asked to participate in decisions by their bosses only to the degree that their participation will really make a difference to the outcome. Over the years, a range of participation has become well recognised amongst researchers and practitioners (see Fig. 16.1). This range begins on the left with less involvement (degree D), simply being informed in a face-to-face setting about a change or development planned by another group of people, usually at a higher level of hierarchy. From there, managers or employees might be consulted about decisions that are being taken by others (degree C). The consultation certainly takes the form of offering reactions, positive and negative, to any proposals. But it might as well take the form of giving data or information prior to or after significant points in decision-making.

Being involved (degree B) is a more engaged form of involvement. Those managers or employees who are involved to this degree actively generate information, analyse those data and develop options or proposals. They may also make recommendations. This work then becomes the input for decisions taken elsewhere.

Finally, taking decisions is the most occupying form of participation (degree A) and shows up on the right hand side of the continuum (see Fig. 16.1) Here there is no difference between participation and taking decisions. People who take decisions usually do so in small to medium-sized groups. Their decision may need to be ratified elsewhere, but the power to conclude a decision lies with the group.

Genuine involvement, therefore, is a process of matching the degree of participation with the actual authority available to groups of managers and employees by virtue of their hierarchical or occupational position, or by

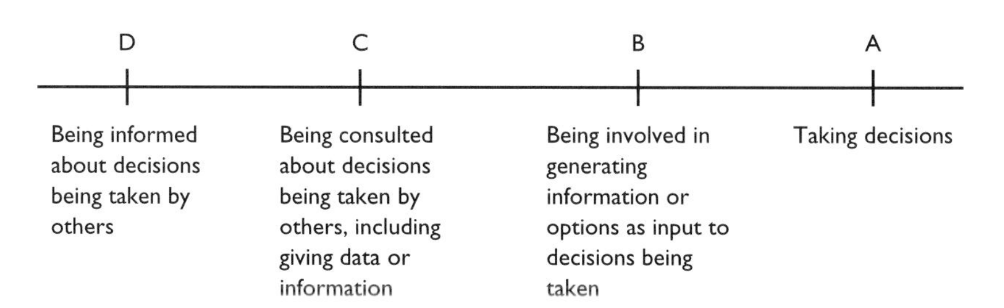

Figure 16.1 Continuum of degrees or range of participation in decision-making

powers invested in them by others whose hierarchical positions makes it possible. One of the main reasons why people do not participate when given the chance (Neumann, 1989) is that the participation on offer is not real. That is, managers or employees feel that they are being asked their opinion but no one will take notice. Within the construction industry, one of the gross examples of this is the so-called consultation that takes place long after the design is finalised. It is too late for managers and employees actually to make a difference with their opinion. Even so, a launch event is held and people are asked to fill out a questionnaire or make comments on a design that is practically written in stone.

Appropriate involvement refers to some degree of participation that makes sense in the light of the content of the decision and its actual relevance to the individuals or groups being invited to participate. Research and practice have demonstrated repeatedly that the majority of people prefer more involvement in decisions that affect them directly (Neumann, 1989). One way of thinking about this is in terms of three concentric circles, with the individual and his or her work at the centre (see Fig. 16.2). Design and construction issues, for example, that are going to affect the individual and his or her work directly are those in which the individual will most probably want to have a say. Theoretically, such decisions are profoundly appropriate for that individual's involvement.

Secondarily, the individual feels concerned about the group or unit in which he or she works. Generally speaking and of relevance to design and build, the more closely one works with others, the more concerned an individual will be with the fate of those individuals. If design options relocate a department or shift the relationship patterns between groups or units that interact continuously, then their involvement to a significant degree on the participation continuum would be seen as appropriate.

Lastly, individuals have some claim on appropriate involvement when their position in the larger organisation is going to be affected by construction. 'Position' would normally refer to issues of occupational identity or status level. Design options can and often do have implications for the 'relatedness' (Miller, 1990) of groups to each other. Relatedness means the

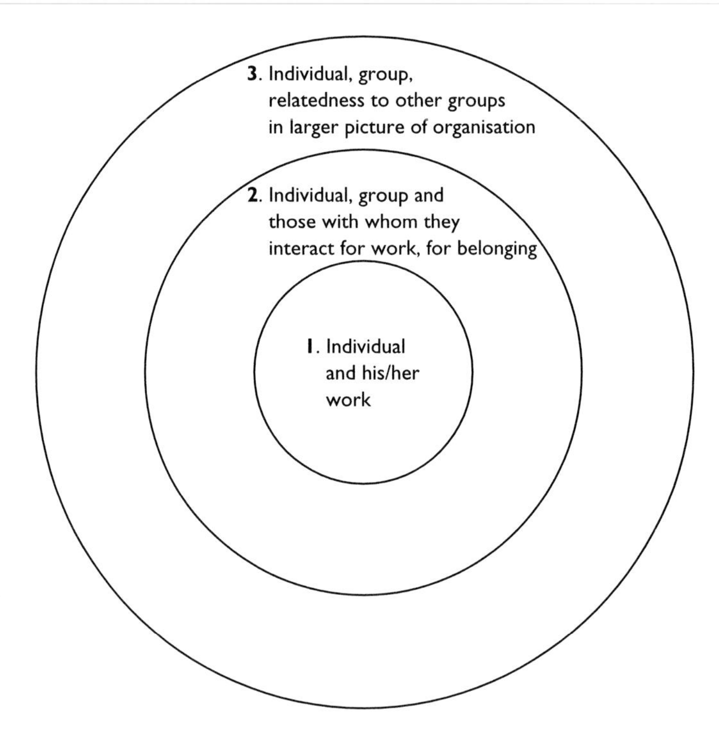

Figure 16.2 Concentric circles of preferred involvement in content issues

fantasies and projections that groups have about each other in a complex social system; those perceptions, feelings and opinions that are not necessarily based on face-to-face interaction but have to do with symbols and stories. Issues like size of offices, differentials in furniture, special meeting rooms and the like often speak to the symbols of relatedness. In the light of how seriously such matters can affect morale, appropriate involvement in the larger picture of the organisation needs to be kept in mind.

Genuine and appropriate involvement of managers and other employees tends to raise important interconnections between construction decisions and other organisational concerns. The sort of issues indicated certainly will relate to the larger organisation as a social system: through involvement, the emerging meanings that managers and employees are placing or might be inclined to place on certain design options become apparent. But more concretely, there may well be implications for job and organisational design in terms of flow of communication and tasks. These may have knock-on effects in terms of training and development for implicated groups. Previous research has identified five substantial interconnections in most comprehensive change projects (Neumann *et al.*, 1995) – which most construction projects can be considered to be.

Points of involvement in design and construction processes

The principle of genuine, appropriate and interconnected involvement can be applied to design and build processes. A generic model of the main stages through which a building goes from design to build to use to eventual demolition offers a basic framework onto which can be added the involvement of users. Such a model can be used to help identify points of involvement that might be beneficial or necessary from a construction perspective.

At The Tavistock Institute, consulting social scientists have been working with construction companies and academics to improve cooperation across, and to reduce adversity in, the construction supply chain. Known as the 'building down barriers' approach to supply chain management, one of the tools that has emerged through this action research is a model of the construction process (see Fig. 16.3). This approach, also called 'prime contracting', replaces short-term, single-project adversarial supply chain relationships with long-term, multiple-project relationships based on trust and cooperation.

The advantage to this discussion of managerial and employee involvement is strong. Based on supply chain relationships, the model incorporates accepted flows of construction work with points of needs assessment related to clients. A new model is not necessary; greater involvement of clients along

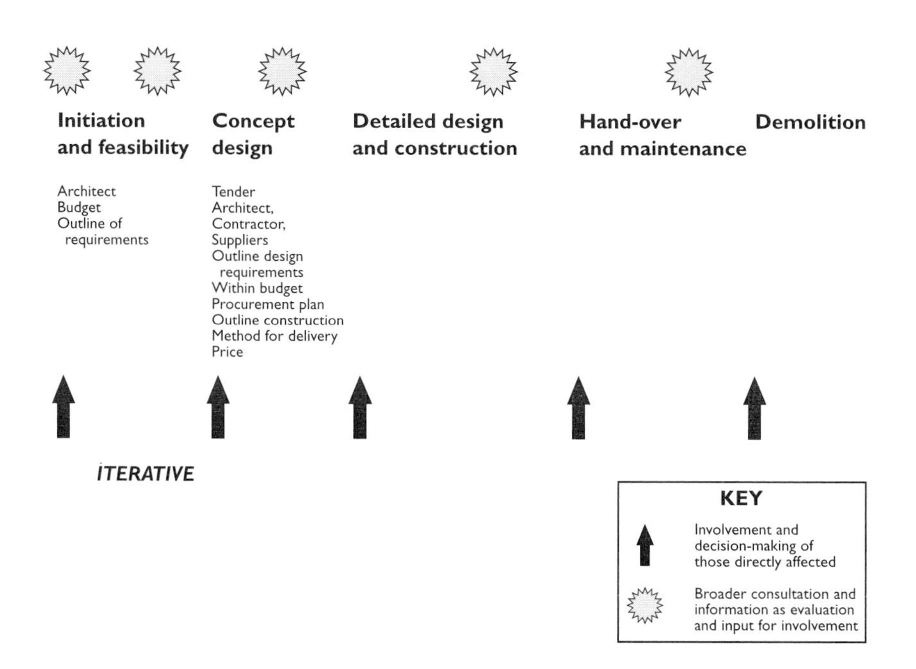

Figure 16.3 Points of potential involvement in design and construction processes

the principle of genuine, appropriate and interconnected involvement can be hypothesised based on notions already developed by the industry.

Five phases have been agreed as common across all construction projects: initiation and feasibility; concept design; detailed design and construction; hand-over and maintenance; and demolition (Holti and Nicolini, 1998). Initiation and feasibility is the phase of inception of a new construction project: the client's needs are established; an internal client-led team is set up; a strategic brief of business requirements is produced; a prime contractor selected; and an outline programme and fee for the overall project through the concept design phase agreed. The second phase, concept design, explores the client's functional requirements in more detail resulting in a project brief: the supply chain will be involved at this point, developing and appraising solutions to problems with an eye on life-costs, risk management and value analysis. A project brief and firm price, signed off by all parties, constitutes the terms of reference for action.

The third phase of detailed design and construction is completed with the involvement of the supply chain to allow construction to happen in the most efficient, cost-effective and risk-neutral manner. At the completion of this phase, the building is handed over to the client for use and maintenance (Holti and Nicolini, 1998). Post-hand-over may involve the prime contractor in monitoring the operation of the building through its life-cycle, to the point of demolition – the final phase.

I have been experimenting in my consultancy practice with the idea that significant points for potential user involvement come at the beginning and end of each of the phases of the design and construction processes. At the beginning of each phase, those managers and employees directly affected by the design and construction need to be given opportunities to be involved (degree B) or to take decisions directly (degree A). During or towards the end of a phase, those who are less affected need to be informed of what is or will be taking place (degree D), or even consulted about proposals and plans (degree C).

As the first two phases are likely to be iterative, clarity as to who is involved when and about what issues is important. In practice, experimenting with this model has meant the establishment of a small steering team made up of those internal clients who will be directly working on the design and construction processes (e.g. facilities managers, information technology engineers, heads of departments directly and powerfully affected, change agents). This team works throughout the design and construction processes to interface with the prime contractor as well as the clients' other contractors and suppliers as are implicated in the change. This approach normally means that the client has a prime contractor who handles much of the construction management process.

In addition to interfacing with the prime contractor, the small steering team ensures that appropriate, genuine and interconnected involvement takes place. This means that executives need only be brought in as appropriate; similarly, managers and employees directly affected can be involved as appropriate. The team also ensures that broader consultation and information is made available to themselves, their bosses and their prime contractor as evaluation of actions and decisions from one stage and as input into the next stage.

Summary

Workplace productivity problems can be caused by design and build processes that inadequately involve users. The chemical plant without windows, the colour coordinated and open-plan scheme for the patents solicitors, and the 'great building' with numerous petty problems are some examples. Perceptions of the multiple actors that are necessary to designing, constructing and using buildings can hinder suitable involvement. Both organisational and social factors stand as crucial in willingness and ability to include the actual users in design processes.

The principle of genuine, appropriate and interconnected involvement of user groups can help to craft an involvement strategy. Mapped onto a model of cooperation across the supply chain network, users may be involved at the beginning and end of each phase of design and construction. An infrastructure within the client system, typically using some sort of steering committee, has proved facilitative for prime contractors and clients alike. Such involvement can help the construction industry to make a positive contribution to workplace productivity.

References

Holti, R. and Nicolini, D. (1998) *Building down Barriers: a case report on an initiative in progress*. London: The Tavistock Institute.

Holti, R. and Standing, H. (1997) *Psychodynamics and Inter-occupational Relations in an Industrial sector – the Case of UK construction*. London: The Tavistock Institute.

Miller, E.J. (1990) Experiential learning in groups, in *The Social Engagement of Social Science, A Tavistock Anthology, Volume 1: The Socio-Psychological Perspective* (eds E. Trist and H. Murray), 165–198. London: Free Association Books.

Neumann, J.E. (1989) Why people don't participate in organisational change, in *Research in Organisational Change and Development*, vol. 3 (eds W.A. Pasmore and R.W. Woodman). Greenwich, CT: JAI Press. pp. 181–212.

Neumann, J.E., Holti, R. and Standing, H. (1995) *Change Everything at Once! The Tavistock Institute's Guide to Developing Teamwork in Manufacturing*. Didcot: Management Books 2000.

Chapter 17

The Intelligent Workplace: a research laboratory

Volker Hartkopf and Marshall Hemphill

The Center for Building Performance and Diagnostics (CBPD) is located within the School of Architecture at Carnegie Mellon University. It was founded by Volker Hartkopf and Vivian Loftness after their experience researching the causes of high levels of occupant discomfort being experienced in Canadian government buildings.

The CBPD emphasises the need for a design team with all the appropriate disciplines in place and functioning at the onset of a building design project. The design team makes trade-offs affecting total building performance with the full knowledge and input of all team members.

Total building performance

Total building performance (Hartkopf *et al.*, 1986) is measured by four metrics: user satisfaction; organisational flexibility; technological adaptability; and environmental and energy effectiveness.

For user satisfaction (the human performer), six elements under the control of the design team are identified: the thermal, acoustical and visual environments; spatial quality; air quality; and building integrity. The requirements of each of these elements must consider the positive or negative effects on the other five.

The human performer has physiological, psychological and sociological needs, while the organisation has economic needs – all of which interact with the six elements of user satisfaction. Table 17.1 is a matrix suggesting issues where each of the elements and needs intersect.

As the concepts expressed in Table 17.1 were taking shape, it became clear that a platform was needed that provided both further input from users, designers and suppliers, and a stage from which to articulate the message. Towards this end, the Advanced Building Systems Integration Consortium (ABSIC) was established at the CBPD in 1988.

Table 17.1 Organising performance criteria for evaluating the integration of systems

	Physiological needs	*Psychological needs*	*Sociological needs*	*Economic needs*
Performance criteria specific to certain human senses, in the integrated system				
1. Spatial	Ergonomic comfort, handicap access, functional servicing	Habitability, beauty, calm, excitement, view	Way-finding, functional adjacencies	Space conservation
2. Thermal	No numbness, frostbite; no drowsiness, heat stroke	Healthy plants, sense of warmth, individual control	Flexibility to dress with the custom	Energy conservation
3. Air quality	Air purity; no lung problems, no rashes, cancers	Healthy plants, not closed in, stuffy; no synthetics	No irritation from neighbours, smoke, smells	Energy conservation
4. Acoustical	No hearing damage, music enjoyment, speech clarity	Quiet, soothing; activity, excitement, 'alive'	Privacy, communication	
5. Visual	No glare, good task illumination, way-finding, no fatigue	Orientation, cheerfulness, calm, intimate, spacious, alive	Status of window, daylit office, 'sense of territory'	Energy conservation
6. Building integrity	Fire safety; structure strength + stability; weathertightness, no outgassing	Durability, sense of stability, image	Status/appearance, quality of construction, 'craftsmanship'	Material/labour conservation
Performance criteria general to all human senses, in the integrated system				
	Physical comfort, health, safety, functional	Psychological comfort, mental health, psychological safety, aesthetics	Privacy, security, community, image/status	Material, time, energy, investment

ABSIC

Initially ABSIC membership was drawn from the private sector and comprised design professionals, construction firms, users and component and service suppliers. ABSIC's mission and its early research findings brought it to the attention of the National Science Foundation (NSF), which awarded it the designation of an 'Industry University Cooperative Research Consortium', the only one of fifty such research consortia at that time to address the needs of the construction industry.

ABSIC was soon joined by four Federal agencies, each acting on its own and for its own needs and missions: Department of Energy, Department of Defense, Environmental Protection Agency and the General Services Agency (the world's largest owner and leaser of building space). The support and input from these Federal partners has greatly strengthened ABSIC, not only since they are large users of office space, with large employee pools, but also because of the great depth of their talents in the very technologies necessary to optimise total building performance. Table 17.2 lists ABSIC members past and present.

As part of its research into issues of building performance and building system integration, ABSIC undertook an in-depth study of a number of leading buildings around the world. The CBPD staff, supplemented by experts in mechanical systems, telecommunications, power and signal distribution and self-assessment surveys, comprised a team that would visit and simultaneously assess each building. The resulting assessment represented a multidimensional evaluation, devoid of the bias which results when the view of only one discipline is reported. Things done well and those not done well were

Table 17.2 ABSIC members

Current members and partners	*Current federal sponsors*
• AMP Incorporated	• National Science Foundation
• Bank of America	• US Department of Energy
• CADSpec Multimedia	• US Department of Defense
• Consolidated Edison	• US Environmental Protection Agency
• Interface, Inc.	• US General Services Administration
• Johnson Controls	
• LG – Honeywell	*Past members*
• LTG – Lufttechnische GmbH	• Armstrong World Industries
• Siemens Energy and Automation	• Bechtel
• Steelcase	• Bell of Pennsylvania
• United Technologies/Carrier	• Duquesne Light
• Zumtobel Staff Licht	• Miles
• Josef Gartner and Company	• PPG
• Mahle GmbH	

identified and catalogued. Buildings in Japan, Germany, France, the UK, Canada and the USA were studied over a span of three years. The critical findings were folded into a number of papers and keynote addresses delivered by the CBPD staff. A book – *Designing the Office of the Future* – published in 1993 details the assessment methodology and the Japan building studies, and summarises finding from the other countries (Harkopf *et al.*, 1993).

As the essential characteristics and issues surrounding the 'Office of the Future' came into focus, the need for a test bed to evaluate and demonstrate new technologies and new design concepts became clear within ABSIC.

Exogenous developments

During this decade of ABSIC's existence, substantial changes were occurring in the outside world. The economic spheres within which organisations function were 'globalising'. Organisational structures were modified (downsizing, outsourcing, etc.) to respond to the macroeconomic upheaval – all with unprecedented speed. The scope of organisational activities has greatly increased and the importance of the individual human performer – the knowledge worker – has eclipsed that of the bureaucratic organisations of earlier times.

The advent of the new tools of work lies at the centre of the new economic and organisational order. Computers, electronic databanks, video conferencing, portable offices, 24 hour commerce, and many other technologies well known to us all are redefining work, on an almost daily basis.

During this same decade, concerns for the finite environment and resources provided to us have risen to positions of significance in governmental agendas. Sustainable architecture has become a focus for the design team.

The appropriate structure to house and support the new organisational forms will be an intelligent building. Intelligent buildings will provide unique and changing assemblies of recent technologies in appropriate physical, environmental and organisational settings to enhance worker effectiveness, communication and overall comfort and satisfaction. They will ensure user health and comfort; organisational flexibility; technological adaptability; energy effectiveness; and environmental sustainability (Hartkopf and Loftness, 1996).

The Intelligent Workplace: a 'lived-in' laboratory

Responding to the need both to demonstrate and to research the implementation of the new building requirements, ABSIC and the CBPD have just completed construction of the Intelligent Workplace (IW). The IW is sited on the rooftop of an existing classroom building of the School of Architecture, Carnegie Mellon University (Fig. 17.1) and will be occupied by the CBPD staff and advanced degree students. It will be a knowledge factory – a

Figure 17.1 The Intelligent Workplace, on the rooftop of the Margaret Morrison Building at Carnegie Mellon

surrogate for most of today's emerging organisational forms, both public and private. We call it a 'lived-in' laboratory, to distinguish it from the more traditional isolated research structure.

The IW will incorporated advanced technologies from around the world, combining them with traditional building assemblies. The building design allows for change-outs of almost all components, including the building skin and mechanical systems. Within the building, sub-assemblies can be compared side by side. Advanced computer technology is being designed to control the building subsystems in anticipation of changing external and internal conditions, instead of in reaction to them (lead, not follow). Additionally, site-specific opportunities are provided by the rooftop location, allowing for technologies such as roof-located daylighting, photovoltaics, and roof ridge ventilation to be explored (Fig. 17.2).

Major innovations

Dynamic layered facade

The facade construction includes light redirection panels to redirect daylight into the interior while minimising glare at the perimeter. These panels will be motorised and controlled in response to both the predicted solar position and the actual conditions, as monitored by an on-site weather

Figure 17.2 Longitudinal section through the Intelligent Workplace

station. The panels are spaced away from the skin, allowing for both a walkway around the perimeter and operable windows. The windows are occupant-activated for natural ventilation, which in turn will save on energy to condition the interior over considerable periods of the year. The windows are double-glazed and have surface coatings to reduce infrared transmittance, while allowing 70 per cent visible light transmittance. The benefit of daylighting for both energy conservation and occupant well-being has been well documented by CBPD research.

On the room side of the windows, mullions are water cooled or heated, to modify the radiant environment and enhance the thermal comfort of occupants located near the windows.

Bolted modular construction

The steel structural components were carefully detailed by computer modelling using bolted connections, both to minimise waste and to assure constructability. As a result, erection of the steel was completed in four days and with no job site waste. By transferring all sizing and detailing to the factory site, the detrimental environmental impact of cutting and fitting at the job site was essentially eliminated. Steel waste was transferred to the factory and was recycled instead of being hauled from the job to a tip site as land fill. Additionally, the bolted structural components can be disassembled and moved to another site, further enhancing the sustainable nature of this architecture.

Open web floor structure

Using an open web floor support structure instead of solid beams allowed integration of services and structure. The top surface is an access (pedestal) floor, supported off the joists (Fig. 17.2). Electrical and signal/data wiring is routed in defined paths by cable trays. Connection nodes are located 6 m on centre, providing plug-in, plug-out connectability within 3 m for all task locations. Similarly, air supply and return ductwork is routed through the joists. The net effect is a reduced depth of this space and reductions in building volume and materials usage, both steps towards a more sustainable architecture.

Floor-based support systems

All workstations are supported from floor-based systems. Conditioned air is supplied through either floor-located discharge grilles or nozzles located on the workstation work surface. This approach, more typical in Europe, is relatively novel in the USA. Choice of velocity and direction of conditioned air is left to the occupant in all cases, and for the workstation-located system, the occupant has the further option of controlling air temperature.

Power and telecommunication links are through plug-ins within the floor boxes and these are, in turn, easily relocated through 'plug-in, plug-out' hardware. Task lighting, used to supplement the daylight and/or the low-level ambient fluorescent lighting, is furniture-located and user-controlled (in both intensity and direction). In some workstations the occupant is given a low-powered masking sound system (beyond that provided by the air supply nozzles). This system allows the worker to adjust the level of background noise to mask nearby conversations and thus enhance concentration by reducing distraction. Furniture and workstations (Fig. 17.3) are intended to be user-movable. Relocation of workspace furniture to support changing work patterns, such as team projects, is within the users' capabilities, especially because of the 'plug and play' telecommunication subsystems.

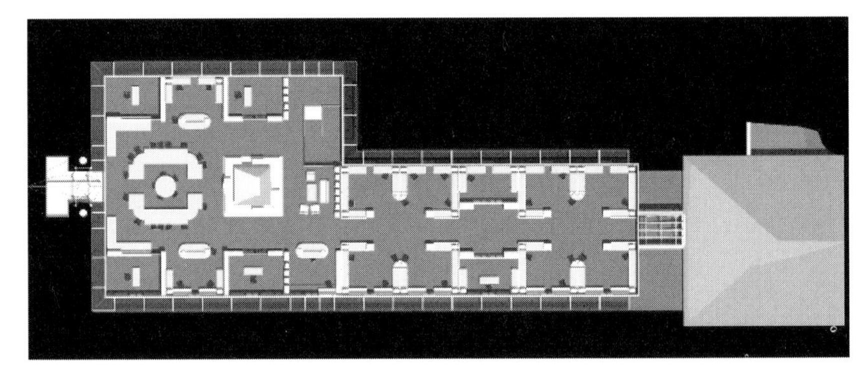

Figure 17.3 Floor plan of the Intelligent Workplace

Intelligent controls with learning systems

Carnegie Mellon has a formidable computer technology strength. This group has been involved with the IW, developing computer software to control building systems in support of both user needs and energy conservation. Rather than controlling in reaction to existing conditions, and thus incurring inefficiency due to building lag, these systems will use predictive algorithms to anticipate change. Working from both known solar positions and historical weather data, supplemented by the on-site weather station, as well as gradually acquired knowledge of individual workers' habits and preferences, the software will anticipate building condition needs. The setting of elements of the dynamic skin, the mechanical system and the lighting system will be integrated by the software in anticipation of user needs. Minimum energy usage will be one of the software's goals, but user override is provided.

Significance of the IW

Using the research capability of the IW, the ABSIC/CBPD agenda will directly address the compelling physiological and social needs of the over 50 million office workers in the US. With certain modifications, the results may be extrapolated to other cultures. This agenda stresses the importance of the environmental quality of the built space to the user. Recent publication of studies on government workers in the UK links low levels of the individual's perception of control of the work environment to higher levels of heart disease (*Wall Street Journal*, 1997). While, in that study, environmental control refers primarily to management-controlled elements, providing the worker with a high level of control over his or her own immediate thermal, acoustic, visual and air quality environments responds, we believe, to the concerns noted in the study.

ABSIC/CBPD will also address the balance of the spatial, thermal, acoustics, visual and air quality elements to the long-term goals of building integrity, energy conservation, and life cycle value.

Anticipated outcomes

In the process of defining the range of user preferences for the various built environments, it is expected that these preference ranges will exceed the values enshrined in building codes and design practices. For example, ASHRAE standards limit local air velocities to 0.25 m s^{-1}. It is obvious that user-controlled systems routinely operate at higher velocities. Similarly, IESNA standards seek to limit brightness ratios in officing spaces to 4:1 or 8:1. Again, daylighted spaces greatly exceed these values, yet are widely preferred by occupants. In fact, proximity to windows has been shown by

ABSIC research to be positively related to user comfort and health (Loftness *et al.*, 1995). Thus, IW research will likely suggest new values for these metrics, as long as they can be user-controlled.

IW research will also challenge the relationship of higher first cost to owning cost by: establishing new building design criteria, documenting increased productivity of the structure itself; reducing energy costs; and raising the level of worker effectiveness (Napoli, 1998).

References

Hartkopf, V. and V. Loftness (1996) 'Global relevance of total building performance. *CIB-ASTM-ISO-RILEM 3rd International Symposium: Applications of the Performance Concept in Building*, organised by National Building Research Institute, Tel-Aviv, December.

Hartkopf, V., V. Loftness, P. Drake, F. Dubin, P. Mill and G. Ziga (1993) *Designing the Office of the Future: The Japanese Approach to Tomorrow's Workplace.* New York: Wiley.

Hartkopf, V., V. Loftness and P. Mill (1986) The concept of total building performance and building diagnostics. ASTM Special Technical Publication, *Building Performance: Function, Preservation, and Rehabilitation*, STP 901, May, 5–22.

Loftness, V., V. Hartkopf, A. Mahdavi, S. Lee, J. Shankavaram and K.J. Tu (1995) The Relationship of environmental quality in buildings to productivity, energy effectiveness, comfort, and health – how much proof do we need? *International Facility Management Association (IFMA) World Workplace Conference*, Miami Beach, FL, 17–20 September.

Napoli, L. (1998) Where every worker is ruler of the thermostat. *New York Times*, 15 February.

Wall Street Journal (1997) Lack of control over job is seen as heart risk, 25 July.

Chapter 18

Air-conditioning systems of the KI Building, Tokyo

Hidetoshi Takenoya

The KI Building in Tokyo, Japan, provides a comfortable, productive environment for occupants while also meeting the requirements of an intelligent building. Many new technologies were successfully implemented to meet these goals.

This chapter describes these technologies and summarises the results of the systems employed in the building. It begins with the planning process and concludes with a physical and psychological survey of the occupants regarding the completed building and its systems.

At the centre of the KI Building is a garden atrium that opens to corridors on the floors above (see Fig. 18.1). The atrium was constructed on a comparatively small scale, with a floor area of 600 m^2 and a height of 15 m. This space is designed to be relaxing and comfortable for occupants.

Natural light filters through the atrium's skylight down to the tropical plants surrounding the ponds. This produces a satisfying environment for workers, particularly those with work areas facing the atrium. However, the atrium and work areas required original, innovative systems designed to provide indoor comfort.

Special systems

Besides systems that control the thermal environment, the KI Building features other special systems. These systems enhance the building's amenities by considering the effects of sight, smell and sound on the human occupants.

Thermal environment

One of the features of the atrium is that there is no glazing and the office space is open to the atrium. Due to a variable and large direct solar gain, and also because of large temperature stratification between the top and bottom of the atrium in summer, some difficulties were envisaged as to whether the thermal environment in the office area would be satisfactory. Due to these concerns, two ideas were introduced.

Figure 18.1 Atrium at the KI Building

The first idea is the extract fan system to remove hot air at the top of the atrium in summer. This fan will be operated automatically by thermostat. The second idea is an air curtain along the office perimeter facing the atrium. This air curtain is provided by low-profile floor-mounted fan coil units (FCUs) and is expected to block hot air into the office. It was anticipated that supplying high-temperature air would result in a major stratification in winter, therefore supply temperature limit control is introduced to minimise stratification, and ceiling-mounted FCUs discharging air downwards are installed around the atrium for heating.

Computer fluid dynamics (CFD) analysis was carried out to assess the atrium environment. Figure 18.2 shows traces of markers in the atrium;

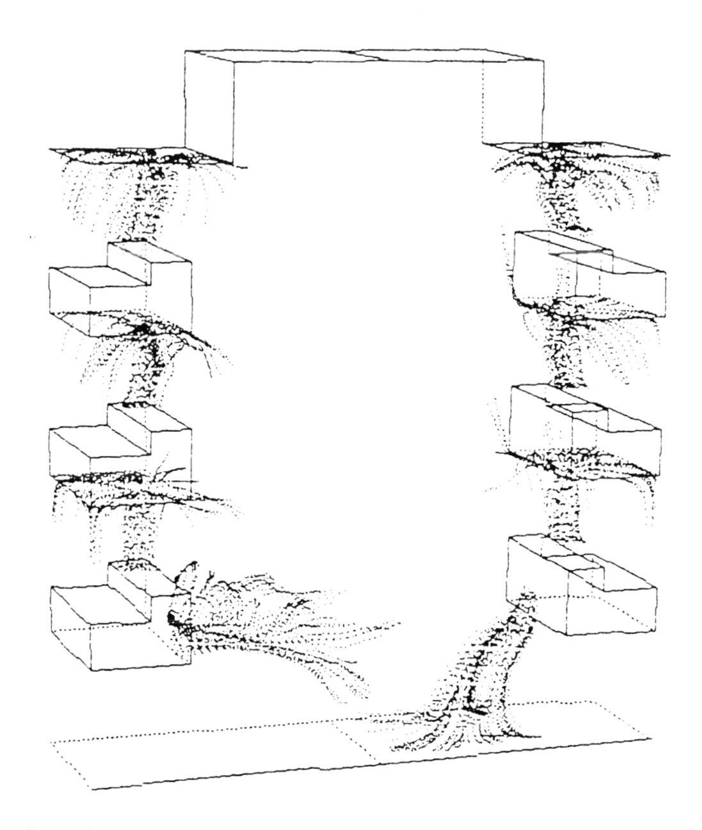

Figure 18.2 Traces of markers in the atrium

Fig. 18.3 shows the air distribution flow to the atrium. Fig. 18.2 shows marker tracing only within a cross-section having a narrow width. The starting points of markers are randomly arranged in the cross-section. The marker which has overshot space is rearranged nearest to the overshooting location. By this method, it is possible to express the secondary flows which occur near the main in relation to airstream.

Site measurement was carried out after practical completion and the result is shown in Fig. 18.4. Occupancy zone temperature in the atrium is relatively low and varied between 24°C and 26°C (setting temperature in cooling mode is 25°C). The stratification in cooling mode is shown in Fig. 18.5 and temperature at fifth floor level is higher than occupancy level by only 1 to 2 K, while maximum temperature at the top of the atrium is 40°C.

The measurements match the results of a survey of worker satisfaction with the atrium environment quite well. Fig. 18.6 shows the survey results for a typical summer day. Satisfaction with the thermal environment was

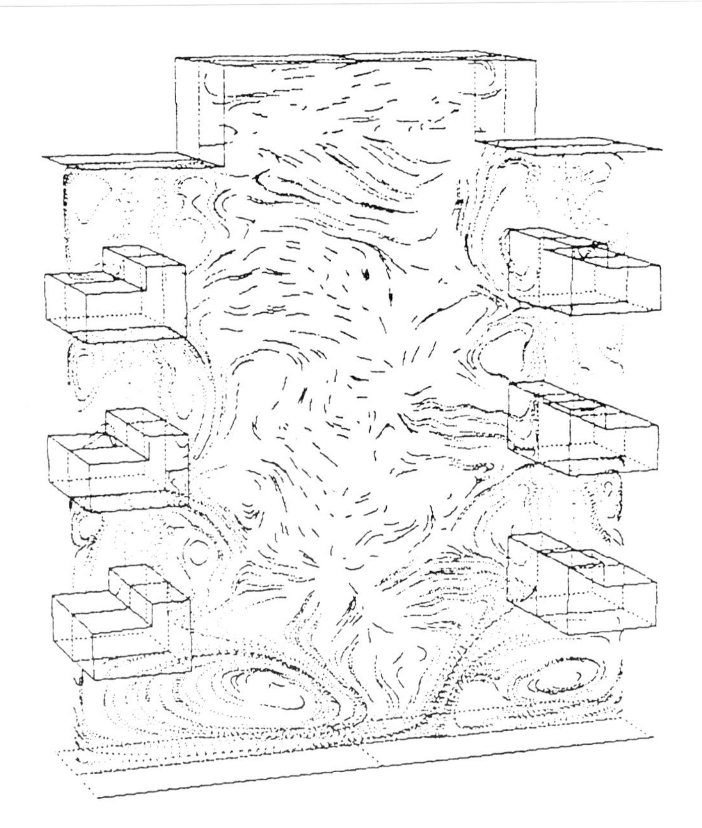

Figure 18.3 Air distribution flow in the atrium

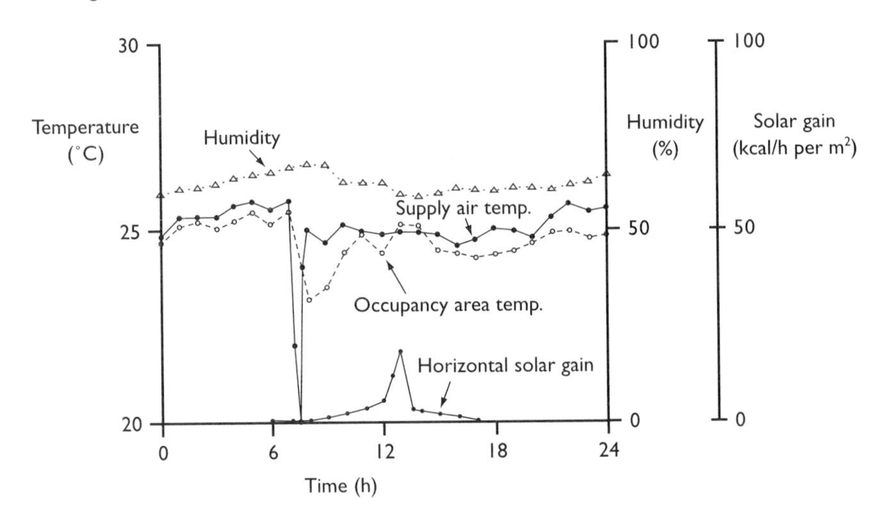

Figure 18.4 Thermal environment in the atrium

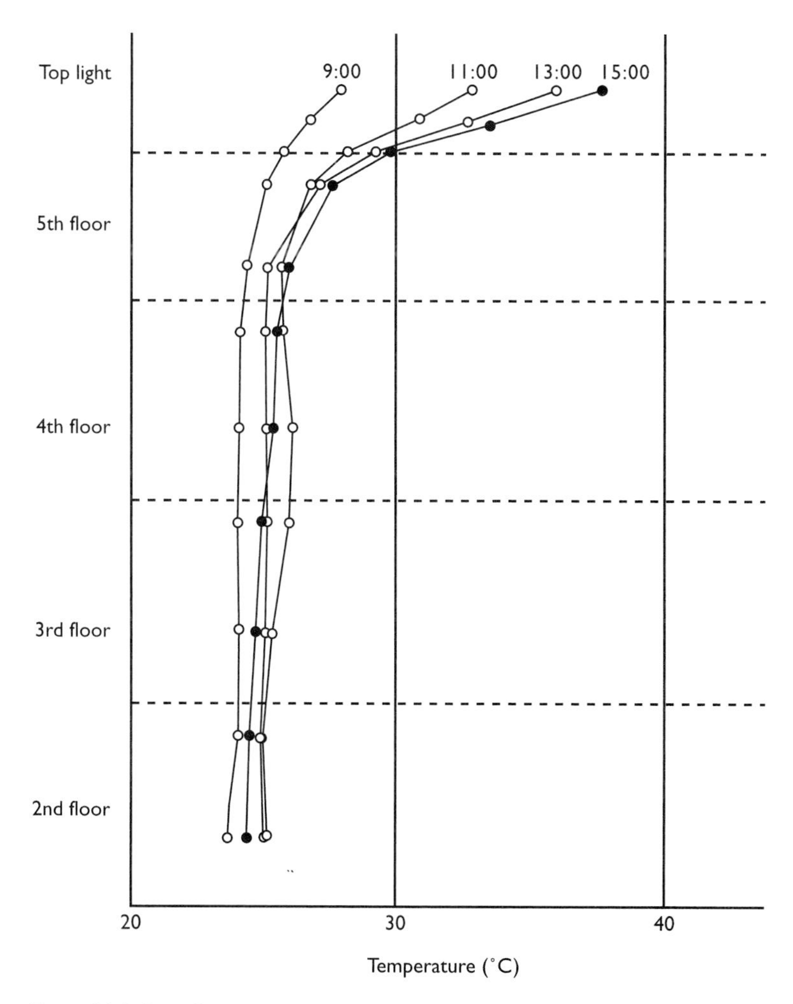

Figure 18.5 Stratification in the atrium: summer

high, although some occupants claimed to be 'a bit hot' when the floor of the atrium had a direct solar gain at around 13:00.

Fragrance environment

Essential oils have a long history in Oriental healing traditions and are popular today in aromatherapy. A recent study shows that 'bathing in woods', meaning walking through a wood with relaxed feelings, improves freshness. Through recent measurements of work efficiency (such as measuring a test group's brain waves, particularly expectation waves), some human physical

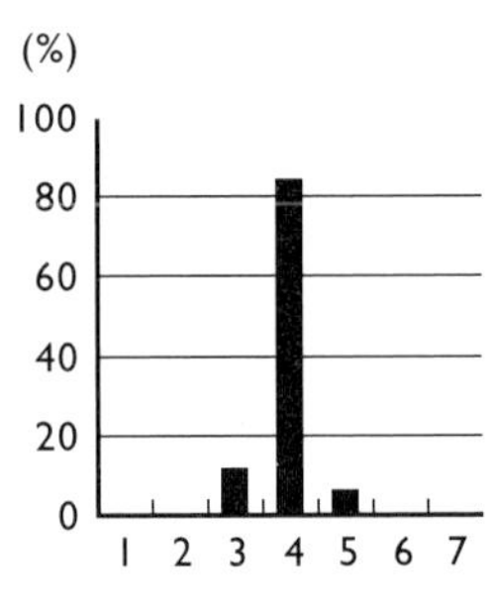

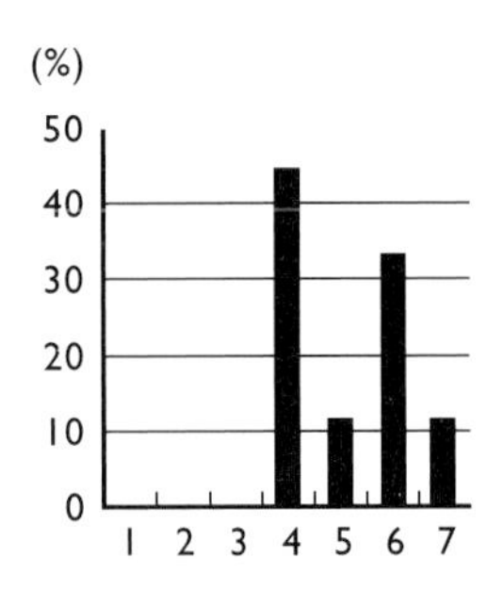

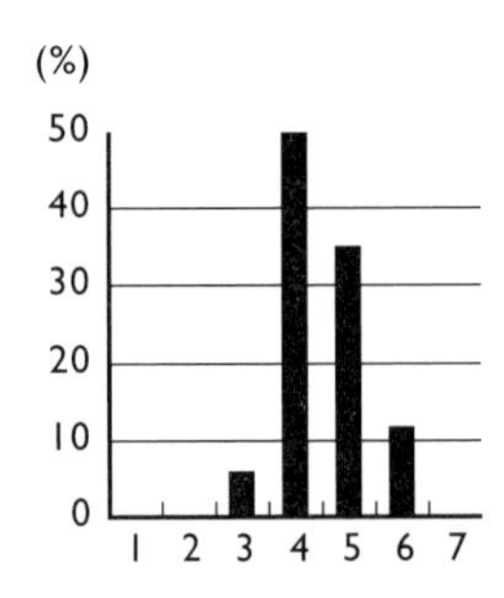

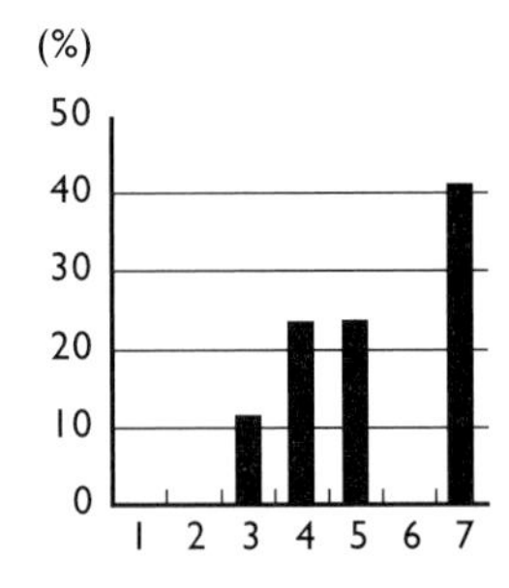

Vertical axes show percentage of vote.
Horizontal axes show occupant's evaluation.
Thermal environment 1, cold; 4, neutral; 7, hot
Comfort feeling 1, uncomfortable; 4, neutral; 7, comfortable

Figure 18.6 Post-occupancy evaluation for thermal environment of the atrium: summer

and psychological responses to fragrances have been documented (Sugano, 1987).

On the other hand, the latest intelligent buildings are fully equipped with office automation equipment, e.g. personal computers, in order to adapt to information technology or borderless business circumstances, and are planned to improve intellectual productivity and creativity of workers. Hence office occupants working in such intelligent buildings are forced to deal with a large amount of information exactly and speedily. This sort of mental tension in the office could result in a physical and psychological strain, so-called **techno-stress** (Shuei-sha, 1991; Industrial Survey Committee, 1987).

This phenomenon indicates that office automation originally intending improvement of the working environment for workers could result in impaired physical and mental health. Intelligent buildings in a real sense have to provide a comfortable workplace in terms of not only physical but also mental aspects, by overcoming harmful effects.

Based on these studies, a fragrance control system was introduced in the KI Building. Its purpose was to create a comfortable environment and combat techno-stress (see Fig. 18.7). The fragrance system consists of an air-handling unit (AHU) for the atrium, a fragrance generator containing three fragrance essences, and a direct digital control unit, with which the air-handling unit controls the dispersal of the fragrances.

The fragrance control scenario (see Fig. 18.8) was based on the human reaction to each aroma and the living pattern of the building occupants. Fluctuation control was introduced to counter the sense of smell becoming dull. Operating status of the fragrance environment system, e.g. status of dispersal, kind of fragrance, and accumulated dispersal volume is monitored by the building management system in real-time mode.

CFD analysis was carried out in order to optimise the volume, pattern and time of fragrance dispersal into the atrium. Figure 18.9 shows the simulated fragrance density distribution in the atrium, and it was expected that the density of fragrance in the occupied zone would reach the specified density in ten minutes of dispersal. The trend graphs showing the fragrance density of supply air from the AHU and return air to the AHU are shown in Figs 18.10 and 18.11.

Post-occupancy evaluation was carried out to prove the effect of the fragrance control system. Figure 18.12 shows occupants' responses. Through careful experimentation, a comfort ratio increase of 24 per cent

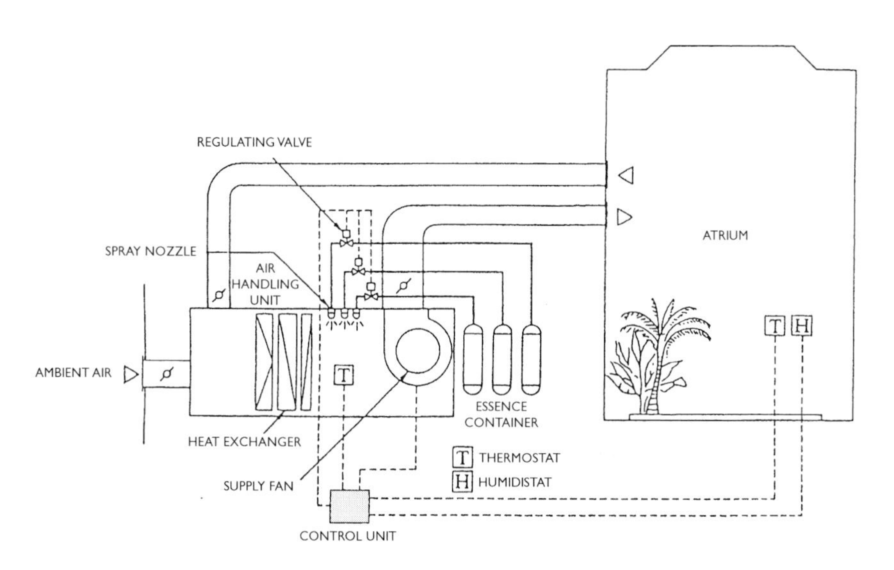

Figure 18.7 Atrium fragrance control system

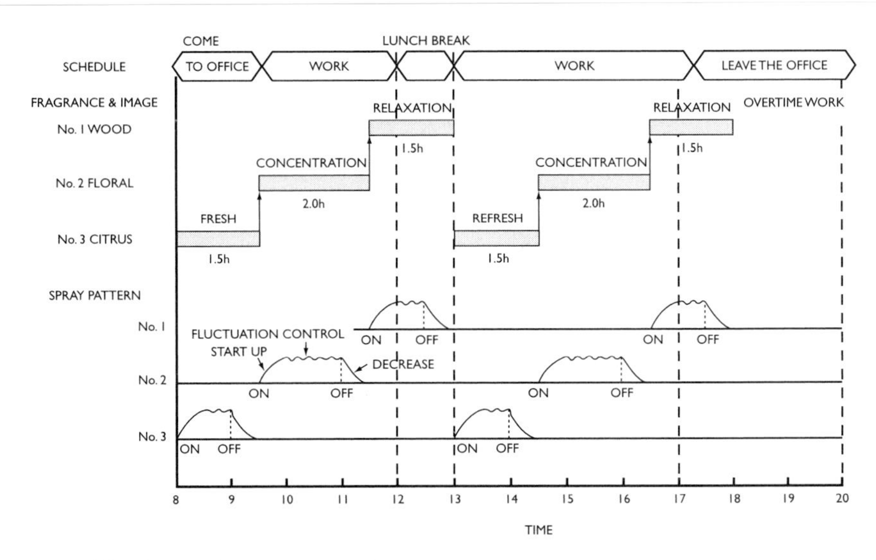

Figure 18.8 Atrium fragrance control scenario

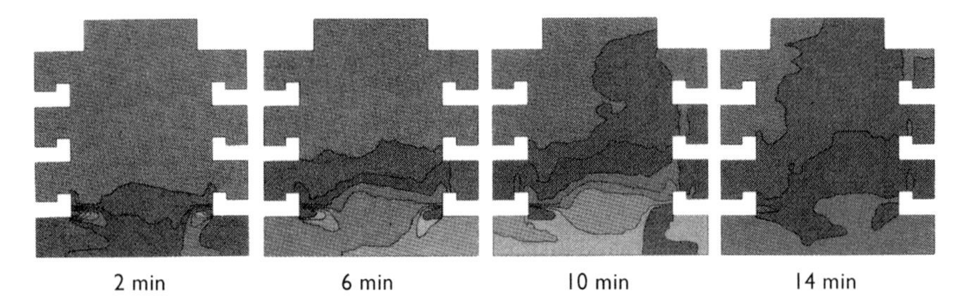

Figure 18.9 Predicted fragrance density distribution in the atrium

(simple mean value) was achieved by implementing the fragrance environment.

Airflow fluctuation control

The pleasurable feelings associated with congeniality and positive attitudes (such as when a person is experiencing a calm breeze at the seashore) can be observed in the heartbeat rate. When spectral analysis with the frequency of f is applied to the intensity of heartbeat fluctuations, the distribution to $1/f$ can be obtained. This is so-called $1/f$ fluctuation. This theory was duplicated and introduced in the atrium's air-conditioning system on a trial basis (Ichinose, 1992).

By loading the software of $1/f$ fluctuation into the building management

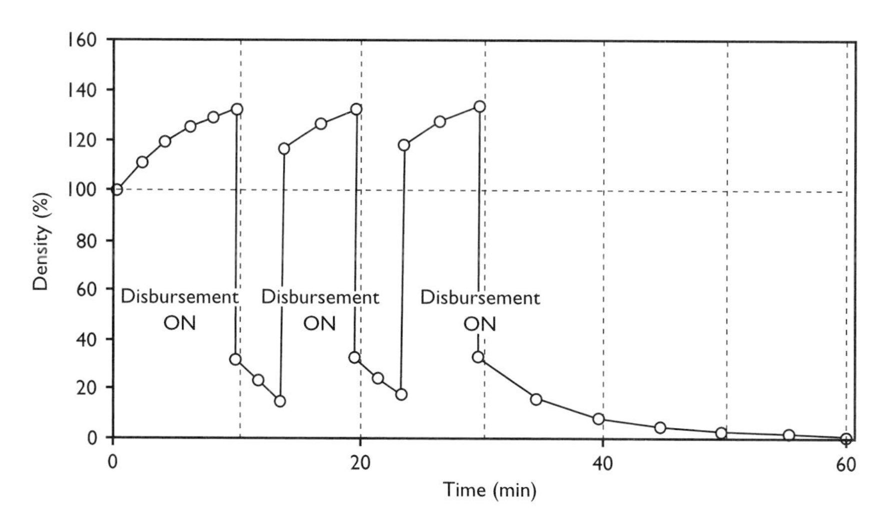

Figure 18.10 Fragrance density of supply air from AHU

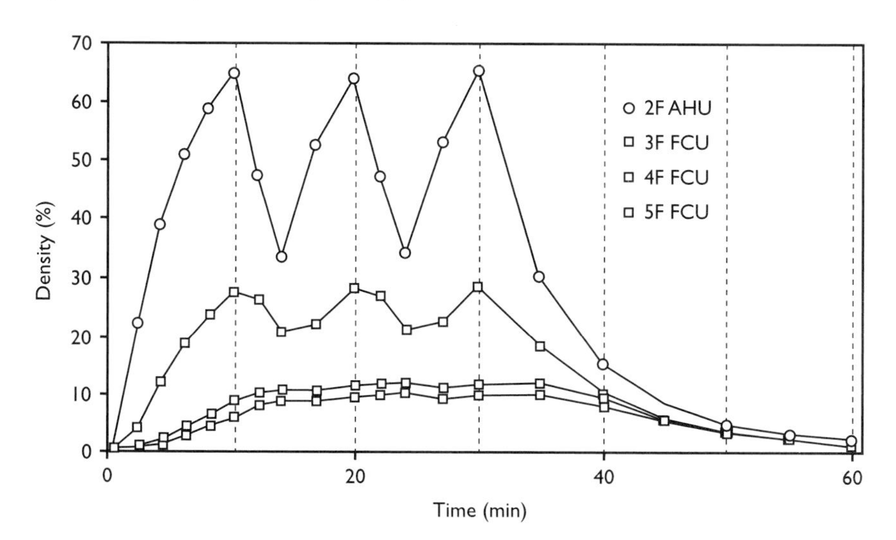

Figure 18.11 Fragrance density of return air to AHU

system, the speed of the inverter (variable voltage and variable frequency) in the atrium's air handler is controlled. The control cycle was set for approximately 10 minutes and spectral analysis of air velocity measurement results shows the distribution of $1/f$ (see Fig. 18.13). $1/f$ fluctuation control was applied for only cooling, and it was set to operate only when the supply air temperature was equal to or less than 24°C (set temperature in summer is 26°C).

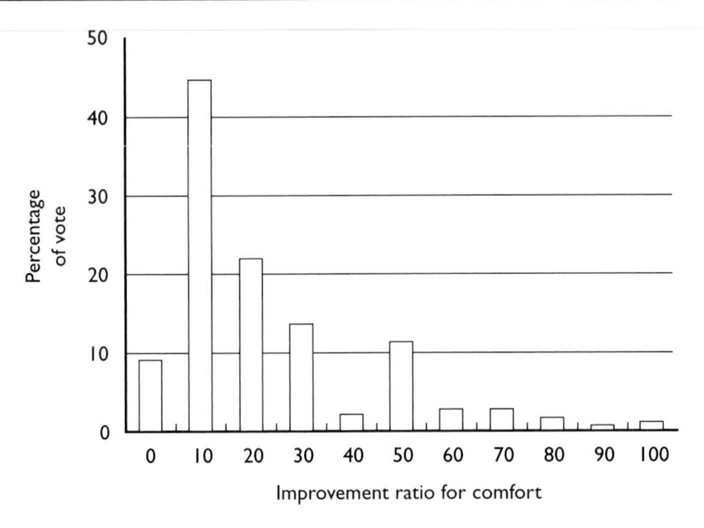

Figure 18.12 Post-occupancy evaluation for fragrance control

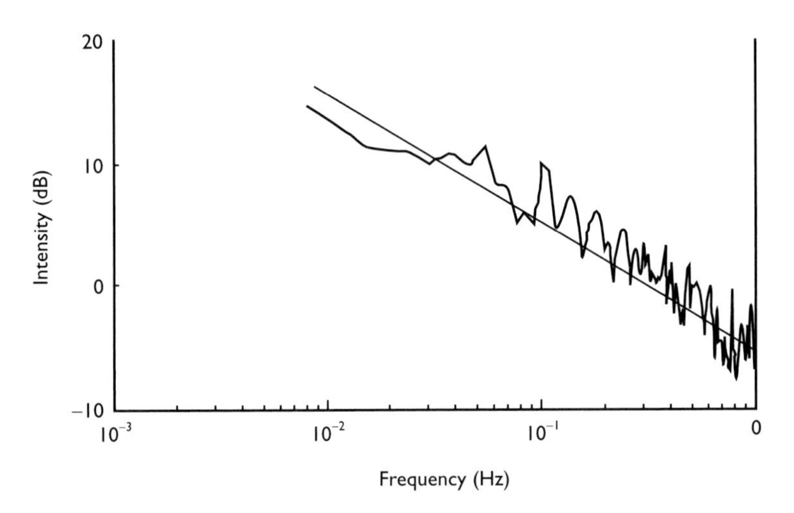

Figure 18.13 1/*f* distribution

A post occupancy evaluation was carried out to prove the effect of airflow fluctuation control. A total of 77 people replied, and we can conclude that a certain level of satisfaction has been achieved. The details are as follows:

- more than half of the occupants that replied felt air fluctuations, but some of them may not have recognised it had they not been asked
- most of the occupants who did recognise the airflow fluctuations considered it preferable and comfortable; some would have preferred it to be stronger

- most of the occupants who were not comfortable felt cold due to the air flow
- male occupants generally recognised the air fluctuations rather than the female occupants; this tendency depended on the location and clothing
- airflow made the occupants feel colder, and sunlight made the occupants hotter; this tendency was dependent on clothing
- most of the occupants who claimed they were not comfortable were those who felt less air flow and more sunlight.

Biomusic

Another innovative amenity of the KI Building was the creation of a musical environment. Music was composed for the building based on analysis of the human brain wave response to sound waves. The so-called biomusic was introduced specifically to upgrade office comfort.

The musical programme alternated between achieving stimulated and relaxed mental states. In a relaxed state, α waves of 8 to 12 Hz are dominant; in a stimulated state, β waves of 14 to 20 Hz are dominant (Nuki, 1990). The different types of biomusic were properly combined for the programme. Fig. 18.14 shows the percentage variance of α and β waves with time when test subjects listened to stimulating music and relaxing music.

The music in the atrium area has had favourable results. Accordingly, a

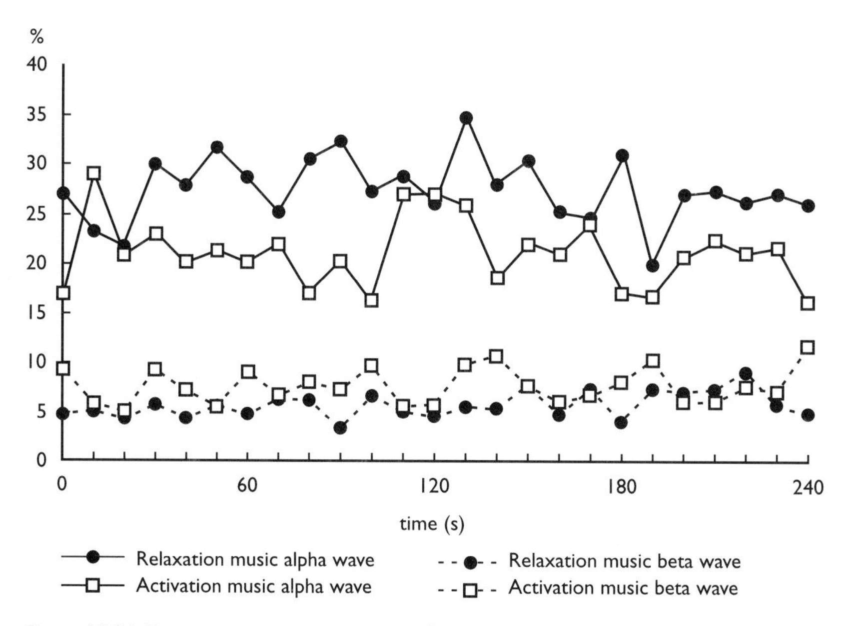

Figure 18.14 Percentage variance of α and β waves

music programme is being planned for work spaces separated from the atrium, to comply with the needs of the occupants in these areas.

Diversifying workspaces

The environmental advantages of intelligent buildings are often degraded by changes in office layout or applications after practical completion of the building. These changes are typically deemed necessary to meet changing demands on corporate resources.

However, these changes can have adverse effects on the indoor environment. Specifically, cooling loads from new office equipment can increase, and these loads can be unevenly distributed.

To prepare for such situations in the KI Building, technologies were introduced to upgrade flexibility and reliability and save energy. The VAV diffuser was one such technology. This new product has three functions: temperature control, temperature detection and air flow rate control. The air flow rate can be controlled for each 3.6 m grid.

As the room temperature begins to rise due to the increasing cooling load, the power element containing wax in the VAV diffuser expands to open the damper, providing a higher air flow rate (see Fig. 18.15). When the room temperature drops, the power element contracts, reducing the airflow rate until the room temperature reaches the set point. According to the test in the mock-up room, it was proved that room air was induced toward the temperature sensor, which detects approximately the room temperature. At the start of heating operations in winter, the diffuser spring of shape-memory alloy fully opens the damper when the supply air temperature is equal to or more than 28°C. This aids in reducing warming-up time. This self-regulating control VAV diffuser works without the building management system, therefore installation costs as well as running energy costs are lower.

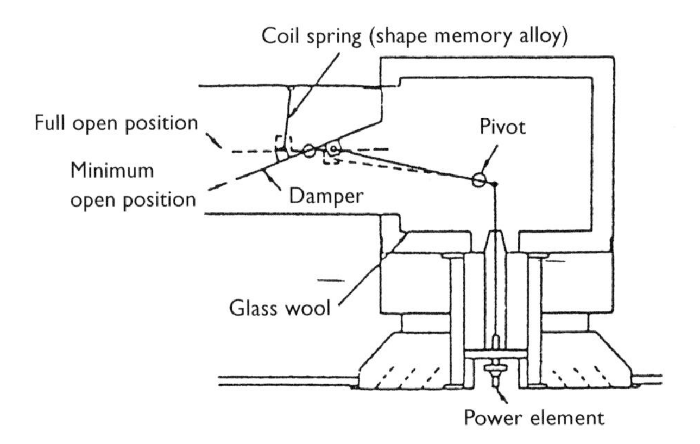

Figure 18.15 Cross-section of the VAV air diffuser

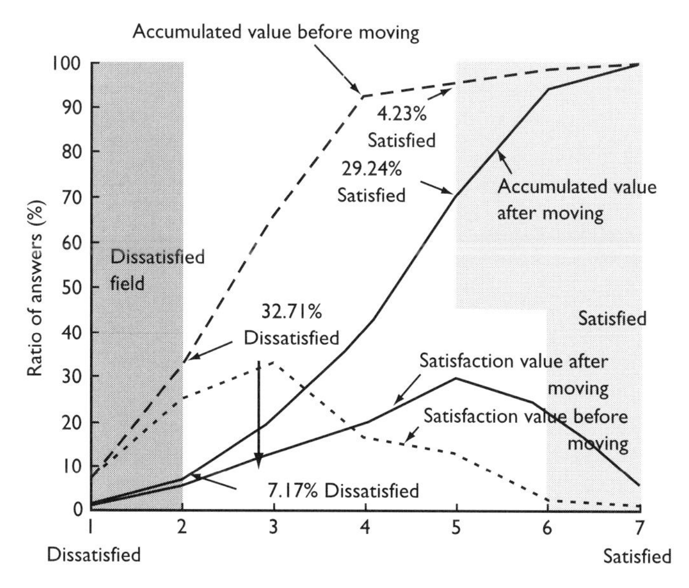

Figure 18.16 Occupant satisfaction with the indoor environment

Sound masking

In an open office area, occupants sometimes complain about noise from the conversation of others, especially telephone conversation. In the KI Building, there is no glazing between the adjacent office space and the atrium, therefore it was expected occupants would suffer from the atrium noise. A sound masking system was introduced for the open office area facing the atrium to combat this noise. According to the questionnaire responses, it had favourable results.

Occupant satisfaction

The creation of a comfortable environment by using several new technologies has been appreciated by KI Building occupants. Figure 18.16 depicts the satisfaction of occupants with their working environment.

Specifically, comfort dissatisfaction decreased to 7.17 per cent. This is approximately one-fifth of what it was before workers moved into the KI Building. The satisfaction level increased seven-fold.

References

Ichinose, H. (1992) *Air Conditioning of 1/f Fluctuation*. National technical report. Matsushita Electric Co., Tokyo **38**(1), Feb.

Industrial Survey Committee (1987) *Ergonomics and Building Management/Office Dictionary*. Tokyo.
Nuki, Y. (1990) Bio-Music for Amenity Formation. *Bio-Musics Seminar*, Tokyo, Japan Management Association, Feb.
Shuei-sha (1991) *Mental Distortion by Stress/Atlas of the Human Body*. Tokyo.
Sugano, H. (1987) Psychophysiological research of odors. *Fragrance J.*, no. 86.

Chapter 19

Employee productivity and the intelligent workplace

Walter M. Kroner

This chapter presents the results of two consecutive research studies related to individual control of the microenvironment and human performance. First, it briefly describes the results of a year-long study of the emerging developments related to individually controlled environmental systems (environmentally responsive workstations, ERWs[1]) and their associated technologies. Second, it presents the results of a year-long field study related to ERWs and worker productivity conducted at an insurance company in the northern Mid-west of the United States. The study revealed a statistically significant positive association between the change in productivity and the change in 'overall satisfaction' with the workspace (Kroner *et al.*, 1992). Researchers found that improved indoor architectural and environmental design contributed to an overall increase in productivity of 16 per cent. ERWs were estimated to increase the level of worker productivity by approximately 3 per cent.

Towards an ERW research agenda: state-of-the-art study background

Three distinct events between 1985 and 1989 predated this study:

1 1985 – the Architectural Research Center Consortium Workshop on the Impact of the Work Environment on Productivity (Dolden and Ward, 1985)
2 1988 – the International Symposium on Advanced Comfort Systems for the Work Environment (Kroner, 1988)
3 1989 – the Office Productivity and Workstation Environment Control Research Planning Workshop (W.I. Whiddon and Associates and Ostgren Associates, 1989).

Two clear research challenges resulted from these three workshops: (1) the ERW concept should be researched and developed, and (2) research related to ERWs' impact on worker productivity was essential.

All three groups emphasised that in the context of continual change, whether those changes are in the building design decision-making process, in the occupant's organisational and managerial processes, or related to the shift towards employing ever-increasing numbers of knowledge workers, expectations are constantly shifting. With such a dynamic work-environment it is difficult to predict needed or desired levels of performance for almost any of the building system and environmental quality categories. As a result, the ERW concept should be explored and developed along with the concept of heterogeneous environmental systems. Such systems include combinations of ERWs, conventional HVAC systems, and other environmental strategies.

In 1990, in direct response to the above research challenges, the authors undertook a state-of-the-art study of ERWs and related technologies in an effort to: (1) pull together information relating to divergent and isolated ERW developments; (2) review critically the technical dimensions of ERW-related developments by industry and their experience; and, (3) identify the key issues and questions which should inform the continued development of this potentially significant research area (Stark-Martin and Kroner, 1991).

The study concluded that ERWs are an emerging technology and represent a new and innovative way of potentially improving worker comfort and productivity and energy efficiency in buildings, and providing new demand-side management strategies to building owners. There is strong evidence to suggest that, if ERWs are integrated with a building's thermal inertia, night-flush cooling potential and similar strategies, energy costs in office-type buildings can be reduced by as much as 40–50 per cent (Von Thiel, 1987).

The impact of ERWs on productivity and satisfaction: an insurance company's office environment

The study was initiated in 1990 in direct response to the research needs identified above. The study's research objectives were to analyse the impact of ERWs on office worker productivity and worker response to individualised environmental conditioning. The study consisted of three distinct surveys: (1) a productivity analysis of individual workers; (2) a comfort and satisfaction survey; and (3) an examination of worker absentee patterns.

The study had several unique strengths. It used an established company-generated productivity monitoring system; combined objective productivity data with multiple subjective assessments of worker satisfaction and comfort; included measurements of three distinct influences on productivity (a major organisational relocation, a new built environment and a new environmental conditioning technology); and included randomised experimental intervention to assure the internal validity of assessments of causal effects.

The insurance company and ERWs

The insurance company occupied its 'old' office building from 1960 through 1991. Expansions to the original structure in 1972 and 1982 resulted in a final gross floor area of 61,800 square feet. The old building had conventional lighting, HVAC, and ventilation systems similar to most office buildings today. In July 1991, the company moved into a newly designed office building, within the same city, encompassing 149,800 square feet of gross floor area. This new building was designed to incorporate 370 ERWs, making it the largest ERW office installation in the world.

The particular ERW involved in this study provides the workstation occupant with individualised control of: the temperature, velocity and direction of air delivered to the desktop through two air diffuser towers; a radiant heat panel located below the desk top; a desk-mounted task light; and a sound-masking device. Each ERW has its own replaceable air filters. An occupancy sensor shuts down the unit if the workstation is not occupied for over ten minutes and returns it to the set levels of operation when the occupant returns.

Methodology

The study began on 1 January 1991. For 27 weeks researchers observed conditions and collected productivity data in the old building. In July 1991, the company employees moved to the new office building, providing the study team with 24 weeks for productivity data collection and observations there. The move to the new building created problems of disentangling the separate impact of the new building and the ERWs, however. The research team's response to this challenge was deliberately to introduce variation in the 'application' of ERWs. This meant that, while we continued to receive information from the insurance company on each worker's productivity, we randomly disabled their ERWs during the last 24 weeks of the study. We selectively 'disabled' the ERW units by disconnecting three features: air temperature and velocity control and the radiant heat panel. All other features were left intact.

The productivity analysis

The productivity analysis focused on 118 workers in the Underwriting Departments. The measure of productivity was the number of files each worker processed during a given week. The productivity assessment method used was company-generated, had been in place for over two years prior to the analysis, and was known to and accepted by the Company's employees. Subjects were not informed that an analysis of their productivity was being conducted by the research team. They were told that two surveys were under way; one which examined their reactions to their work environment through a questionnaire, and the second an energy study which required the random disabling of some ERWs for a period of time. Since the company's productivity measurements were ongoing and were not specifically noted by the employees, the effect, productivity, was not affected by the subjects' knowledge of being under study.

The Tenant Questionnaire Survey Assessment Method

To measure worker comfort and satisfaction, the study team utilised the Tenant Questionnaire Survey Assessment Method (TQSAM) (Dillon and Vischer, 1987). It is based on occupant surveys using a standardised questionnaire. The TQSAM provides a quick and efficient method for assessing office buildings from the user's point of view. The results of the survey are used to generate a profile of the tested building, which can then be compared to a normative profile and/or a previously occupied building's ratings. The decision to use the TQSAM was based on the following: (1) it was a readily available 'tool'; (2) its use allowed the study team to compare the company's office building with other office buildings in the TQSAM database. The study team realises that there may be some inherent problems with the TQSAM due to the statistical techniques used in its development. However, because the emphasis of this project was on the productivity study, the TQSAM survey form was adequate for our use.

The TQSAM was administered to the company's employees on three separate occasions: in March, while the company was still in its old building; in July, soon after the company moved into its new building; and in November, after occupants had a chance to acclimatise to the new building.

Study results

The productivity study

We studied changes in productivity associated with three different causes. Table 19.1 shows the results of the significance tests. Figure 19.1 presents a schematic view of the changes in productivity over the course of the study. The greatest effect was the drop in productivity caused by the disruption of the move between the old and new buildings. The second, and perhaps most significant, effect was the improvement in productivity that resulted from the move to the new building with its innovative environmental and architectural design concepts. The third most notable effect was the increase in productivity associated with the new ERW technology.

The most subtle effect was the drop in productivity associated with temporarily disabling the ERWs. The size of this effect was attenuated by noncompliance with the experimental protocol: workers frequently insisted that

Table 19.1 Estimated median changes in productivity by cause (95 per cent confidence interval for median percentage change)

Cause of change	*Low*	*Mid*	*High*
Disabling the ERWs			
unadjusted	−18.3%	−11.8%	−1.7%
adjusted	−22.9%	−12.8%	−2.7%
New building with ERWs	4.4%	15.7%	28.4%
Move between buildings	−39.7%	−31.7%	−21.4%

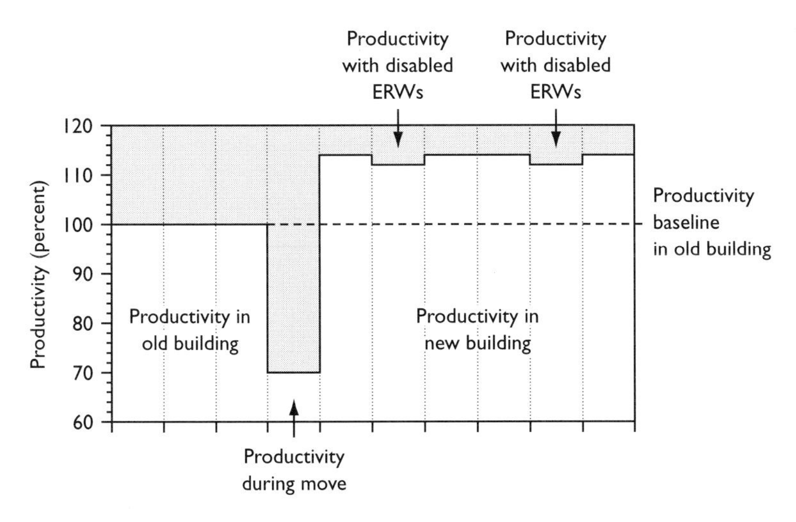

Figure 19.1 Changes in productivity over the course of the study

the disabled ERWs be reconnected. When we plotted a worker's average percentage change in productivity against the average fraction of full ERW function during the experimental weeks, we found a weak ($R^2 = 4$ per cent) but statistically significant ($p = .04$) positive association. Therefore, we extrapolated to full compliance with the experimental protocol and used the regression intercept as an adjusted estimate of the experimental impact. After this adjustment, the drop in productivity associated with disabling the ERWs was even more significant. Furthermore, the fact that there was a correlation between the extent of ERW disabling and the extent of the drop reinforces the case that it was indeed the experimental manipulation that caused the change in productivity.

In round numbers, we can summarise Table 19.1 as follows. The move between buildings temporarily reduced productivity by about 32 per cent. In the longer run, the combination of new building and ERWs increased productivity by about 16 per cent over the baseline in the old building. Disabling an ERW for one week reduced productivity by about 13 per cent relative to the new baseline in the new building.

These figures apply to the median worker, so half the workers had larger changes and half had smaller. We note that the figures, while quite certain as to sign, are rather uncertain as to magnitude. Thus, the productivity drop associated with the move could actually be anywhere in the range from 21.4 per cent to 39.7 per cent, with 31.7 per cent as the most likely estimate. The increase associated with the combination of new building plus ERWs could be anywhere in the range from 4.4 per cent to 28.4 per cent, with 15.7 per cent most likely. Likewise, the decrease associated with disabling the ERWs could be anywhere from 2.7 per cent to 22.9 per cent, with 12.8 per cent most likely. These uncertainties are rather large but accurately reflect the extent of variability within and across workers and the relatively small numbers of person-weeks in the dataset.

Finally, it appears that disabling the ERWs resulted in something like a 2 per cent overall drop in productivity. We arrived at this figure by multiplying the median 12.8 per cent drop in productivity when the ERWs were disabled by the median 15.7 per cent increase in productivity achieved in the new building [(+ 15.7% change from new building + ERWs) × (– 12.8% change from disabling ERWs) ≈ – 3 per cent impact on productivity associated with ERWs alone] (Fig. 19.2) Given the uncertainties in the components, the actual figure could easily fall anywhere between a fraction of 1 per cent and 7 per cent, with 2 per cent a reasonable point estimate.

Stepping back from the details, we believe these findings are a major step in the study of the link between work environment and productivity. We

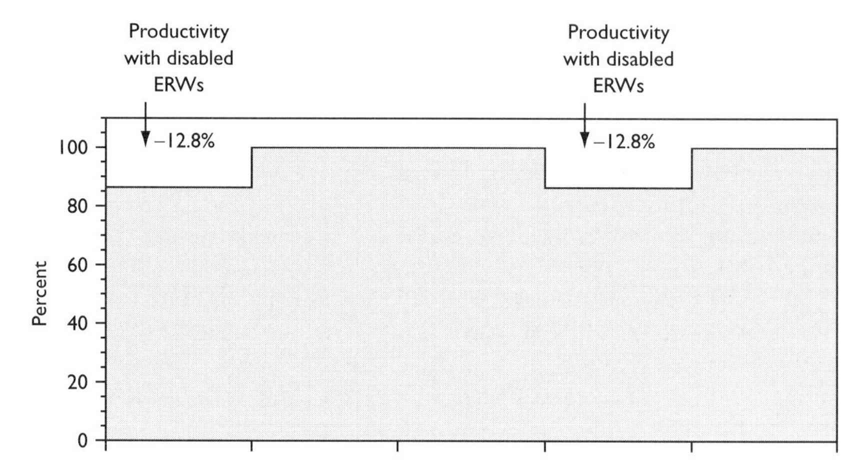

Figure 19.2 Total increase in productivity in the new building with ERWs

have solid statistical evidence that objectively computed measures of productivity increased for underwriting personnel in the new, ERW-equipped building. Because of our experimental manipulations, we have solid statistical evidence that the ERWs played a significant part in the increase, because temporarily disabling the ERWs caused statistically significant decreases. Finally, to give some scale to these changes, we showed that the combined positive effect of the new building and ERWs was about half the size of the transient negative effect caused by the move between buildings.

As previously mentioned, the 16 per cent increase in productivity associated with the move to the new building is perhaps the most significant finding of the study. For the first time we have a quantitative measure of the significance of a high-quality indoor environment that has been architecturally and environmentally designed to facilitate work-flow management and maximise occupant comfort and satisfaction. The fact that this study was carried out in a real office building, as opposed to a laboratory, and used an established employer-generated productivity measure is equally significant.

While the results of the productivity study provide useful and interesting productivity data as mentioned above, they also raise several additional questions. For example, though we know that productivity dropped when the ERWs were disabled, we cannot be certain whether this effect was caused solely by a reduction in comfort, by the loss of individual control associated with the disabled ERW, or because individuals were frustrated due to being inconvenienced. Additional questions remain as to whether the results would have been similar had we disconnected the task-lighting and/or some other ERW function. Further study in these areas is indicated.

Conclusions

A number of factors contribute to making this study a unique contribution to the field. The impact of quality architectural design has often been debated, but never quantified. The quantifying of productivity as it relates to architectural design and/or a particular environmental technology satisfies a long-standing need for a basis for decision-making related to building design, innovation in environmental technology and energy, and other considerations.

Through an almost uncanny series of coincidences, we were able to take advantage of a rare opportunity that involved the largest installation of environmentally responsive workstations (ERWs) in the world, in a company that had a tested productivity measure in place, and at a time when the company was moving from a conventional office building to a modern and innovative architectural design. We were handed, in other words, a clear-cut before-and-after scenario. Up to now, a significant amount of analysis has been performed on the impact of environmental quality on productivity. This work, however, has either been performed in laboratories or in field studies that attempted to measure the impact of isolated environmental factors on productivity using measures meaningful to scientists, but not to managers in the workplace. The study was designed to focus on knowledge workers and provide information pertinent to management as well as the research community.

Out of our good fortune came a series of findings, of which the most significant are the following.

- We can claim with certainty that individual control over one's environment yields a 2 per cent improvement in productivity. The actual improvement in productivity produced by the ERWs is undoubtedly greater, but given the relatively short time span (two weeks) in which we were able to disable the ERWs and study the results, this is what we can claim conservatively. However, the Senior Vice President of the Insurance Company, states that 'By our results, productivity went up from 4 to 6 per cent, and with a payroll of over $10 million a year, that's a substantial return on investment'.
- In the TQSAM study of the individual workers, the tremendous transformation of the workers' response to the perceived quality of their environment is equally significant.
- In a holistic sense, these findings allow us to quantify the impact of a well-designed, high-quality environment on productivity.

What our findings do **not** tell us is what particular aspect of environmental quality is most important to an individual worker. We do not know, for instance, the relative importance of the individual's control of air velocity versus air temperature. We do not know whether a direct flow of air is preferable to an indirect flow of air. And we do not know the circumstances under which a worker would be willing to trade a movable radiant heat panel below the desk for one at a different location or with a different configuration.

However, it **is** important to be able to establish quantitatively and without doubt that the ability of an individual to control his or her own work environment in response to personal needs makes an appreciable difference. In fact, we have anecdotal evidence that individual control is such a powerful factor that workers have refused offers of higher-paying jobs elsewhere once they realised that the new position would not include individual environmental controls.

As is invariably the case, arriving at answers to pressing questions has uncovered a whole new set of concerns, for example:

- the availability of individual controls appears to contribute to lower energy consumption – but the extent of the reduction needs to be quantified

- the availability of these controls leads to a significant drop in worker complaints – but that, too, needs to be quantified
- economic impacts result from worker satisfaction and retention – employers have fewer people to replace and train – but these impacts remain to be measured.

Clearly, the study justifies additional studies of this type. We feel that it is well worth investing R&D funds to gain a better understanding of individually controlled environments in both economic and environmental terms.

References

Dillon, R. and Vischer, J.C. (1987), *Derivation of the Tenant Survey Assessment Method: Office Building Occupant Survey Data Analysis.* Public Works Canada.

Dolden, M.E. and Ward, R. (eds) (1985) *Proc. ARCC Workshop on the Impact of the Work Environment on Productivity*, April, AIA, Washington, DC.

Kroner, W. (ed.) (1988) A new frontier: environments for innovation. *Proc. International Symposium on Advanced Comfort Systems for the Work Environment*, May. Center for Architectural Research, Rensselaer, Troy, NY.

Kroner, W.M., Stark-Martin, J.A. and Willemain, T. (1992) *Using Advanced Office Technology to Increase Productivity: The Impact of Environmentally Responsive Workstations (ERWs) on Productivity and Worker Attitude.* Troy, NY: Center for Architectural Research, Rensselaer.

Stark-Martin, J.A. and Kroner, W.M. (1991) *Environmentally Responsive Workstations (ERWs): Toward A Research Agenda.* Troy, NY: Center for Architectural Research, Rensselaer.

Von Thiel, D. (1987) *Vergleich verschiedener Heiz-und Luftungssysteme mit einer Klimaanlage.* Cologne: Schmidt-Reuter.

W.I. Whiddon and Associates and Ostgren Associates (1989) *Proc. Office Productivity and Workstation Environment Control Research Planning Workshop*, Electric Power Research Institute (EPRI). Palo Alto, CA, USA.

Note

Environmentally responsive workstations (ERWs) integrate and provide heating, cooling, lighting, ventilation and other environmental qualities directly to the occupants of workstations. Additional integrated components may include: communication and information systems, electric power service, optical view panels, and fragrance options. The key feature of an ERW is that the occupant controls, modulates, and maintains the environmental conditions. ERWs are designed to operate when the workstation is occupied. ERWs are at their best if they are integrated with an environmentally responsive architecture.

Chapter 20

Future design – guidelines and tools

John Doggart

Designing buildings is a complex task, requiring many judgements and priorities to be made. Costs, spatial layout, services and aesthetic issues are just four of the factors which the design team must reconcile, and the good designer is the one who achieves the right balance.

One priority is starting to stand out above all others: to produce a high-performance building that outperforms its competitors by contributing to better occupant health and well-being. By raising the importance of these factors, productivity gains can be made which overwhelm most other considerations. For example, a 1 per cent gain in productivity is typically worth more than saving the entire fuel bill, while a 5–10 per cent loss in productivity can wipe out a company's profit and send it to the wall.

So how can the designer contribute to increased productivity? Until recently information has been scanty and uncertain, but there is a growing body of knowledge that allows design performance to be predicted with some confidence. Even when benefits cannot as yet be quantified, and there are many areas where this is true, the evidence shows that one direction is better than another, i.e. which factors are beneficial and which are to be avoided.

Productivity is affected by a huge range of design factors:

- comfort temperatures
- cooling systems
- thermal mass
- window design
- day and night light
- controls
- finishes
- building sickness
- built form.

All of these are under the control of the designer. So what should the design team aim for, and what tools and guidelines are available? These notes aim to provide some guidance, while recognising that these are early days and much more remains to be investigated before a full and stable picture emerges.

Comfort temperatures

In the 1930s the idea took hold that people were like machines, requiring certain average operating conditions to function satisfactorily. For the designer this manifested itself in simple guidelines, for example that temperatures should be 21°C in winter and 22°C ± 1 K in summer. Clearly this assumes that all people want the same temperature, all the time. This over-simple view of human needs is being supplanted by more subtle understanding of people and their needs.

One of the most interesting challenges to the mechanistic view is: why do most people in the UK prefer naturally ventilated buildings, when these plainly don't meet the standard comfort criteria? This was shown in a survey for agents Richard Ellis, where 89 per cent of the sample said that they preferred buildings without air conditioning (Harris Research Centre, 1994). Another survey indicated that this was only true if severe overheating was avoided in summer.

These results require that comfort definitions are revisited. It now appears that occupants accept, and perhaps even like, variations of temperature over time, provided that they remain within overall limits. In particular occupants seem content to accept higher temperatures for short periods of time: the higher the temperature the shorter the time. These times and temperatures vary between different sources, but a coherent view emerges when these sources are aggregated (Table 20.1).

The values in Table 20.1 provide a coherent time–temperature profile which can be specified by the client and used by the design team to test their design. This is carried out using a detailed energy and temperature analysis program such as TAS, ESP, or APACHE. Predicted performance is then compared with the time–temperature profile. Because of program

Table 20.1 Acceptable temperatures and time periods

Temperature	*Maximum time for temperature to be exceeded*	*Source*
21°C	100%	All
25°C	5%	Cohen (1993)
27°C	2.5%	CIBSE (1986)
28°C	1%	Cohen (1993)
28°C	<30 degree hours	Cohen (1993)
29°C	0%	Petherbridge *et al.* (1988)

uncertainties, we like to use a safety factor of two, i.e. that predicted temperatures never exceed half those allowable by the time–temperature criteria.

By using this technique, severe summer overheating is avoided. At last a client can get a building that will perform well and is likely to be preferred by occupants, without summer overheating which would erode productivity gains.

Comfort cooling

In virtually all offices, some degree of comfort cooling will be required at some time of the year. This can be provided by conventional air-conditioning, but this route is now being severely questioned. Air-conditioning is an expensive user of fuel, with consequent effects on increasing greenhouse gas emissions. Fuel costs can be £8/m^2 per year more than natural ventilation, and extra maintenance costs can increase the cost difference to £13/m^2 per year. Air-conditioning is also frequently cited as a cause of sick building syndrome, and a paper at the 1997 European Respiratory Society Conference reported that people working in air-conditioned offices are almost two and a half times more likely to suffer from respiratory infections than those in naturally ventilated buildings. These conditions accounted for 17 per cent of days off work among staff in air-conditioned offices, compared with 9 per cent in other buildings (Teculescu, 1997). Coupled with the preferences previously stated for naturally ventilated buildings, there now must be a serious question mark on the continued use of air-conditioning, except where necessary. Certainly air-conditioning is no longer a safe haven for designers and occupants, but must be judged alongside alternatives which often appear less risky and more healthy.

Obtaining reliable cooling from natural sources requires a different approach to design, using many elements of the building in combination to obtain the desired effect. The techniques vary according to climate. In northern temperate climates the most common method is to design with more heavy materials to make the building into a giant thermal store, which absorbs temperature rises during the day. At night the building is cooled down by cool air, leaving it ready for the next day. In Mediterranean countries this strategy is reinforced by more active techniques such as pre-cooling the air through pipes embedded in the ground. In hot countries air movement can be used to provide comfort conditions (ECD Energy and Environment, 1998). Detailed modelling and air flow analysis is needed to predict performance with confidence.

Window design and daylight

Windows are vital to the comfort and well-being of the occupants, but their benefits have been neglected for many years. On apparently reasonable economic grounds, floor plates have been getting deeper and windows often cannot be opened. Yet this goes completely against occupants' desires. When Longman's new head office was being designed, the staff were asked what features they would prefer. Unprompted, 80 per cent said that they wanted an openable window, far higher than any other wish (Standley, 1994).

In another example when the Open University was upgrading its premises, a trial was made with retrofitting air-conditioning on one floor, as an alternative to improved windows and thermal mass to another. In a subsequent survey occupants overall said they preferred the naturally ventilated floor (BRECSU, 1997). To produce this result, designers needed to design the windows carefully to avoid draughts, make sure that blinds did not clash with opening windows, and that the manual controls were easy to use and immediately understandable. Programs such as N-Light and Daylight provide the designer with the means of predicting adequate daylight (ECD Energy and Environment, 1997, 1999).

The growing importance of windows has resulted in Dutch Health Codes requiring that no one should sit more than 6 m away from a window for any extended period. This requires buildings to be narrow plan, typically 12 m wide.

Nightlight

In the absence of daylight, tasks still need to be well lit. Six per cent productivity increases have been reported by the Rocky Mountain Institute at a Post Office at Reno, and 15 per cent gains at Lockheed (Romm and Browning, 1994). The tools for achieving this improvement are well known, taught to engineers, and easy to apply – there really is no excuse for poor design.

The quality of light is also very important. The BRE has shown that high-frequency lighting cuts headaches and eyestrain by 50 per cent in offices (Wilkins *et al.*, 1989), worth about 0.5 per cent in productivity improvements through reduced absenteeism. High-frequency lighting should therefore be fitted as standard.

Controls

Personal control is greatly prized by occupants. There can be problems between occupants when they need to share controls; to avoid problems the number sharing them should be kept to seven or less. Controls which switch on lights automatically are disliked intensely; the general rule is to switch on manually, and have absence detection to switch off when people leave their room or space. Controls should result in fast reaction response, either directly by occupants operating their own controls or through a facilities manager, the latter requiring excellent organisation if delays, dissatisfaction, and consequent productivity losses are to be avoided.

Finishes

Allergic reactions to office materials are becoming increasingly common, due to a cocktail of problems. Two main issues stand out. The first problem is off-gassing from materials and finishes. Volatile organic compounds (VOCs) result from many materials, such as solvent glues used to fix down carpets, solvents used in paint, and formaldehyde used in chipboard and many manmade fibre boards. Some of these have short-term effects, but others can last for months and even years. Careful specification and preference for natural materials can avoid the problem at source. The likelihood of off-gassing of suspect materials can be checked using a method developed at SBI, the Danish Building Institute (BRECSU, 1997).

Dustmites are the second major problem. As buildings have become sealed up and temperatures have risen, conditions for dustmites have improved. To combat the problem, materials such as carpets, soft materials and open shelves which act as breeding grounds, should be avoided.

Building sickness

Sick building syndrome is not yet understood, but its effect on absenteeism is undoubted. Figure 20.1 shows that the reduction of one building sickness score reduces absenteeism by around five days per person per year, or about 2.5 per cent (Jones *et al.*, 1995). Avoidance is more difficult, but guidance is given in the Building Research Establishment Environmental Assessment Method (BREEAM) for existing offices (BRE, 1993). This lists the main features of a building that correlate to building sickness and their relative weighting. One BREEAM credit is given for 30 points or more, two for 45 points or more and a maximum three credits for 60 points or more (Fig. 20.2).

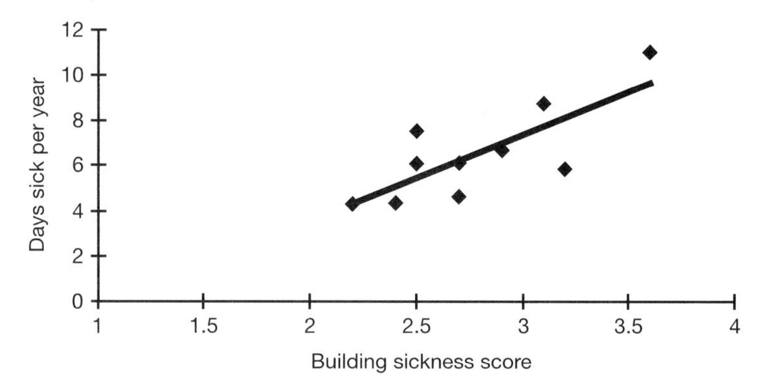

Figure 20.1 Health and productivity in the office. Each diamond represents the data from one building

Building form

The requirements listed in Table 20.2 suggest a particular building form for temperate climates, with floor plates about 12 m wide, maximum of seven people sharing controls (preferably individual control), $\geqslant$3 m high ceilings to aid natural ventilation, and cross-ventilation designed in. Air-conditioning is not needed in most temperate climates, and this can be extended to Mediterranean climates with enhanced natural cooling systems.

Table 20.2 Indicators of healthy buildings

Points	*Issue*
	Heating, ventilating and air-conditioning
5	no air-conditioning (except in computer suites, secure rooms and other special high heat load situations) and building designed to avoid overheating
3	openable windows or mechanical ventilation with individual control
3	air intakes (a) designed to ensure exhaust air does not re-enter, (b) located away from sources of outdoor pollution and (c) protected by suitable filters (one point for each)
2	steam or no humidification*
3	no recirculation of used air*
1	recirculation with adequate particulate filtration
3	systems designed and installed for easy maintenance and cleanliness, with filter media, thermal and acoustic insulation prevented from releasing fibres into the airstream*
3	extract ventilation to areas used as toilets, kitchens, and for smoking, photocopying and other polluting activities
2	occupants provided with local control of temperature, e.g. by thermostatic radiator valves, and, where appropriate, information on the use of these controls
3	commissioning complete or system recommissioned in past five years, including check procedure on the commissioning data*
3	all systems thoroughly cleaned before hand-over or at least within the past five years*
2	no collection of stagnant water or dirt within ventilation system*
	Building and furnishing materials
3	all furnishings thoroughly cleaned within the past year or shown to be clean
1	use of non-static carpets
	Lighting and solar control
2	no tinted windows
2	solar control blinds (external or inter-pane) on all windows oriented more southerly than NE or NW
2	lighting able to meet CIBSE standards with provision for visual display unit work where necessary
2	use of high-frequency ballasts for fluorescent lighting
2	artificial lighting with local controls plus task lighting and with a view out of a window from each workstation

Table 20.2 continued

	Noise
2	for meeting the BREEAM indoor noise requirement
	Layout
3	cellular rather than open plan, i.e. at least 90% of rooms designed for ten or fewer occupants
3	shallow plan: maximum distance of occupants from a window of 7 m
	Operation and maintenance issues
3	hygiene and maintenance schedule including: all air intakes checked and cleaned regularly, filters replaced, wet regions in air-conditioning systems cleaned and sterilised and regular checks for and cleaning of dirt accumulations in system*
3	operational maintenance schedule including: automatic control system (e.g. local thermostats), humidification units
3	operational maintenance schedule including: control point settings, a requirement for recommissioning on repartitioning and a check for blocked air supply grilles*
3	carpet cleaning specification requiring high performance, regularly maintained vacuum cleaners with high efficiency, hot water extraction (steam) cleaning (with minimum operating temperatures of 70°C) or liquid nitrogen treatment at least once a year and, where papers are stored for more than two years, cleaning them
3	questionnaire-based staff survey to determine prevalence of sick building syndrome symptoms one year after occupation and then every two years
3	policy to minimise the use of polluting processes, equipment and materials including adhesives, floor waxes, stains, polishes, spray cans, deodorisers, detergents, etc.
3	smoking ban or smoking allowed only in designated and separately ventilated rooms which make up less than 5% of the floor space
2	lighting levels checked and light fittings cleaned on a regular basis

Note
* Automatic credit for natural ventilation.

The benefits of good design

The combination of techniques leads to an impressive array of benefits (Table 20.3). Well designed buildings can dramatically improve the organisation's bottom line. Savings of 3 per cent to 15 per cent can be made, worth £600 to £3,000 per person per year at typical salary of £20,000 including overheads. These savings are all bottom-line benefits, and can be made by good design, with short payback times.

Evaluation

Assessment and evaluation of performance can be complex and time-consuming. To overcome these difficulties, ECD has developed two programs, AssessA and PerformA, which simplify evaluation. AssessA is for new office designs, while PerformA is used when buildings are occupied.

Table 20.3 Benefits of good design

Issue	*Guidelines*	*Benefits*
• Comfort temperature	5% over 25°C 2.5% over 27°C 1% over 28°C 0% over 29°C	89% of people prefer natural ventilation, provided high summer overheating is avoided
• Comfort cooling	Avoid mechanical cooling where possible	Staff preference, also reduced operating costs of around £13/m^2 per year
• Window design and daylight	Openable, maximum 6 m distance to window	80% of people want openable windows
• Nightlight	Normal lighting guidelines, max 400 lux, avoid glare, personal control	6–15% work improvement
• Light quality	High-frequency lights	0.5% absenteeism improvement
• Controls	Quick response, understandable, personal control	Reduced complaints, increased staff satisfaction
• Finishes	Avoid VOCs, fabrics and carpets, dust collectors	Reduced allergic reactions
• Building sickness	Follow BREEAM guidelines	2.5% absenteeism improvement

The programs report on what has been achieved so far. They then outline where further improvements can be made, and provide an action plan for achievement.

PerformA has been developed to assess the likelihood of sick buildings and to report on the financial benefits of avoidance. It also generates financial improvements through better lighting, as well as reporting on health, legal and other financial benefits of high performance buildings. PerformA has a module to allow whole buildings portfolios to be inventoried and managed.

Conclusions

There has been a step increase in knowledge on how buildings can be designed to improve productivity. Designers can now predict how to improve productivity, and offer their clients a real improvement to their bottom-line performance. Tools such as PerformA and AssessA exist so that current performance can be quickly assessed and areas for further achievement identified. The potential savings are vast – between £600 and £3,000 per person per year.

This allows designers to establish their position, to offer buildings which provide real tangible benefits to their clients. A client in turn will demand not only a beautiful building but also one that helps their company to be more productive than their competitor. Good designers will be able to respond positively. Good design really is good for business.

References

Anglia Polytechnic University (1998) *Daylight* (Daylight Calculation Program) Chelmsford.

AssessA and PerformA evaluations programs, ECD Energy and Environment Ltd. Tel. +44 171 405 3121.

BRE (1993) *BREEAM/Existing Offices Version 4/93. An Environmental Assessment Method For Existing Office Designs.* BRE Report BR 240, Building Research Establishment. Garston.

BRECSU (1995) *Environmental Assessment of Buildings.* THERMIE Programme Action No. B108, Garston.

BRECSU (1997) Good Practice Case Study 308. *Naturally Comfortable Offices – A Refurbishment Project.*

CIBSE (1986) *CIBSE Guide*, vol. A8. Chartered Institute of Building Services Engineers. London.

Cohen, R. (1993) A comforting future. *Building Services*, Sept., 35–36.

ECD Energy & Environment Ltd (1998) *EC2000 Project.* Rue Abbe Cuypers 3, Bruxelles 1040, Belgium.

ECD Energy & Environment Ltd (1999) *N-Light. Daylight Factor calculation program.* 11–15 Emerald Street, London WC1N 3QL.

Harris Research Centre (1994) *Occupiers' Preferences.* A survey for Richard Ellis by the Harris Research Centre. Part 2 – The Performance of the Workplace. Available from Richard Ellis, London, Tel. +44 171 256 6411.

Jones, P., Vaughan, N., Grajewski, T., Jenkins, H.G., O'Sullivan, P., Hillier, W., Young, A. and Patel, A. (1995) *New Guidelines for the Design of Healthy Office Environments.* University College Cardiff and University College London.

Petherbridge, P., Milbank, N. and Harrington Lynn, J. (1988) *Environmental Design Manual – Summer Conditions in Naturally-Ventilated Offices.* Report BR 86, Building Research Establishment, Garston.

Romm, J. and Browning, W.D. (1994) *Greening the Building and the Bottom Line – Increasing Productivity Through Energy Efficient Design.* Snowmass, CO: Rocky Mountain Institute.

Standley, Marilyn of Longman addressing the *Green Buildings Turning The Tide* seminar held at the Royal Society of Arts 27 October 1994. Details from ECD Energy & Environment, London, Tel. +44 171 405 3121.

Teculescu, D. (1997) Respiratory and irritant symptoms in French office workers. *European Respiratory Society Conference*, Berlin.

Wilkins, A.J, Nimmo-Smith, I., Slater, A.I. and Bedocs, L. (1989) Fluorescent lighting, headaches and eyestrain. *Lighting Res. Technol.* **21**(1) 11–18.

Chapter 21

Optimising the working environment

John H. Jukes

Two of the great mysteries of the 1990s are as follows.

1. Why do 90 per cent of computer installations fail to meet their original productivity performance specifications? (Recently published figures show IT investment in the USA increasing by 700 per cent and labour productivity plummeting by 80 per cent.)
2. Why do over 80 per cent of staff in ordinary offices suffer from the classic list of sick building syndrome symptoms? (Headaches, tiredness, dry/itchy eyes, sore/dry throat, cough, cold/flu symptoms, irritability, skin rashes/itches, pains in the neck, shoulders and back, etc.)

An involvement with the first problem, as we shall see, led to the second one. The solution to both questions has evolved over several years of practical development in a number of different organisations.

In 1988, after 25 years in office productivity measurement and improvement programmes, we found an increasing demand to investigate new computer systems that were failing to meet their performance standards.

For example, a new PC network system for a major financial services organisation was planned for forty staff and was failing to cope with eighty staff. Checking the workload measurement standards showed that forty staff should handle the work easily. Observing key tasks in operation showed that staff were able to do them in the time planned. However, at the end of the day their output was only some 50 per cent of what it should be. Attempts to improve performance met with unusually emotional response, so they were not pursued. Staff on non-computerised tasks using the same basis for setting time standards had no problem in meeting the daily target. The observed situation was that sitting someone in front of a PC all day resulted in their personal performance dropping by around 50 per cent without them being aware of it. The problem was eventually resolved by making conventional improvements in the system and organisation. Eventually forty staff were coping with the workload. However, the new systems meant the planned staffing should be twenty. The productivity gap remained. Management felt that enough was enough, and were happy to let sleeping dogs lie.

After several PC network projects with similar results, we decided to investigate this mysterious productivity gap. The question was: why did the same staff working in the same office seem to work at a much slower pace using a PC all day than they did doing non-PC tasks?

The opportunity to find the answers came in the form of a project involving some 250 design engineers and support staff. The first twenty CAD systems were not performing to target and could not justify any further investment in CAD. Here was a powerful incentive, since all the younger engineers felt that CAD experience was necessary for their future as engineers.

This time we made the conventional improvements and set targets including the mystery gap. Within 6–8 weeks four staff reached them and the rest went sick – with a bewildering array of different symptoms. Our medical advisers said it was stress: we were making them work too hard.

We took the twenty off the CAD machines to train another group while we sorted out the problem. The original twenty, once back on the drawing boards, recovered their normal productivity levels and stopped going sick.

An article on sick building syndrome listed symptoms which coincided with our list of symptoms experienced by staff. Further investigation showed that most staff experienced them to some degree, but usually in a mild form that did not prevent them from coming to work. They tended to go critical under external stress.

We reasoned that an anti-sick building syndrome pill would solve all our problems. Further investigation revealed that, although there seemed to be general agreement on the symptoms, there was less agreement on the causes, other than they were likely to be many different ones.

We therefore decided to see if we could find our own solution, so we worked back from the symptoms. The common major complaint was tired eyes. Experimenting with a variety of VDU filters eventually produced one or two that had an effect in reducing eye symptoms but had little effect on productivity. The type of filters that we used also seemed to give protection from the various more controversial variety of non-ionising radiation that the VDUs seemed to generate. This is everything from soft X-rays through ultraviolet, infrared, microwaves, radio frequencies, very low frequencies, extra low frequencies and static.

We tried the NASA solution of one large specimen plant per person. This had no measurable effect other than the experimental group asking for them to be taken away because other staff kept offering them bananas and nuts.

From one eminent university professor we had our first workable clue. 'People sitting in front of a VDU all day tend to sit still; they sit badly; they become upper chest breathers; they don't get enough oxygen to the brain and they slow down; their immune system gets impaired; the slightest stress and they go sick.'

The next clue was from a well known consultant optician: 'Staff are sitting gazing at brightly coloured images at a fixed focal length. Their blink rate drops, their eyes dry out. The eye muscles get tired from being locked in one position all the time. When they try to look at a more distant object, the eye muscles won't relax.' Another source of advice suggested that looking at a bright image at a fixed focal length for extended periods induced a light hypnotic trance.

So the solution was – get them to sit right, blink, look away from time to time to change their focal length and have regular breaks.

The problem of what was right sitting took us through the world of ergonomics, orthopaedics, chiropractors, osteopaths, the Alexander System and A.C. Mandal. After much conflicting advice and experiment we came to the conclusion that we had to be able to adjust the height of the desk for the individual. We found a 15 inch differential in the sitting height between our tallest and shortest staff member. A stress-free active sitting position meant getting the chair and desk height right for the individual and supporting the pelvis rather than the lumbar region. It also meant rethinking the shape and dimensions of the desk to accommodate VDU, keyboard, etc. and leave enough room for paperwork.

With some Alexander tuition on correct sitting posture for the staff, most back, neck and shoulder aches went, as did some headaches. Productivity started to improve noticeably. We waited for the gap to close, but it didn't – it only closed about 25 per cent.

Since everyone complained about the air-conditioning system we looked at that next. There were enough data to show that productivity falls when it is too hot or too cold. However, we finally gave up trying to find an air temperature that everyone agreed was comfortable. We decided that there was some work to be done on high air velocities which made a reasonable temperature feel too cool. We found air turbulence created by the hot air rising from groups of computers reacting with the cool air coming in from ceiling slots. We found people sensitive to radiant cold from windows. We decided that the ideal thing would be to give everyone their own mini system to control air temperature and velocity, but at the time we could not find one. We found that putting the air intake into the floor got rid of turbulence and created a more even temperature with lower air velocities.

We found that the humidity was too low, at 15–20 per cent. The main building humidifiers had been turned off due to a legionella scare some three years before. Some staff were drinking thirteen cups of tea and coffee a day and were showing dehydration symptoms.

A small army of portable humidifiers and gallons and gallons of water later the humidity was up to 50 per cent – and everyone hated it. It felt clammy and thundery and stuffy and close and oppressive and headachy, etc.

This time another well known university came to the rescue. 'It's negative ions that you need. Air-conditioning strips them out of the outside air and the computers' static charge soaks up the rest. Positively charged air feels stuffy and thundery.'

It took some time to find ionisers that really did work and did not wipe your credit card or computer disk, or plaster the place with black goo, or make people feel woozy (from too much ozone). We found one type which combined ionisation with air filtration.

The ioniser filters soon filled with large amounts of black stuff. It was carbon – apparently derived mainly from the skin cells. We all shed skin cells at the rate of several million a day. Eighty per cent of household dust is skin cells but they are grey. The PC and associated electrical equipment provide the electromagnetic fields which cold-cook the cells down to carbon, oxygen and hydrogen.

Airborne carbon particulates, it seems, have some unpleasant characteristics in that they become positively charged; they are brittle and break into finer and finer pieces; they get down to below 0.1 of a micrometre (millionth of a metre); they go through the body's filtration defences; they get into the bloodstream and wreak havoc. This huge volume of tiny particles comprising several million a cubic foot provides a vast absorbent surface area which is home to volatile organic compounds, bacteria, viruses, fungi spores, etc. The body's immune system gets rid of them but gets tired and stressed in the process. One theory is that 70–80 per cent of SBS symptoms may be attributed to this one factor.

The air now seemed fresh and sharp. In fact the building manager accused the staff of opening the windows. 'You can tell when they have opened the windows, you know; the air smells different.' Staff seemed more cheerful and alert. Most of the respiratory problems and eye problems seemed to vanish.

The output took a sharp upward swing and we waited for the gap to close this time. It did not. It stuck at about two-thirds of the way. This was pretty good, but there was obviously something else we had not found.

We noticed that despite all the complaints about screen reflections and glare, the output from the staff near the windows was always higher than that of those in the centre of the office. There was something about daylight, but what?

We seriously contemplated putting light wells through the upper floors, using optical fibres or even mirrors. Someone then said that his uncle used some special lights to grow his prize orchids which imitated natural daylight. They turned out to be daylight full-spectrum lighting. The facilities manager nearly had a heart attack when he saw the price of the tubes. We decided to invest in another pilot project.

Some staff did not like the new lights and complained of flicker. We found later that full-spectrum lighting increases visual acuity, which made some staff aware of the flicker that was always there. Fluorescent lighting is flashing on and off at 50 times a second: the brain only processes at twenty to thirty times a second, so it seems like continuous light. An electronic ballast puts the frequency up to 30,000 cycles, which is virtually continuous light. We added electronic ballasts.

Some staff still did not like the light and found it too harsh. We then found a polarising filter which got rid of reflected glare and covered the bare tube. Since daylight is polarised by the atmosphere, we now had our goal of virtual daylight. Polarised light, it seems, is largely absorbed by the objects it fall on.

In a day or so the hard core of headaches had vanished. We had now not only closed the productivity gap but passed it, and symptoms were down by about 80 per cent.

We still had some skin trouble and one or two staff experiencing nausea. Although the ionisers neutralised most of the static in the environment, we adopted a policy of earthing everything we could to prevent any build-up on chairs, desks, floors and other surfaces.

We had a rash of respiratory symptoms which we found related to fungus spores breeding in the carpets. We swopped the carpet tiles for ones that had received anti-fungal, anti-bacterial, anti-microbial treatment.

By now we regularly monitored symptoms as well as productivity.

Staff now complained that the office was too noisy. We got in a sound engineer and sound-absorbed everything in sight – ceiling, floor, walls and screen panels. Everyone then said it was too quiet. We experimented with the wall and screen panels until (nearly) everyone said that it was OK, but – they could now hear other people's conversations more clearly. Sound masking provided the answer to this by increasing the ambient noise level.

Rather to our surprise, the productivity curve took another sharp upward lift. We had not expected any significant effect from sound suppression. It was some time later that we realised the reasons. Apart from distracting conversations, the different wave form patterns, particularly saw-tooth from continuous machine noise, can create muscle tension and stress. Also, reflected sound creates confusion and stress in our sound-direction sensing mechanism.

We now realised that what we were doing was recreating the outdoors, indoors. In hindsight this is, of course, the sort of environment that the human body has evolved to cope with over the past 273,000 years, compared with the office environment which has only existed for the past 100 years.

If the great outdoors is good for the body – what about plants again? We found that by using a different mix of hydroponic plants researched for NASA, we could add to cooling and humidity and get rid of volatile organic compounds, which are an unavoidable component of all plastics. More importantly, they could add to the oxygen levels and reduce carbon dioxide. The result was a marked reduction in respiratory symptoms and got total symptom reduction to around 90 per cent.

What we also decided we were attempting to do was to get people to work in a poised, relaxed, alert manner free of undue physical stress. In this way they could handle and recover from external psychological stress which is the inevitable part of any job. Some staff were now getting sick when they went home. We now seemed to have a well building syndrome situation.

This raised the question of how far can we go with this. Can we create an environment that does not just not make people tired and sick? Can we create an environment which positively supports the mind and body in achieving a maximum performance all the time? Can we leave staff at the end of the day with energy in hand to enjoy their leisure? What about geopathic stress, what about Feng Shui, what about the positive psychological effects of colour and smell, what about even more user-friendly desks and chairs, what about reclining chairs, what about the effect of aesthetics (can offices be beautiful instead of bland, boring and messy?), what about flat screens, what about what people eat? Why not treat staff like athletes who could be coached, groomed and supported by an environment that could provide super performance? All this we are working on and testing now.

So what was the problem? The evidence indicates that a body working in an office with other people is subjected to a number of ergonomic and environmental stressors for which the body has not yet evolved an adequate adaptive mechanism. Electricity has only been around for the past 100 years, and personal computers for the past ten years. All this is in the context of 273,000 years of evolution. The PC itself is not the problem, it is just one more additional source of stress. Air-conditioning is not itself the problem, since the problem is observable in non-air-conditioned offices.

It seems that the body responds to the right light, air quality, working position and neutralising the effects of electromagnetic radiation (EMR), static, sound, bacteria and fungi spores. The observed effect is an automatic improvement of around 32 per cent in personal productivity from sharper reaction times; a reduction in errors from better visual acuity and a marked improvement in cheerfulness and morale.

Can an employer afford to provide such an environment? We suggest that the cost of not investing in optimising the working environment is already being paid in the form of additional staffing and occupancy costs. The improvement in personal productivity can be translated into tangible staff savings in the order of 12–16 per cent. The cost of implementing the improvement of the working environment is equivalent to around 4 per cent. This is a return on investment of 3:1 to 4:1. Carefully phased, the project can be virtually self-funding, with an additional profit contribution of around £2,500 per employee.

All this depends on being able to measure productivity. Contrary to popular belief, there are many well-tried productivity measurement systems. Alexander Proudfoot's short interval scheduling system has been in use for forty years and has been applied successfully to almost every human activity in thousands of organisations.

So the solution to the two mysteries is that working with a PC in an ordinary office subjects the body to a complex cocktail of ergonomic and environmental stressors. The body, adapting to outside stress, uses energy in the adaptation process. The body slows down to conserve energy. As more energy is consumed the body gets tired and reaction times increase. Eye fatigue results in impairment of visual acuity. As the adaptive process falters, so other parts of the body malfunction and manifest what are called sick building syndrome symptoms.

An optimum workplace environment reduces external stress. Normal response times return, productivity is automatically enhanced, visual acuity is recovered and enhanced. Environmental health symptoms just disappear. The working environment becomes a pleasure to be in and enables staff to cope with the inevitable psychological stress that work and life bring to all of us. Now stress can be dissipated by well-spent leisure and a good night's sleep without slowly accumulating.

There are some 100 million people working with computers around the world that are costing their employers some £250 billion in lost productivity, and some 60 million staff are suffering physical discomfort unnecessarily. Here we have a workable affordable solution that is applicable to everyone. The computer and the office are with us. Let us learn to live with them for the benefits they bring and avoid the 'disbenefits'. These are some of the answers to the problem; there are bound to be others. Sick building syndrome need be no longer a mysterious complex complaint but something that can be solved logically and systematically.

Part 6

The future

Chapter 22

New ways of working: a vision of the future

Francis Duffy

A legacy of resentment

Why doesn't the office – where all the clever people we need to run an information economy are housed – get the management attention it deserves? Perhaps it is because many managers have inherited the idea that the office is a minor appendage to real production, something that has to be put up with if not bitterly resented. If so, they are wrong. Often today the office *is* the business, certainly offices have become much more central to business performance. Business can now use office space aggressively and imaginatively as a major factor of production as well as a powerful way of expressing values.

What has changed? Information technology has now become so powerful and reliable that all office processes are being completely rethought. As office technology develops exponentially, new kinds of office culture are being invented on the run – plural, fluid, responsive, knowledge-based – that are in complete contrast to conventional forms of office work. The old, top-down, hierarchical structures are dissolving. Conventional boundaries are disappearing. New processes are being invented. Clerical work is being automated or exported to economies where labour is cheaper. The chronology and the geography of newer kinds of office work are being redrawn as IT allows more and more people to control, and indeed redesign, the ways in which they manage their energy and intelligence, shape their working days and reconfigure the connections between home, work and leisure.

In this period of rapid transition conventional office locations, buildings and interiors are by no means neutral. They can be disastrous. Managers, particularly in old countries like Britain and France, but also in North America, where the Taylorist legacy in office architecture is physically still so strong, should worry about them as potentially formidable obstacles to business change. The outmoded fabric of the office has been shaped by decades of managerial neglect as well as by the supply-side bias of developers, architects, engineers and furniture salesmen. What has suited the property, construction and furniture industries turns out to be the unwitting preservation, in concrete and glass instead of aspic, of the top-down, divide-and-rule, mechanistic values of Taylorism. Frederick Taylor's *Scientific Management* was, of course, the predominant managerial style of the early decades of this century, when the conventional form of the office crystallised.

The catalytic function of design in a time of change

Today many clients want to understand how to make office buildings relate more closely to organisational performance.

For decades in office design, architects have voluntarily taken a back seat. This was certainly not the case at the turn of the century, when architects such as Sullivan and Wright were hammering out the norms that determined the shape of what became the conventional twentieth century office building. Subsequently architects stopped innovating in office design quite simply because they no longer had to do so. Everyone knew what an office was like – a successful formula had been worked out relating buildings and interiors to a particular, and very limited, notion of tenant and user demand. For a long time the norms and conventions of office design remained as stable as the simple technology of the typewriter and the straightforward habits and behaviour of clerical organisations.

Scott Adams' cartoons, featuring the hapless Dilbert in his hopeless labyrinth of endless carrels, capture the worst side of modern enterprise. Nothing can be achieved. Everyone and every thing is blocked. Millions of Americans laugh daily at these caricatures – the conventional office is easy to mock and the jokes may make the horror and the waste tolerable for a moment – but the symbolic connection between the futility of the organisation and the banality of its environment is not funny. *Buildings can kill.* They can kill easily enough through poisonous air-conditioning and lethal materials. But, much more lethally at this particular point in the development of organisational ideas, office buildings can kill through their amazingly accurate capacity to express and even exaggerate whatever is wrong, backward and inhuman about organisations. This is because the iconography of the conventional office is deeply rooted in the old-fashioned, mechanistic values of Taylorism. These values are exactly what advanced managers today are struggling to escape from. It is ironic that the architects' habit of treating office space as a supply-side commodity has dulled management as well as themselves to the significance of the design of the working environment.

This is the negative side. A contemporary organisation that wishes to change its culture – to abandon hierarchy, to encourage interaction, to stimulate creativity, to accelerate innovation, to break across previously impenetrable organisational silos – would be foolish to attempt such changes while persisting with an office environment that expresses, through inertia, exactly the opposite values. Such messages, communicated all too eloquently by the old office environment, fundamentally and fatally contradict the objectives of any contemporary change management programme.

The positive side of the same argument is this. Managers who genuinely wish to change their organisation's culture have an immensely powerful tool available to them in the re-design of their office environment. The values that are essential to the development of new ways of doing business – egalitarianism, transparency, stimulus, creativity, lateral thinking, accelerated responsiveness – all have very concrete equivalents in the powerful language of design. The task for both managers and architects today is to unlock the eloquence of innovative design to reinforce business performance by expressing business ideas for business purposes.

Innovation in office design and using office design to improve business performance are closely related – as the next three examples illustrate.

A supply-side innovator

'Good design is good business' is one of the favourite sayings of the most brilliant of all contemporary British property developers, Stuart Lipton. Lipton made his name by creating in the early 1980s the most innovative business park in Europe – Stockley Park, near Heathrow. He was at the same time the genius behind the Broadgate office complex in the City of London, a huge inner city development for the financial services sector which has had roughly the same impact on the fabric of the City of London as the Rockefeller Center had on midtown Manhattan fifty years earlier. Commercial property developers have not usually been noted in the UK for fine discrimination in the commissioning of architectural talent. Lipton is an exception – not only through using such excellent and well established practices as Arup Associates, Richard Rogers, Sir Norman Foster and SOM, but in giving such relatively unknown but equally talented architects as Eric Parry and Nicholas Hare their first chances to build major communal buildings.

What is so remarkable about Stuart Lipton as a developer is that he is totally convinced that *design really does matter to business*. Rapid construction, excellent facilities management, rational office floor plates that tenants can quickly plan and re-plan, intelligent environmental services that can cope with changing technologies and new patterns of use, the provision of such amenities between buildings as the fine new urban squares at Broadgate and the elegant, park-like landscaping at Stockley – these are all examples of Lipton's imaginative use of physical resources to anticipate and satisfy emerging user needs. Only by challenging the best architects with the toughest briefs based on the most thorough user research can the developer escape from the stereotypes that have been the curse of the real estate industry for decades. At both Stockley Park and Broadgate sectoral requirements of different types of office user have been carefully studied. *Market and user research has at last reached the world of property development.* Lipton is a developer whose commercial success comes from testing intuition against empirical reality. He then uses design invention to provide office organisations with the physical infrastructure that gives them the capacity, in all their varied ways, to do their work as efficiently and effectively as possible.

Two demand-side innovators

Robert Ayling, the Chief Executive of British Airways, is an equally interesting innovator in the creative use of design, this time from the demand rather than the supply side of the great equation that links people and buildings. Ayling and his Project Director, Chris Byron, have led the creation of a new headquarters near Heathrow – designed inside and out by the Norwegian architect, Niels Torp – which is explicitly and purposefully intended to change the culture of the airline. BA was still a nationalised industry a decade and a half ago when the threat – and the opportunity – of sophisticated international competition galvanised it into becoming the success it is today. The first improvements in attitude and service were achieved naturally enough on the tense frontline where the airline connects with its customers at the service counters and in the aircraft. The inertia that was holding the airline back from even more commercial progress was all too physically apparent in the endless brown and beige corridors in the ramshackle collection of buildings that previous decades of a bureaucratic and quasi-military regime had bequeathed as headquarters accommodation to BA.

Today everything has changed. The brown and the beige have gone. The long corridors are no more. Fragmentation is over. Waterside, BA's new headquarters building, is one of the most advanced examples in the world of how to use architecture in a programmatic and carefully managed way to achieve business goals. Waterside is a low building, consisting of six four storied, open plan, office pavilions, each a substantial building in its own right, connected by what would have been called, in the nineteenth century, an arcade – a continuous architectural device that is far more like a street than an atrium. This cobbled street undulates up and down and from end to end of the complex. It is glass roofed, crossed by bridges that link the pavilions to each other, and is immensely rich and varied in the amenities that it offers – shops, restaurants, conference and meeting rooms, sunny cafés as well as centres for information for office staff and centrally located, highly visible training areas for flight crews.

Two thousand eight hundred people work permanently at Waterside, practically all at fixed workstations. Hundreds more BA staff and others flood in every day – from flight crew to consultants. The street is as lively and interactive as Covent Garden. Given the elaborate infrastructure of information technology that has been installed, including cordless phones, there is no doubt that the capacity of the building could be pushed even higher. The barriers have come down, not because openness looks nicer but because BA, as a modern service organisation, has decided that it must use the building to reinforce transparency as an agent of cultural change. Ayling knows that communication, teamworking and fluidity are necessary for survival, let alone success. Consequently the building is deliberately designed to broadcast the critical importance of togetherness, of interaction, and of accessibility and transparency. Ayling also knows that British Airways has no choice today but to provide the kinds of environment that attract and retain not old-fashioned, time-serving clerks but the brightest, the best, the most ambitious and the most discriminating people in the job market.

All these policies are deliberately built into Waterside. The building is programmed to achieve management goals at two levels – directly, by design features that work efficiently, and indirectly, by using architecture as a highly effective way of expressing values and ideals that could not be transmitted more powerfully or more consistently in any other way.

Stuart Lipton and Robert Ayling are excellent but by no means unique examples of the kind of thinking businessmen who are rapidly bringing the UK economy up to date – not least by design. The third example of innovative use of design for business purposes is American. This is John Lewis, formerly CFO of Andersen Worldwide, the Chicago headquarters of the two recently separated components – Andersen Consulting and Arthur Andersen – the twin epitomes of successful, rapidly growing, knowledge-based organisations.

John Lewis's contribution, from 1995 to 1996, was to use Andersen Worldwide's move within downtown Chicago of some 1200 staff to revolutionise the culture of the Andersen headquarters. The opportunity, which he took hold of so vigorously, was to use the move to promote interaction between all levels and parts of Andersen Worldwide and, at the same time, to simplify and delayer what had been a very hierarchical organisation. SOM, and particularly Neil Frankel, were responsible for the interior design. Identical 'home base' workstations have been provided for practically everyone, whatever their seniority. Trappings of status have been swept away. The proportion of space dedicated to common facilities – particularly project and meeting spaces as well as quiet rooms for concentrated individual work – has been substantially increased. Bands of common facilities alternate with home bases so that none of the predominantly open plan, individual workplaces is further than a few feet away from a substantial array of supplementary shared accommodation. Cordless telephony, as at BA, allows people to plan their use of space and time in a much more fluid and intelligent way than in the conventional office. The impossible conundrum of the conventional office, i.e. to create individual workplaces that are equally good at accommodating both interaction and concentrated work in the same spot, is solved by zoning. Mobile workers have access to whatever environment is appropriate for the task at hand.

Three features of John Lewis's leadership of the Andersen Worldwide project deserve special attention. The first is the economic basis of the project. Because of the radical rationalisation described above, 1,600 people have now been accommodated in two-thirds of the rentable area previously occupied. What this means to Andersen financially is that what is an elaborate fit-out, substantial by any standards, has paid for itself in rather less than four years, so significant has been the reduction in the total annual rent. The second key feature is the enormous care taken in this project to establish a business case for each design innovation – as thorough and as well based on data as for any other key business decision. The third feature has been the involvement of large numbers of staff in the programming and design process, not just through data collection – important as that is – but through feedback of observations and data, seminars to discuss issues, focus groups to determine collective priorities, the use of 1:1 mock-ups, visits, and, of course, rigorous post-occupancy evaluation. All three features are critical to using design as an agent in the change management process – which, of course, was what the Andersen Worldwide project and, indeed, the British Airways project described above, were all about.

New ways of working – the price of relevance

For architects, however, there is a cost in helping to achieve such successes – which might partly explain their tardiness to accept the challenge to innovate. Architects and designers, confronted by fee cutting as well as marginalisation, have often preferred to retreat into their own private worlds accepting, like nineteenth century Bohemians shivering in their Parisian garrets, the inevitability of being misunderstood and rejected. This is nonsense. Design must be relevant. Architects must involve themselves in the issues of the day. Invention can and must be used to facilitate change.

Integration is everything. *Systems thinking*, i.e. *linking the design of the physical environment with the design of information technology with the design of the use of human resources, is the secret of success.* An autonomous designer can no more achieve change by issuing unilateral orders from on high than politically astute managers like Robert Ayling or John Lewis. Neither, of course, would dream of such an approach. Architects and designers, if they wish to unleash the potential of design, in the context of the change management that is now so vital for organisational success, must learn to be humble enough to embrace three practical conditions:

- a willing acceptance of the need for overall business leadership, setting targets, articulating values, insisting on performance
- an absolute equivalence in the design process between the design of IT, the design of the use of human resources and the design of the physical environment

- an enthusiastic involvement in the total design process of the clever, demanding, sometimes unreliable end-users on whom business success undoubtedly depends.

Business, particularly in a time of rapid change, depends heavily on design. Architects and designers must learn to relate design systematically to business and societal goals. The corollary is that successful design depends on understanding business and society.

The equation linking supply and demand

The equation between supply and demand has been referred to more than once in this chapter. Architects and designers – and, of course, engineers of every kind – have the ability to straddle that equation. They can easily, if they want, manipulate it for the benefit of their clients and also to advance the potential of architecture and design. Equations, of course, are useless without measures, and measures are useless without values. What is absolutely necessary, if office architecture is to become truly relevant to business, is to have reliable measures that link business performance with the capacity of buildings, environmental systems and interiors to accommodate and enhance that performance. Both sets of measures must be expressed in the same terms if the equation is to work. Both sets of measures must relate to organisational values and organisational purpose.

One powerful and systemic way of doing this is to follow the line of thought first developed in *The Responsible Workplace* (Duffy *et al.*, 1993) and taken further in Chapter 5 of *The New Office* (Duffy, 1997).

The gist of this argument is as follows. When it comes to matters of space use, all office organisations, in an increasingly competitive world, must obey two iron laws to achieve their commercial goals and, indeed, to stay in business. They must simultaneously *drive down occupancy costs* – which in offices are often higher than IT and second only to the costs of salaries – and *use the physical environment to attract, retain, stimulate and inform the increasingly valuable people who work for them*. The first dimension, restraining occupancy costs, is primarily a matter of efficiency, i.e. doing the most with the least resources. The great management theorist, Peter Drucker, calls this 'doing things right'. The second dimension is much more open-ended. It is to do with effectiveness, i.e. establishing values, achieving results by escaping from constraints, inventing unanticipated solutions, reframing the problem, getting out of the box. This is what Peter Drucker calls 'doing the right thing'.

Efficiency and effectiveness may seem to be contradictory. Working together they produce a simple – but dynamic and purposeful – model in which a vector represents the resolution of the two forces (Fig. 22.1).

Office organisations that do not feel themselves under any particular

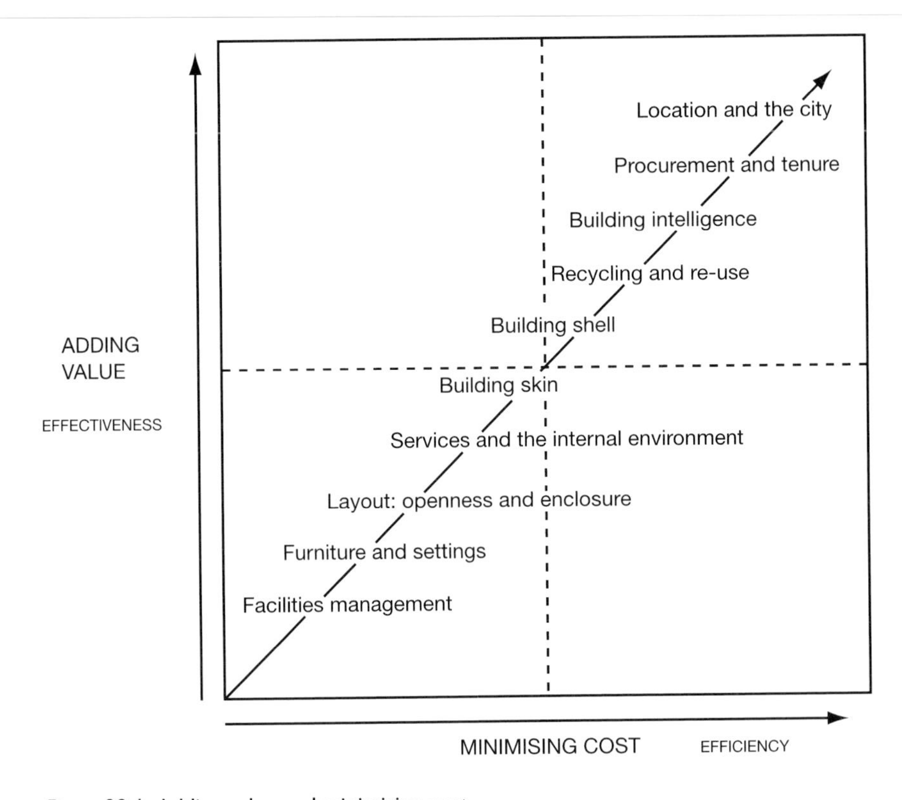

Figure 22.1 Adding value and minimising cost

pressure to achieve greater efficiency or effectiveness are likely to be willing to tolerate conventional office environments in which space is allocated, as it has been for decades, crudely by grade and status. The more the pressure grows to improve business performance on both dimensions, the more likely organisations are to allocate space by more rational criteria. When both pressures become extreme, organisations are most likely to innovate. In today's terms, this means taking advantage of space use intensification. For example, some office organisations may only be able to afford the quality of environment they need to attract the best possible staff by sharing workplaces. Only by achieving greater efficiency through sharing are they able to afford a richer, wider range of more effective work settings.

Putting the same argument in another way, it is possible to explain with the same logic (Fig. 22.2) why conventional North American and Northern European offices have become the way they are. North American offices of the Dilbert variety neither are particularly efficient (because they consume so much space and so much energy) nor can they remotely be called effective (because they are so crushingly unattractive and unstimulating). Northern

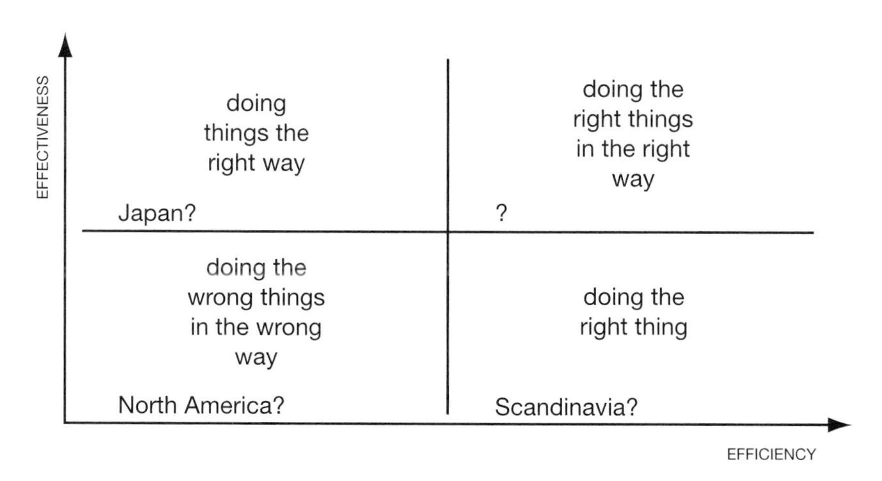

Figure 22.2 The efficiency/effectiveness balance

European offices are not efficient but often they are humane, attractive, lively, pleasant to work in – much more so than most offices in the US. Japanese offices customarily are very densely occupied (thus maximising efficiency) but are generally not the kinds of places that are likely to stimulate knowledge workers (thus minimising effectiveness). The kinds of environments advanced organisations are crying for – hence the impetus described above for design invention at Broadgate, British Airways, Andersen Worldwide – are ones that maximise efficiency and effectiveness simultaneously and in the same place. This is the goal for the future.

Measuring efficiency and effectiveness

Efficiency and effectiveness can both be measured, but in two very different ways. The measurement of efficiency is direct – for example, floor plates that maximise the ratio of usable to lettable area, increased density of occupation, reduction in the cost of churn, increase in space sharing. Each of these brings a direct financial benefit to the bottom line which can be measured in gains of so many per cent compared with other buildings.

The measurement of effectiveness, on the other hand, is indirect. The potential to enhance the effective use of office space can be installed, but results cannot be guaranteed. There is a big gap between the existence of a physical resource and the willingness and the ability of the people who inhabit it to exploit it. Nevertheless, not providing or taking away the same resource has a quite disproportionate effect. Any possibility of exploitation is removed. This is exactly analogous to what happens when a decision has to be made about the installation of new software in PCs. The possession of the advanced software does not mean that it will be used properly. However, not having the appropriate software rules out, quite simply, the chance of successful exploitation.

It is exactly the same with office buildings. They can be designed like BA's Waterside in such a way as to increase the possibility of interaction. Like Andersen Worldwide they can be crammed with opportunities to accommodate a wide range of different tasks. Both these offices have been designed to add fresh resources and to express positive messages to staff. In both cases enormous care has been taken to train everyone involved to understand management's message and to take advantage of the new resources – exactly as one would with new software. The consequence in both cases has been to increase the likelihood of beneficial use by a multiplying factor perhaps of three or four. Post-occupancy studies will no doubt eventually reveal what has actually happened.

An important point has to be made here about the burden of proof. Most management involves judgement. Most managerial decisions are made taking many parallel factors into account. Most management depends upon complex and changing external factors. The design of the working environment is no different. How things are done is often as important as what is done. Certainty is rarely available. There is no point in demanding – as so often happens when people want to block innovation in office design – higher (or indeed lower) precision of prediction, proof or subsequent verification than would be expected in the preparation of any other kind of business case for change.

A vision of the future

Managers should understand that design is much more than what it has often been reduced to – a fashion accessory or a technical appendage. Both are far too easily brushed aside. Design can instead be seen as a factor of production, given its capacity as a powerful catalyst of change. The physical design of the office has an enormous potential to broadcast positive as well as negative messages about what is and what is not important for management purposes. The one big feature that design and management share – and both modes of thinking are more similar than many people think – is organisational intent. Design is a vitally important way of getting things done. It is the skill of using the physical resource of buildings, space and furniture to solve the problems of users, clients and society in ways that help them achieve better, and otherwise unattainable, futures.

Innovative organisations, especially those that are determined to manage the process of change to their advantage, have begun to learn, once again, how to use the power of design.

The design of office space has been static for far too long. The inventive use of design to increase efficiency and effectiveness, i.e. *to drive down costs and to stimulate creativity simultaneously*, is revitalising not just how office organisations perform but also what offices look like. That design is now coming to be recognised as a critical factor of production as well as a powerful means of expressing management's intentions will revitalise not just the future of office architecture but the future of the design professions themselves.

References

Duffy, F. (1997) *The New Office*. London: Conran Octopus.

Duffy, F., Laing, A. and Crisp, V. (1993) *The Responsible Workplace*. Oxford: Butterworth Architecture.

Appendix

Creating the Productive Workplace: Summary of Conference held at Westminster Central Hall, London, 29–30 October 1997

Derek Clements-Croome

Can we design for productivity?

Buildings affect our existence. There is an interaction between our human senses and our surroundings. Productivity is related to the morale of the people working for the organisation as well as the attention level of the senses. It is fairly easy to tell when the productivity is lower than it might be because absenteeism, medical records and complaints become prevalent and work output is affected. There is a 'hum' in a place where productivity is high. Individuals too, know when they are working effectively.

Productivity can be measured in absolute or comparative terms. Research on productivity looks at **individual** needs rather than the responses of the **group** as a whole. There is overwhelming evidence that personal control of the environment is highly significant together with job satisfaction, effective work organisation and social ambience.

There are a cluster of factors that are conducive to high productivity. They include well-being, health and comfort. Conveniently, these are more likely to occur in a building which is well managed and has a low energy consumption. The act of producing quality work gives a person some self-fulfilment and satisfaction. Informed organisations are just beginning to understand the benefits of good workplace design and to acknowledge the notion that a **healthy workforce means a healthy organisation**.

Valerie Sutherland and Cary Cooper made the point that life within organisations is one of constant change, with endless modifications to work structure and climate fuelled by rapid technological and social changes. Identification of sources of stress using an **occupational stress indicator** prevents problems, and indicates the need to devise curative strategies for stress management in the workplace. Some degree of pressure is an inevitable part of living in a constantly changing work environment and this makes the audit and the identification of stress highly important.

Charles and Chad Dorgan presented some convincing data about productivity benefits. Their emphasis is on indoor air quality, which if not satisfactory affects the health of the building occupants and results in reduced productivity. Their analysis shows that the total productivity benefits amount to some $55 billion per year. Taking into account the cost of installing indoor air quality improvements, this gain results in an average simple payback time of just over half a year. The study was carried out by focusing on the compilation of some 500 reports of published and unpublished work which link indoor air quality and productivity in offices. Building wellness and employee inventory, health and medical effects, health cost benefits, productivity benefits and recommended improvements are all part of the methodology. Building wellness is defined in terms of the rate of complaints, the degree of sick building syndrome, and comfort.

The majority of studies indicated an average productivity loss of 10 per cent due to poor indoor air quality. The study was concentrated on non-industrial buildings and concluded that a one-time upgrade of these would cost almost $90 billion, with an annual operating cost of nearly $5 billion, giving a total annual benefit of nearly $63 billion and resulting in a simple payback period of 1.4 years. The benefit was based on a combination of productivity and health benefits.

David Mudarri from the United States Environment Protection Agency considered the economics of installing enhanced quality environmental services in buildings and highlighted the disparity between the great economic loss that society suffers from poor indoor environment and the cost necessary to improve it. He believes that this in some measure is due to an imbalance in the market place, because private entities that want improved environments have been unable to translate this desire into an overt expression of market demand that would justify the expenditure and risk that the improvements require. He believes that some of the ultimate goals are good building practice; a rational integration of energy and indoor environmental polices; guidance and software packages for building owners.

While acknowledging that economic losses from poor indoor environments have not been rigorously studied, it is contended that there is good evidence that the economic losses sustained by industrialised nations are substantial. The United States Environmental Protection Agency estimates that a total annual cost of indoor air pollution in the United States is approximately $6 billion due to cancer and heart disease, but approximately $60 billion due to reductions in productivity. The productivity loss was derived from self-reported survey data, and represents an average productivity loss of approximately 3 per cent for all office workers, or approximately twice that level for all those workers actually reporting some loss.

Other evidence is quoted which shows that the annual savings and productivity gains from buildings improvements in the USA range from $12 to $125 billion for improved worker performance, compared with $17 to $33 billion for improvements in health. It is reckoned that the benefits that would result from improvements vary from 18 to 47 times the cost of the improvements. The work of Dorgan and Dorgan has already been referred to, which established a productivity gain worth $63 billion per year, with a $90 billion initial investment giving a payback of 1.4 years.

It is reasonable to conclude that modest improvements in the quality of environment inside buildings can result in very large social benefits. Perhaps building owners do not appreciate this assessment, or maybe the required changes in the patterns of behaviour are not easy to come to terms with. Tenants and owners market their building space according to its location, appearance, parking and other items that are visible and tangible, in spite of the fact that issues of environmental comfort and building services rate high on the list of major complaints by tenants. The effort in time and money for commissioning, writing maintenance manuals and performing maintenance are not considered as being an investment.

Enhanced environmental systems offer a saving due to reduced maintenance as well as from energy conservation, but these savings are insignificant when compared to the potential health, comfort and productivity gains. The 30 per cent saving in energy, and in the maintenance of building services systems, corresponds to a cost of about $2.5 to $3.5 per square metre, but a 3 per cent loss of productivity associated with poor environmental quality corresponds to about $45 to $60 per square metre; in addition there are the impacts of discomfort and poor health on the occupants.

In order to make these issues marketable there is an urgent need for indoor environmental quality protocols in the form of guidelines and standards. In these there needs to be a set of procedures that integrate the needs of energy cost reduction with the needs for good indoor environmental quality. The US Department of Energy, in the US Environmental Protection Agency, has begun to develop a protocol in conjunction with a wide variety of stakeholders which will be part of the International Performance and Verification Protocol.

Demand side management reflects awareness of the utilities to encourage energy efficiency, rather than supporting energy growth patterns. **Performance contracting** provides the mechanism by which energy savings from energy-efficient investments can be used to pay for better environmental systems. One objective of the international protocol is to facilitate these mechanisms by standardising measurement and verification procedures.

Owners should market and advertise the environmental quality of the spaces they provide, and there would then be a good chance of the building owner's net revenue potential being enhanced. It is also important to emphasise that the term **building performance** does not just refer to energy and operating costs, but also to how well the building services provide the occupants of the space with healthy conditions which help to maximise productivity.

Audrey Kaplan described the barriers to productivity as being interruptions; discomfort; illness symptoms; a lack of privacy; resources constraints and the corporate culture. She went on to describe productivity as depending on:

- the physical facilities and infrastructure
- facilities management and practices
- occupant behaviour
- the management approach.

The environment is not neutral and can enhance or hinder the work output of an organisation. People react to the environment as a whole, and it is essential to remove any obstacles that impede the effort in their work. Well-being depends on a healthy mind and a healthy body. Research has shown that where an environment exhibits more than two building sickness syndrome symptoms the productivity tends to decrease.

John Jukes gave a very cogent account of how people working with computers can have an optimum healthy workplace environment to reduce stress and hence improve productivity. He estimates that there are some 100 million people working with computers around the world, and employers lose £250 billion per year in reduced productivity mainly because some 60 million people are suffering physical discomfort. In his experience, sitting in front of a PC all day can result in a person's performance dropping off by about 50 per cent without their being aware of it. His work has established that the major sources of workplace health problems are: tired eyes; bad posture; airborne carbon particulates; low humidity; flickering lights; quality of light; noise; and fungus spores which breed in carpets.

These symptoms have an impact on the immune system, which means the body's defences are low and the slightest stress can cause sickness. For example, people who sit badly do not breathe properly, so there is not enough oxygen circulating to the brain and they slow down, which results in a lowering of the immune system. Airborne carbon particles become positively charged; because of their small size they overcome the body's filtration system and enter the bloodstream. Again, the immune system suffers.

There is no magic solution and each case requires a fresh approach. Good posture, outside views where the eyes can relax looking at more distant objects, plants to decrease the CO_2 and lower the temperature, careful choice of furnishings, are all important issues. Many questions remain unanswered. We do not know enough about geopathetic stress or about Feng Shui and many other factors.

John Jukes considers that it is not a matter of whether an employer can afford to provide such an environment, but rather that the cost of not investing in a quality environment is paid for in the form of additional staffing and occupancy costs. Improvement in personal productivity can be translated into tangible staff savings of the order of 12 to 16 per cent. The cost of implementing improvements to the working environment is equivalent to around 4 per cent.

People, concentration and work

To be productive, one's concentration has to be sustained at a high level and remain uninterrupted. How do we think when we are working? How does our conscious mind work when it is being bombarded by all sorts of inputs from the environment around us?

Liam Hudson is satisfied that people are affected by their surroundings but is more concerned about the effect on occupants of half-hidden psychological needs. He comes to an interesting conclusion from his life's experience. Work production seems to have gone well in temporary or converted types of buildings, whereas productivity has often been difficult to achieve in custom-built buildings. This idea is described in Stewart Brand's book entitled *How Buildings Learn*. There are, however, good modern as well as good old simple buildings. Perhaps it is inconsistency that remains a problem. The Latham Report (*Constructing the Team*, 1994) acknowledges this. To probe the matter further Liam Hudson concludes that there needs to be a resonance between the imagination and our material environment; there has to be a parallel which allows our buildings to act as both vehicle and metaphor for our states of mind.

He goes on to make another point: most of us think poorly most of the time. The assumption is made that we know how to think, and is treated as a kind of automatic process. Edward de Bono has been trying to instil good thinking in school children. It is suggested that it would be a good idea to foster a new habit of conversation between architects, psychologists and engineers. A common language might evolve that would allow working environments to be designed which yield to conditions, rather than dominate them. Creative tension is valuable.

David Schwartz and Stephen Kaplan lead us into new directions in the theory and assessment of attention and concentration. Mental effort is expended with the objective of reaching some goal. Without a goal or objective we let the outer world's impressions, and the constantly changing currents of memory and imagination, carry us where they will. There are many obstacles in conserving the mental effort required for concentration. There are multiple sources of information around us; only a small part of this information is likely to be relevant to any particular goal. Effective thinking therefore requires the ability to select, both from our knowledge and from the world, the information required to attain the objective.

There are certain limits in our capacity to concentrate. A distinction has to be made between jobs that are repetitive and vigilant in nature and those that are creative and open-ended. Intrinsically, demanding tasks require one to think. When we remove opportunities for exploration, for creativity and experimentation and for wrestling with uncertainty, we take away the very conditions that nurture excellence and remove the features which make intellectually demanding work so satisfying. Extraneous interference such as random noise, interruptions, lack of privacy, poor lighting, stuffy atmospheres all interfere with the capacity to think and fatigue our cognitive ability. It is simply that more mental effort is expended in managing the extraneous interference, which leaves less energy for the demands of the task in hand. The result is that people become ineffective and productivity declines. There is also the possibility of increased irritability, which will magnify the rate of decline.

Many simple pencil-and-paper tasks can be used to measure concentration. The point is made that the effect of some workplace intervention on workers' ability to concentrate can be studied, but it is important to administer several different measures to produce reliable results. In this way we can study the independent contribution of concentration ability to the performance of various tasks, and also study how this ability changes in response to our attempts to enhance it.

Roy Davis considered the characteristics of skilled tasks from the point of view of their potential demands and gave a few examples to show how better design can lead to improved comfort and efficiency. There is no reason why work environments should not be as satisfying as those that surround us when we enjoy leisure. Work can be comfortable but it can also be fulfilling and enjoyable. Again the point is made that skills involve coordinated goal-directed activities. Productivity is high when relevant information is filtered from the environment and used to achieve the aims of the goal-directed activities. The control of attention can move between levels. One can imagine that as a result of training, actions at lower levels become almost automatic, whereas organisational design issues take place at a higher level and much attention is needed to achieve this. Roy Davis went on to describe the problem of mapping physical variables onto psychological ones, and how it is important to take account of the user's experience. However, people do not always say what they mean. Furthermore, many workplace skills are based on implicit, procedural knowledge which is difficult, if not impossible, to put into words. Hence it is important to watch what people do and analyse their behaviour as well as listen to what they are saying.

Jean Neumann brought some important matters to our attention using three vivid case studies. Buildings are designed, constructed and managed by a variety of people who perceive different priorities and needs as to what constitutes a productive and healthy workplace. Often there can be inadequate consultation with a broad sample of the user population. This means there will be a mismatch between users' perceptions and needs and those of the building design team. The productivity and well-being of the building design and construction team tends to take priority over the productivity and well-being of the users in some cases. Effective participation of the various user groups is vital. Negotiation and collaboration are highly important.

Michel Cabanac believes that human liberty is the freedom to choose one's own way to maximise pleasure and joy, which although transient give contrast and variety to what might be described as stable, neutral or indifferent conditions. In the context of the workplace this means that buildings need to respond to individual human needs; people need to be adaptable too.

Best practice in gaining a competitive edge

Philip Ward described a £600 billion opportunity for CO_2 reduction brought about by energy savings. Comfort, low energy and high productivity go together. All are aspects of sustainability.

Oseland and Williams considered how best practice can improve productivity and, in particular, the relationship between energy efficiency and staff productivity. Total energy costs of a typical office can be offset by an increase in productivity in the order of 1 per cent. They maintain that an increase in productivity of up to 8 per cent is needed to offset running and installation costs of heating, ventilating and air-conditioning systems. Their research establishes that energy-efficient lighting and the provision of individual control increases productivity, while at the same time saving energy. Good design, installation, commissioning and maintenance ensure that energy efficiency gains are continual, and this ensures staff satisfaction and improved productivity.

Adrian Leaman and Bill Bordass illustrated the question of productivity in buildings in terms of the **killer variables**. In other words, the variables that have a critical influence on the overall behaviour of the buildings systems. The present state of knowledge suggests that losses or gains are up to 15 per cent of turnover in a typical office organisation, and these might be attributable to design management and the indoor environment. Uncomfortable staff tend to show consistently lower productivity. There are also associations with related factors such as perceived health, management, design and use characteristics which improve perceptions of individual welfare, and also contribute towards better energy efficiency. There is also evidence to suggest that people need some variation in the levels of environmental factors. These variations can stimulate the arousal system, a fact which was demonstrated by the early work of Pepler, which indicated that for short periods it was better for people to work in slightly cooler, rather than warmer conditions relative to the neutral zone.

The killer variables are described as **personal control**, **responsiveness**, **building depth** and **work groups**. Personal control includes heating, cooling, lighting, ventilation and noise. Building depth is very much concerned with the daylighting characteristics of the building. Deeper buildings may, but do not always, result in lower satisfaction and productivity. But depth of space is also a correlate for other variables which affect human performance. For example, it relates to complexity, because services systems have to cope with perimeter zones as well as central zones which have quite different thermal characteristics.

Perceptions of productivity appear to be higher in smaller and more integrated work groups.

Leaman and Bordass conclude that system control, rapid response to the environment, shallow plan forms and the services selected to match activities are particularly important issues with regard to productivity. There is a need to improve the feedback process, the integrated design process, the care of the occupants, the formulation of the brief. There is a disturbing tendency to forget conveniently the bad news which can help to rectify situations. It is important to understand the contexts, the risks and the manageability of systems by treating the perceptions of the occupants about their environment seriously.

John Worthington made the point that the measure of productivity can be both the effectiveness of the process and the performance of the building for its occupants in supporting business objectives. The briefing process is important in establishing clients' needs, sets out the process of procurement and establishes the measures against which performance can be evaluated. In North America there is now emerging a new professional role for a **design brief manager**, who should understand the language and expectations of both user and hence, the business, and the design and construction team. The wording used by John Worthington in describing design brief management is interesting. He writes that it provides 'a framework in which the design team can **elegantly** allocate the available resources to maximise the clients' needs'. The brief, therefore, is concerned with defining the problem and identifying the solution area in order to find a balance between maximising effectiveness (i.e. productivity) and minimising costs (i.e. efficiency). This work clearly needs to be integrated with the process of identifying the various user groups and ensuring that there is sufficient dialogue between them and the design and construction team.

The Kajima Building in Tokyo (KI building) is an example of a sensitive environment design where an attempt is made to address the occupants' multi-sensory experience of their environment. This is a reference to the fact that workers exposed to intelligent buildings for extended periods of time can suffer a physical and psychological strain called **technostress**. A fragrance control system is installed based on the human reaction to various aromas which causes a pattern of freshening the environment, sustaining concentration and coping with fatigue. This recognises people's ultradian rhythms which pulse every 1.5 to 2 hours and are characterised by attaining optimum arousal allowing peak performance, followed by stress. Ultradian thrythms are a cyclical pattern of rest and activity.

Biomusic has been composed for the building based on an analysis of how human brain waves can be used in the auditory response process. The music programme comprises a relaxed state denoted by alpha waves of 8–12 Hz, and beta waves denoting stimulating states of 14–20 Hz. Fluctuations in air flow reflect the heartbeat rate, again relating the rhythms of our environment to those in the body.

The result of these measures was that comfort dissatisfaction was decreased to a fifth of what it was before the workers moved in to the new building. The greatly increased satisfaction level, it is claimed, resulted in an increase of productivity. Potential clients are invited to come to the KI Building so they can experience the environment, which also results in a saving of travel time for the employees of the company.

The work by Walter Kroner on environmentally responsive workstations is well known. This is a classic study which shows that improving architectural and environmental design contributes to an overall increase of productivity of 16 per cent. Environmentally responsive workstations increased the level of work and productivity by nearly 3 per cent. Again this emphasises the need for the individual to control his or her own work environment. The work also supports findings of other research workers that increases in satisfaction generally mean an increase in productivity.

The future for workplace and environmental design

Andrew Carter believes that there is significant evidence to show that physical environments enhance productivity; however, it is difficult to quantify this. He calls for further research to examine a range of models of productive behaviour and to evaluate the changes in revenue or output that can result. He demonstrated this by using the model of productive behaviour known as **high-performance teams**. The idea of introducing competitive teams working within organisations has recently been used very effectively. Very simple high-productive strategies have been produced, generated by the people themselves. The drive for customer focus is very important as a vision for an organisation or a profession. Concentration is then given to all of the issues that can help to raise productivity in quality and quantity terms. Productivity is affected and influenced by the response of people working in their physical workplace.

He went on to describe a particular case study involving the facilities management function. The model used was based on understanding the requirements of the customer; using valid data collection and statistical analysis techniques; analysing the business processes to identify opportunities for improvement; involving a wide cross-section of departments to derive the understanding about the options needed for improvement; setting up and managing of user groups to evaluate and participate in using the services.

Andrew Carter continued to describe some particular examples covering hot-desking; open plan working; a centrally managed records service. These projects achieved 22 per cent reduction in space; a 7 per cent increase in people accommodated; 25 per cent reduction of operating costs; and an increase in customer satisfaction.

Max Fordham contends that the human species has not adapted to meet the extremes of environment, but accepts that we do adapt quite easily in the common middle ground usually referred to as the neutral zone. Does a reduction in the environmental stress result in an increase in the likelihood of being more productive? Is the internal state of mind and the impact of fellow workers on the individual important? The human side to the story is that productivity improves when people know they are loved and are given special attention; this is a kind of gratitude expressed as a reward by an improvement in productivity. Perhaps this is a lesson for facilities managers, or sends a message to those building owners who do not employ facilities managers. If buildings are poorly maintained and not cared for, and occupants not listened to, then productivity will fall.

Environmental conditions do not need to be tightly specified, as the human body trades one sense off with another one, but also each sense has a considerable range of adaptability. Satisfaction with thermal environment can easily be achieved by altering our clothing. Attention was drawn to the fact that the design of clothes can allow free and easy movement and comfort at quite low air temperatures, as has been done in previous centuries. On this basis, buildings could be designed which did not require heating.

Every building has particular requirements and so it is important to clarify the context of each project. In general, building mass, building form, the use of cool night air and the requirements for sound and lighting can be achieved by combining the best of modern technology with simple things such as curtains, shutters and blinds. The added advantage of using these simple things is that they offer the individual personal control.

David Wyon discussed individual control at the workplace. He described how the environmental stimuli provided by visual, auditory or thermal information can be a distraction as well as a pleasure. The more degrees of freedom that can be designed into the workplace, the better, so that the user has personal control. The point is made again that optimising uniform conditions to accord with group average requirements ignores individual choice. Environmentally responsive workstations offer the user control and choice. Walter Kroner also referred to these systems.

Future workplaces will contain many, or perhaps all, of the degrees of freedom that have been identified. Building control systems in the future will have access to many more sources of information than they have at present. Sensors will be placed at every desk and will act as a personal environmental diary. Users will be able to access the building control management system at any time to obtain online information.

The work of Volker Hartkopf and Marshall Hemphill at Carnegie Mellon

University demonstrates the concept of **total building performance** which is a distinctive feature of intelligent buildings. User satisfaction, organisational flexibility, technological adaptability, environment and energy effectiveness are key issues. User satisfaction needs to be a consideration of thermal, acoustic, and visual environments; spatial quality; air quality; and building integrity. Human physiological, psychological, and sociological needs have to be offset against the economic needs of the organisation. Experience was gained by studying buildings in Japan, Germany, France, Canada, the USA and the UK over a period of three years. The pattern of organising performance criteria for evaluating the integration of systems was drawn up using the network of companies in the Advanced Building Systems Integration Consortium which was established at the Centre for Building Performance and Diagnostics at Carnegie Mellon University in 1998.

The intelligent workplace was demonstrated by the laboratory recently built on the rooftop of the Margaret Morrison building at Carnegie Mellon University. The design and construction includes many innovations such as a dynamic layered facade; a bolted module construction; an open web floor structure; floor-based support systems; intelligent controls with learning systems. The laboratory provides a catalogue of user preferences and experiences. It is already becoming apparent that some standards do not reflect what users prefer. Examples are quoted of preferred higher air velocities and higher brightness ratios for lighting.

John Doggart set out the guidelines and tools available which will influence future design. Designers have to reconcile costs, spatial layout, services and aesthetic issues, and when this is successful, buildings are generally healthier and productivity gains are more likely to be evident. Again it was pointed out that the environmental standards often used do not reflect user preferences. However, it is important to take each case study and relate the design to the context of that particular building. Not all buildings should be naturally ventilated, nor should all buildings be air-conditioned. Evidence is quoted from the 1997 European Respiratory Society Conference which reports that people working in air-conditioned offices are almost two and a half times more likely to suffer from respiratory infections than those working in naturally ventilated buildings. This resulted in twice as many days being taken off by the staff in air-conditioned buildings. However, there may be a very good case for having an air-conditioned building in some situations. Again the need for good-quality daylight, personal controls, office building materials and building form were emphasised.

John Doggart showed evidence relating building health to days taken off by employees. The BRE Environmental Assessment Method was seen as being helpful in reducing problems, mainly because sustainability means low-energy and low-polluting buildings; generally these buildings are healthier and people are more satisfied, hence more productive.

Frank Duffy asked why we bothered with offices. Physical resources drive down costs (efficiency) and should stimulate creativity (effectiveness). Business and architecture are interrelated, and intelligent buildings are ones that are adaptable in the sense that they have responsive building management, space planning and business organisation systems in place. A process cannot be carried out without understanding in detail what people do and what their aspirations are. Rapidly involving information technology and communication systems are killing conventions. People are working more to their natural rhythms and therefore there is a great need to understand what these rhythms are.

New ways of working link the nature of work; the space for people to work individually or in groups; and the systems selected to control the building environment as three principal interacting attributes.

Productivity of buildings in use requires new thinking and decision-making systems. This involves integrating **change management, facilities** and **technology**. Targets need to be set for improvement and they have to be evaluated.

Conclusions

The Conference established several common themes, as follows.

- We need to cut across the grain of traditional practice from time to time. We need to question and renew our thinking. This means being open to inputs from other professionals such as occupational psychologists.
- It is clear that **productivity is influenced by a number of factors including the environment**, and that **considerable economic and health benefits can be achieved**. It should not be a matter of whether an employer can afford to provide a good-quality environment, but rather that the cost of not investing in this will be paid for by additional staffing and increased occupancy costs.
- It is possible to **assess and measure productivity in absolute or comparative terms**. There is some ambiguity as to whether productivity means an increase in the work output of the people working in the offices or an increase in productivity by the design and construction team. There is no reason why it should not apply in both circumstances. The emphasis of this Conference has been on increasing the productivity of people working in offices.
- **It is possible to identify sources of stress**, which helps to prevent problems. Where this is not done, **stress management** in the workplace is very important. It has been mentioned that sitting in front of a PC all day can result in personal performance dropping off by some 50 per cent. It is said that the average person in Britain watches television 5 hours day in their leisure time, and if we add this to the work time occupation, it makes one wonder what quality of life we are seeking.

- It has been suggested that **most of us think poorly most of the time.** Effective thinking requires us to have the ability to select both from our knowledge and from the world around us. There are limits on our capacity to concentrate and we need to understand these.
- The variables that have a critical influence on the overall behaviour of building systems have been described as being **personal control, responsiveness, building depth** and **work groups.** Staff that are uncomfortable tend to show consistently lower productivity. Losses or gains of up to 15 per cent of turnover can be attributable to design management and the indoor environment.
- There is a need for brief **quality management.** This involves identifying the various user groups and ensuring that there is sufficient dialogue between them and the design and construction team.
- Buildings should be a **multi-sensory experience** and should engage the basic human senses. Buildings that do not do this give rise to inhuman environments and dissatisfaction.
- **Personal control** has been referred to by many speakers and a particular example of the environmental responsive workstation has been described. Future workplaces probably will have many degrees of freedom built into them. This is only part of effective total building performance, which is a distinctive feature of intelligent buildings.
- Designers have to reconcile **costs, spatial layouts, services** and **aesthetic issues** and when this is successfully achieved, buildings are generally healthier and productivity gains are more likely to be evident.
- **Facilities managers have a responsibility for the various user groups.** The quality of human nature is that whilst being very adaptable, they want to know that their environment is being cared for with their interests in mind.
- **There is a great disparity between the great economic loss that society suffers from poor indoor environment and the cost saved by improving it.** Modest improvements in the quality of the environment can result in very large social benefits, as we have already said. Tenants and owners acknowledge this but continue to market their building space according to location, appearance, parking and other **tangible factors.** In order to make productivity-dependent issues marketable there is a need to consider an **indoor environmental quality protocol** in a similar way to what the US is proposing, and to effect **demand-side management and performance contracting.**
- Although it is important that clients understand the ways of human behaviour in their organisations, and this should be recognised in the formulation of briefs for building design and management, **the individual also has a responsibility to understand and manage his or her daily rhythms. In this way productivity will rise and people will be more fulfilled.**

This summary refers to presentations made at the Conference. Not all presenters wrote a chapter for this book, and some authors did not attend the Conference. My own contributions are written – chapters 1, 3 and 10.

Derek Clements-Croome
Reading
September 1998

Index